Chaotic Behavior in Quantum Systems

Theory and Applications

NATO ASI Series

Advanced Science Institutes Series

A series presenting the results of activities sponsored by the NATO Science Committee, which aims at the dissemination of advanced scientific and technological knowledge, with a view to strengthening links between scientific communities

The series is published by an international board of publishers in conjunction with the NATO Scientific Affairs Division

A Life Sciences **B Physics**	Plenum Publishing Corporation New York and London
C Mathematical and Physical Sciences	D. Reidel Publishing Company Dordrecht, Boston, and Lancaster
D Behavioral and Social Sciences **E Engineering and Materials Sciences**	Martinus Nijhoff Publishers The Hague, Boston, and Lancaster
F Computer and Systems Sciences **G Ecological Sciences**	Springer-Verlag Berlin, Heidelberg, New York, and Tokyo

Recent Volumes in this Series

Volume 114—Energy Transfer Processes in Condensed Matter
edited by Baldassare Di Bartolo

Volume 115—Progress in Gauge Field Theory
edited by G. 't Hooft, A. Jaffe, H. Lehmann, P. K. Mitter, I. M. Singer, and R. Stora

Volume 116—Nonequilibrium Cooperative Phenomena in Physics and Related Fields
edited by Manuel G. Velarde

Volume 117—Moment Formation in Solids
edited by W. J. L. Buyers

Volume 118—Regular and Chaotic Motions in Dynamic Systems
edited by G. Velo and A. S. Wightman

Volume 119—Analytical Laser Spectroscopy
edited by S. Martellucci and A. N. Chester

Volume 120—Chaotic Behavior in Quantum Systems: Theory and Applications
edited by Giulio Casati

Series B: Physics

Chaotic Behavior in Quantum Systems

Theory and Applications

Edited by

Giulio Casati
Alessandro Volta Center for Scientific Culture
Como, Italy

Plenum Press
New York and London
Published in cooperation with NATO Scientific Affairs Division

Proceedings of a NATO Advanced Research Workshop on
Quantum Chaos: Chaotic Behavior in Quantum Systems, Theory and Applications,
held June 20-25, 1983,
in Como, Italy

Library of Congress Cataloging in Publication Data

NATO Advanced Research Workshop on Quantum Chaos: Chaotic Behavior in Quantum Systems, Theory and Applications (1983: Como, Italy)
Chaotic behavior in quantum systems.

(NATO ASI series. Series B, Physics; v. 120)
Proceedings of a NATO Advanced Research Workshop on Quantum Chaos: Chaotic Behavior in Quantum Systems, Theory and Applications, held 6/20-25/83 in Como, Italy.
"Published in cooperation with NATO Scientific Affairs Division."
Bibliography: p.
Includes index.
1. Chaotic behavior in systems—Congresses. 2. Quantum theory—Congresses. I. Casati, Giulio, 1942- . II. Title. III. Series.
QC174.84.N37 1983 530.1'2 84-26351

ISBN-13: 978-1-4612-9485-6 e-ISBN-13: 978-1-4613-2443-0
DOI: 10.1007/978-1-4613-2443-0

Softcover reprint of the hardcover 1st edition 1985

A Division of Plenum Publishing Corporation
233 Spring Street, New York, N.Y. 10013

PREFACE

Six years ago, in June 1977, the first international conference on chaos in classical dynamical systems took place here in Como. For the first time, physicists, mathematicians, biologists, chemists, economists, and others got together to discuss the relevance of the recent progress in nonlinear classical dynamics for their own research field. Immediately after, publication of "Nonlinear Science Abstracts" started, which, in turn, led to the Physica D Journal and to a rapid increase of the research activity in the whole area with the creation of numerous "Nonlinear Centers" around the world.

During these years great progress has been made in understanding the qualitative behavior of classical dynamical systems and now we can appreciate the beautiful complexity and variety of their motion.

Meanwhile, an increasing number of scientists began to wonder whether and how such beautiful structures would persist in quantum motion. Indeed, mainly integrable systems have been previously considered by Quantum Mechanics and therefore the problem is open how to describe the qualitative behavior of systems whose classical limit is non-integrable.

The present meeting was organized in view of the fact that scientists working in different fields - mathematicians, theoretical physicists, solid state physicists, nuclear physicists, chemists and others - had common problems. Moreover, we felt that it was necessary to clarify some fundamental questions concerning the logical basis for the discussion including the very definition of chaos in Quantum Mechanics.

Therefore the purpose of the meeting was on one hand to clarify, as much as possible, the rigorous mathematical aspects of the problem and, on the other hand, to discuss the possible applications in nuclear physics, solid state physics, chemistry, etc.

We hope that the meeting has met the above purposes and that the present book will serve as guide for those willing to enter the field.

The present conference, as well as that of 1977, has been organized in collaboration with Prof. Joseph Ford. His personal contribution was very essential for the success of both meetings. As a matter of fact, Joe Ford began the work in nonlinear dynamics near the end of the fifties, alone, without the comfort and the support of fashionable science and therefore with no help in getting through the various difficulties, scientific or not: he was not even aware that another person was following the same road and working on the same problems in buildings just under construction in a far off Siberian forest.

Now, the importance of the studies in nonlinear dynamics, in chaos and in the various applications, need not to be emphasized. Over more than the two last decades Joe Ford has contributed in a decisive and unstinting way to the development of this field and I am sure that the international scientific community is very grateful to him. Personally, I take this opportunity to express to him my warmest gratitude.

Giulic Casati

CONTENTS

MATHEMATICAL AND PHYSICAL PROPERTIES OF SCHRODINGER EQUATIONS WITH RANDOM AND ALMOST-PERIODIC POTENTIALS*

Bernard Souillard

Centre de Physique Théorique
Ecole Polytechnique
91128 Palaiseau Cedex France

ABSTRACT

A review of results concerning the Schrödinger equation with random and almost periodic potentials is given with special emphasis on rigorous results ; some more speculative discussions about the upper critical dimension of the localization problem is however developped at the end. Applications of similar ideas to other problems of wave propagation in inhomogeneous media are also mentionned.

The question of electron propagation in disordered systems or more generally the problem of wave propagation in inhomogeneous media is an old one, which has been strongly revived by the concept of localization introduced by Anderson and by the works of Mott, Thouless and many others. On the other hand electron propagation in incommensurate structures has appeared as a very attractive subject with the works of Azbel and Aubry.

---

*This text is included in the Proceedings for the sake of completeness, under the kind pressure of Giulio Casati. It is the summary of a review I gave at the Les Houches workshop "Common trends in condensed matter and particle physics" (March 1983) the proceedings of which will appear in Physics Reports. The reader will find here most of the results and references quoted in my lectures at Como.

For simplicity, most of this talk will be concerned with the tight-binding models which are commonly used in condensed matter physics when describing non interacting electrons at zero temperature. The equation reads

$$i \frac{\partial}{\partial_t} \psi_t(x) = [H\psi_t]\ (x) = \sum_{\substack{y \in \mathbb{Z}^d \\ |y-x| = 1}} \psi_t(y) + \lambda V(x)\ \psi_t(x) \qquad (1)$$

Nevertheless most of the results presented here are also valid for a continuous Schrödinger equation as well as for other wave equations as will be indicated in the conclusion.

For pedagogical purposes we will also mainly restrict to two opposite cases, namely :

a) the potential $\{V(x)\}_{x \in \mathbb{Z}^d}$ is a set of independant identically distributed random variables with a density r(.).

b) the potential $\{V(n)\}_{n \in \mathbb{Z}}$ (we restrict here to d = 1) is given as $V(n) = \cos(\alpha n+\Theta)$ where $\Theta \in [0,2\pi[$ is a given parameter.

Case a) appears in the modelisation of disordered systems,

case b) appears in the modelisation of incommensurate one-dimensional systems, and also in the study of an electron in a two-dimensional systems, and also in the study of an electron in a two-dimensional crystal with an external orthogonal magnetic field in which case the parameter λ in equation (1) has to be $\lambda = 2$.

As a matter of fact these two cases can be considered as special cases of potentials given as a general ergodic process, case a) corresponding to the most mixing class of such process whereas case b) corresponds to the most deterministic one.

Three main groups of questions are asked about such systems :

I - What is the spectrum of such operators H, i.e. what are the energies accessible to the electron or the frequencies accessible to the wave, and what are the properties of the density of states.

II - What are the properties of the stationary states or

proper modes, namely what are the spatial properties of the physical solutions of the equation $H\psi = E\psi$, what is the nature of the spectrum of H, what are the time evolution properties of the system and its transport properties.

III - What are the universal properties of such systems near transition points if any.

This talk is organized around these three groups of questions.

I - SPECTRUM AND DENSITY OF STATES

1) The spectrum of H

In case a) of a random potential, the spectrum $\sigma(H)$ of the operator H is exactly known : it is proven that for almost all realization of the potential V,

$$\sigma(H) = [-2d,+2d] + \lambda \text{ Supp } r(.)$$

that is $\sigma(H)$ is exactly the set of energies $E = \alpha + \lambda\beta$ where $\alpha \in [-2d, +2d]$, interval which is the spectrum of the non-diagonal part of (1), and β is in the support of the density function r(.) of the distribution of the potential. This theorem was first proven in Ref. [1] but in fact the main ideas had previously appeared in Ref. [2] .

In case b) it is known since an argument of Pastur [3] that, for α irrational ($\alpha \in \mathbb{R} \setminus \mathbb{Q}$), the spectrum of H is independant of Θ. It has been proven recently [4] that there exists a generic* set of pairs (λ,α) such that $\sigma(H)$ is a Cantor set. By this we mean first that it has no isolated point, but more interestingly that its complement in $\mathbb{R}$ is a dense open set, or in other words, in any arbitrarily small neighbourhood of any point of the spectrum there is an open gap. We emphasize the fact that this does not imply the spectrum to be of zero Lebesgue measure and it is in fact known to be of positive Lebesgue measure for λ small.

* i.e. a countable intersection of dense open sets.

2) The density of states

The integrated density of states k(E) can be defined for example as

$$k(E) = \lim \frac{1}{\Lambda} \# \{\text{eigenvalues of } H^{\Lambda} \leqslant E\}$$

where H^{Λ} denotes the restriction of the Hamiltonian H to a finite box Λ. By methods of Ref. [5], k(E) is known to exist for a large class of potentials and in particular for cases a) and b). It is also known to be a continuous function of E [3], constant on gaps and strictly increasing on the spectrum [6].

In the almost-periodic case, e.g. in case b), the integrated density of states k(E) has the following remarkable property : let E be in a gap, then k(E) belongs to the frequency module of V. In other words, if E is in a gap, $k(E) = m_o + \alpha m_1$, where m_o and m_1 are relative integers (in the case of a general almost periodic potential, one has several frequencies $\alpha_1, \ldots, \alpha_p$, and for E in any gap $k(E) = m_o + \sum_{i=1}^{p} m_i \alpha_i$). These integers are constant on the gap and are different for different gaps. Hence this theorem [7] provides a labelling of the gaps by a set of integers, although the spectrum is "often" a Cantor set.

The case where $\lambda = 2$ is particularly interesting since the integer m_1 found here is exactly the integer appearing in the quantized Hall conductivity which on a plateau is exactly equal to $m_1 \frac{e^2}{h}$, and it follows then from the approach of reference [8] that the quantized Hall conductance is an homotopic invariant, which explains its robustness to variations of the potential for example.

The behaviour of the integrated density of states k(E) near the bottom E_o of the spectrum is also a question of interest. For the case a) of a random potential it has been proved [9] that

$$k(E) \underset{E \downarrow E_o}{\sim} \exp\{-c(E-E_o)^{-d/2}\}$$

which is the singularity predicted by Lifshitz.

3) Open questions concerning the spectrum and the density of states

. is the spectrum a Cantor set for $\lambda = 2$ in case b). This does not follow from the quoted results. It is believed to be in this case a Cantor set with zero measure.

. is the spectrum still a Cantor set if the potential is now a more general sequence with more frequencies, e.g. V(n) =

$\lambda_1 \cos(\alpha_1 n_1+\Theta_1) + \lambda_2 \cos(\alpha_2 n_2+\Theta_2)$.

. are all gaps opened ? This question makes sense since we have, as shown above, a labelling of all possible gaps.

. what would be a description of the behaviour of the density of states k(E) near the bottom of the spectrum for an almost - periodic potential, e.g. for case b).

II - NATURE OF THE SPECTRUM AND OF STATIONARY STATES, TIME EVOLUTION AND TRANSPORT PROPERTIES

The physical solutions of the stationary state equation $H\psi=E\psi$ can be square integrable, in which case they are usually exponentially decreasing, or they can be non square integrable. The first case should correspond to localized solutions remaining mainly in finite regions during time evolution and mathematically speaking are associated to pure point spectrum ; the system should be an insulator for Fermi level near E. The second case should correspond to initial states going to infinity during time evolution, and mathematically speaking are associated to continuous spectrum, which may be singular continuous or absolutely continuous ; the system may be a conductor for Fermi level near E.

Let us first consider the random case a). In dimension $d = 1$, for almost all configurations of the potential and any disorder (i.e. any $\lambda \neq 0$) the spectrum of H is a pure point spectrum, with exponentially decaying eigenfunctions [10] [1] [11], the static conductivity as given by the Kubo formula, and the diffusion constant are zero [1]. The spectrum also has recently been proven to be almost surely pure point for quasi-one dimensional random systems, i.e. systems in a strip [12] . Finally for dimensions larger than 1, it has been proven [13] that, for large disorder (i.e. large $\lambda|$), almost surely the diffusion constant is zero.

As for the Anderson transition, the only situation where such a phenomenon is proven, is the case of the Bethe lattice, for which the lattice $\mathbb{Z}^d$ in equation (1) is replaced by a Bethe tree (i.e. a graph without closed loops and with a constant coordination number). It is there proved [14] that for large disorder the spectrum of H is pure point with exponentially decaying eigenfunctions, and the static conductivity is zero, whereas for small disorder, the spectrum is pure point near the edges of the band and purely absolutely continuous in the middle.

External electric fields lead also to interesting questions. Since the spectrum is always a discrete one if one adds to a tight-binding model a discrete version of an electric field [15], one has to study the continuous case for which the Schrödinger operator reads

$$-\frac{d^2}{dx^2} + V(x) + F.x$$

F denoting the strenght of the electric field. For d = 1, it has been proved [15] for a large class of realistic disordered systems that the spectrum is purely absolutely continuous for any $F \neq 0$. In particular there are no localized states (L^2 solutions) for any $F \neq 0$, in contrast to what is sometimes written in the litterature ; but also there is no singular continuous spectrum (sometimes called exotic spectrum).

For the almost periodic case b), one has the following results : if α is poorly approximated by rationals namely if there exist c and k such that for all p and q,

$$\left| \alpha - \frac{p}{q} \right| > c q^{-\kappa-1}$$

then there exists an absolutely continuous component in the spectrum for λ small [16] [17], whereas there exists a pure point component in the spectrum for large λ [18] [17]. In contrast if α is a Liouville number, i.e. is well approximated by rationals, namely there exist $c, p_n \to \infty$, $q_n \to \infty$ such that

$$\left| \alpha - \frac{p_n}{q_n} \right| \leqslant c n^{-q_n}$$

Then by a result of Gordon [19] there are no L^2 solutions of the equation $H\psi = E\psi$, and furthermore for $\lambda > 2$ the Lyapunov exponent associate to the equation is non-zero and there is no absolutely continuous spectrum, [18] [20] [21], the spectrum for $\lambda > 2$ being hence purely singular continuous. It is funny to note that the first set of α has measure one, whereas the second is generic, and hence pure mathematics have the wisdom not to choose between these two situations !

Finally, let us remark that the results mentioned above concerning systems with an external electric field also apply to incommensurate systems, namely in the continuous case, for any $F \neq 0$ the spectrum is purely absolutely continuous.

Opened questions on this section :

. In the random case a), is there an absolutely continuous component in the middle of the spectrum for d large enough and small λ ? Is there a non zero conductivity for Fermi levels lying in that region of energy if any ?

. In the presence of an electric field F, what is the behaviour of the resonances when $F \to 0$ and can we recover the result that at $F = 0$ the static conductivity as given by the Kubo formula vanishes ? This seems an interesting but difficult problem in view of the very complicate behaviour of resonances even in the a priori simpler case of a periodic potential [22].

. Is there a pure point component in the spectrum of a continuous Schrödinger equation with a strong almost-periodic potential. Some attempts to prove the opposite have failed; a pure point component is to be expected but there is no rigorously known example.

III - UNIVERSAL PROPERTIES NEAR AN ANDERSON TRANSITION

The Bethe lattice model is the only one exhibiting an Anderson transition as far as rigorous results are concerned. There, it is proved [14] for a large class of distributions of potential that the density of states is analytic at the mobility edges and that the localization length $\xi(E)$ diverges as

$$\xi(E) \sim (E - E_c)^{-1} \tag{2}$$

at the critical energy E_c, where $\xi(E)$ is defined through

$$\xi(E) = \frac{\sum_x |x| \bar{\rho}(0,x;\, E)}{\sum_x \bar{\rho}(0,x;\, E)}$$

and

$$\bar{\rho}(0,x;\, E) = \left< \sum_{E_\alpha} \delta(E - E_\alpha) |\psi_\alpha(0)\psi_\alpha(x)| \right> \tag{3}$$

E_α and ψ_α denoting respective eigenvalues and normalized eigenfunctions.

Let us now describe some suggestions [23] concerning the upper critical dimension and the critical exponents for the localization problem. First let us note that in order to compare the Bethe lattice, which is usuallly thought as an infinite dimensional system, with d-dimensional lattices for d large, one has to reinterpret the above result (2) as $\nu = \frac{1}{2}$ for $d = \infty$, for the same reasons that one has to do it in the case of the Ising model or for percolation [24]. If one compares this $\nu = \frac{1}{2}$ with the recent, non rigorous but plausible, result of Shapiro [25] namely $s = 1$ on the Bethe lattice, where s is the exponent governing the vanishing of the conductivity at the mobility edge, and if one uses the now well established scaling law of Wegner [26] $s = (d - 2)\nu$ we are led to on upper critical dimension $d_c = 6, 8$ or ∞.

On the other hand we can compare this remark with the (d-2)-expansion of the critical exponents obtained by transforming the localization problem into a non-linear σ-model [27] which yields

$$\nu = \frac{1}{d-2} + O((d-2)^2) \quad \text{and} \quad s = 1 + O((d-2)^3)$$

the first two corrections being known to vanish [28]. It is then tempting to think and in fact we conjectured it, that all corrections should vanish, which would yield $s = 1$, compatible with Shapiro's result and $\nu = \frac{1}{d-2}$ yielding again $d_c = 4$. Let us note that the vanishing of all correction corresponds to the "triviality" of the renormalization function; this is strongly reminiscent of what happens for supersymmetric models. It would be extremely interesting if one could use supersymmetry type of arguments in order to prove our conjecture or if one could negate it. As an additional remark, an attractive feature of our approach is that the critical exponents s and ν would be exactly known and in particular one would get $s = \nu = 1$ for a three-dimensional system. The main assumption here is the hypothesis that the localization length as defined through an average of type (3) has the same singular behaviour as the localization length of a typical system. If this was false, the critical dimension d_c would be different; this would nevertheless not necessarily contradict the conjecture of thevanishing of all corrections in the expansion of s and ν.

CONCLUSION

As a conclusion, I would like to stress that such ideas apply also to other wave equations and it is an attractive feature of these theories that they can be applied with appropriate adaptation to other problems. One interesting question is certainly the one of obtaining a macroscopic modelization and manifestation of the localization phenomenon. Such a nice transposition could be the one to hydrodynamics through the shallow water theory [29]. But many other microscopic or macroscopic applications are possible ; as a matter of fact some of the mathematics needed in the works presented here have their origin in very early attempts of soviet mathematicians to describe radio wave guides and other optical problems with disorder [30]. On the other hand possible applications to acoustic [31] or to plasma physics [32] have also been suggested.

Wave propagation in inhomogeneous media seems to be a subject with a nice future....

REFERENCES

1. H. Kunz and B. Souillard, Commun. Maths Phys. 78, 201 (1980).
2. D. J. Thouless, Phys. Rep. 13, 93 (1974).
3. L. Pastur, Spectrum of random Jacobi matrices and Schrödinger equations with random potentials in the whole axis, preprint (1974), unpublished, and Commun. Math. Phys. 75, 179 (1980).
4. J. Bellisard and B. Simon, J. Funct. Anal. 48, 408 (1982). Previous results concerning Cantor spectrum had been obtained for the case of potentials which are limit-periodic functions by J. Moser, Comment. Math. Helv. 56, 198 (1981).
   J. Avron and B. Simon, Commun. Math. Phys. 82, 101 (1981).
5. L. Pastur, Russian Math. Surveys 28, 1 (1973).
6. F. Wegner, Z. Phys. B 44, 9 (1981).
   J. Avron and B. Simon, to appear in Duke Math. J. (1983).
7. This result was first obtained in the continuous case by R. Johnson and J. Moser, Commun. Math. Phys. 84, 403 (1982). The discrete case was handled in some cases by C* algebra technics in J. Bellissard, R. Lima and D. Testard, in preparation.
   A complete solution through the study of a rotation number for discrete systems was obtained in reference [8] below.
8. F. Delyon and B. Souillard, The rotation number for finite difference operators and its properties, Comm. Math. Phys. 89, 415 (1983).
9. M. Donsker and S. Varadhan, Functional Integration and its Applications, Clarendon Press Oxford (1975).
10. I. Gold'sheid, S. Molcanov and L. Pastur, Funkt. Anal. Prilozhen 11, 1 (1977).
    S. Molčanov, Math. USSR Isvestija 42, transl. 12, 69.
11. R. Carmona, Duke Math. J. 49, 191 (1982).
    F. Delyon, H. Kunz and B. Souillard, J. Phys. A 16, 25 (1983).
    G. Royer, Bull. Soc. Math. France, 110, 27 (1982).
12. J. Lacroix, Localisation pour l'opérateur de Schrödinger aléatoire dans un ruban (preprint).

13. J. Fröhlich and T. Spencer, to appear in Commun. Math. Phys. (1983).
14. H. Kunz and B. Souillard, The localization transition on the Bethe lattice, Journal de Physique - Lettres, 44, 411 (1983).
15. F. Bentosela, R. Carmona, P. Duclos, B. Simon, B. Souillard and R. Weder, Schrodinger operators with electric field and random or deterministic potential, Comm. Math. Phys. 88, 387 (1983).
16. E. Dynaburg and Ya. Sinai, Funct. Anal. Appl. 9, 279 (1976).
17. J. Bellissard, R. Lima and D. Testard, Commun. Math. Phys. 88, 207 (1983).
18. G. André and S. Aubry, Ann. Israël Phys. Soc. 3, 133 (1980).
19. A. Gordon, Usp. Math. Nauk. 31, 257 (1976).
20. J. Avron and B. Simon, Bull. Am. Math. Soc. 6, 81 (1982).
21. M. Herman, Une méthode pour minorer les exposants de Lyapunov, preprint.
22 F. Bentosela, V. Grecchi, F. Zironi, J. Phys. C.15, 7119 (1982)
23. H. Kunz and B. Souillard, On the critical dimension and the critical exponents of the localization transition, Journal de Physique - Lettres 44, 503 (1983).
24. J. Vannimenus, Z. Phys. B.43, 141 (1981).
J. P. Straley, J. Phys. C.15, 2333 (1982).
25. B. Shapiro, Quantum conduction on a Cayley tree, preprint
26. F. Wegner, Z. Phys. B.25, 327 (1976).
27. F. Wegner, Z. Phys. B.35, 207 (1979).
28. E. Brezin, S. Hikami and J. Zinn-Justin, Nucl. Phys. B.165, 528 (1980).
S. Hikami, Phys. Rev. B.24, 2671 (1981).
29. E. Guazelli, E. Guyon and B. Souillard, Localization of shallow water waves by a disordered botton, to appear in Journal de Physique - Lettres (1983).
30. V.A. Tutubalin, Theory of prob. and its applic. 15, 282 (1970) 16, 631 (1971).
31. C. Hodges, J. Sound and Vibrations 82, 411 (1982).
32. D. Escande and B. Souillard, Localization of waves in a fluctuating plasma.

SMALL DIVISORS IN QUANTUM MECHANICS*

*Lecture given at the Como Conference on Quantum Chaos June 1983

Jean Bellissard

Princeton University and Université de Provence Marseille
Mathematics Department
Princeton, New Jersey 08544

It is today a well known fact that the occurence of small divisors in the perturbation theory for a classical system, is the source of instabilities for the motion of a classical system. These instabilities actually have a physical existence since numerical calculations as well as experiments exhibit chaotic behavior for an orbit in the vicinity of an unstable periodic one.

On the other hand, although these orbits are usually dense in the phase space, there are "many more" stable orbits, mainly supported by invariant torii, on which they have a quasiperiodic variation in time. This is the content of the Kolmogoroff-Arnold-Moser theorem, where "many more" must be understood in the sense of the Liouville measure on the energy shell.[1]

Systems with two degrees of freedom have been the focus of attention from the beginning of our century because they are the simplest non integrable systems. Even for such cases, the structure of the orbits is far from being understood. However, much progress has been made especially during the last decades since the contribution of Birkhoff [2]. The K. A. M. theorem, which in this case turns out to be the work of Moser [3], gives only a perturbative result. Even with the improvement of Rüssmann and M. Herman, the results stay far away from the domain observed numerically [4a)b)c)]. The semi-rigorous approach of Doveil and Escande gives results in much better agreement.[5]

The more recent rigorous contribution of Aubry and Mather [6, 7] allows enormous extension of the knowledge of the orbits by the proof of the existence of "Cantorii" sets, which are also called Mather sets by mathematicians, or broken torii by Rheinhardt in his lecture [8]. On the other hand, the vanishing of the "Peierls barrier" gives a test for the existence of invariant torii [6,7].

The instabilities of these "Cantorii," are perhaps at the origin of the chaotic behavior of a typical unstable orbit. As shown by Rheinhardt [8b],the orbit seems to be trapped by one cantorus for a while, then it seems to jump, then it is trapped by another one for another while, and so on and so forth. The succession of cantorii, which are labelled by a rotation number and the periods of trapping could give a coding, describing the orbit.

Aside these orbits, we find stochastic orbits as described by Smale [9] in the vicinity of an hyperbolic point. However, the "Smale horseshoe" phenomena occurs only on a set of zero Liouville measure, and does not provide information on the typical unstable orbit.

The aim of the present contribution is to convince the audience that a similar situation can be described in quantum mechanics in several models. However, we shall not speak about the semi-classical limit for an hamiltonian with discrete spectrum. What we propose as a quantum analog will obviously not be observed in the latter case.

There are several models in quantum mechanics leading to small divisors in the perturbation theory. This is the case for a Schrödinger operator with a quasi-periodic potential. Such systems describe crystal modulated by an incommensurate force. It is the case for conducting organic chains, where the extra modulation is caused by the so-called "Peierls instability." The rigorous study of these systems has been expanding rapidly since 1980. [10]. Even though a qualitative picture is emerging, giving the generic properties of these models, we are far from an exhaustive set of results.

Small divisors occur also in time dependent periodic quantum systems. The first order character of the time dependent Schrödinger equation allows strong resonnances between energy levels and the oscillation of the external forces. Among them we shall present here, the case of a quantum electron in a crystal submitted to an alternate electric field. During this conference, several speakers have also referred to the kicked Rotator model of Casati et al. [11], which is certainly a striking example to look at.As it has been pointed out by the Maryland group, [12] some of these models can be seen as Schrödinger operators for disordered systems. A review of the problem has been done by B. Souillard [13]. In the complete random case, the small divisor problem can be solved, but the case where the disorder is less random, the problem can produce the same effects as for incommensurate systems.

The main feature of all these models is that in some representations, they can be seen as the perturbation of an hamiltonian which is diagonal and possesses a dense set of eigenvalues. We can therefore compare this situation with the case of a perturbation of a completely integrable system. Complete integrability means that in certain canonical frames, the hamiltonian does not depend on the angle variables.

The label of eigenvalues corresponds to the values of action variables in classical mechanics.

On the other hand, the perturbation gives rise to non-diagonal terms, which correspond to angle-dependent terms in the classical case. More precisely if

$$H(A,\phi) = \sum_{\underline{n}\in\mathbb{Z}^v} h_n(A)e^{i\underline{n}\phi} \qquad (1)$$

the numbers $\underline{n} = (n_1,\ldots,n_v)$ are the classical analog of the difference between the quantum numbers of a non-diagonal term in the quantum case.

In this set-up, the stability problem, solved by K.A.M. in the classical case, turns out to answer the question of stability of the point spectrum after perturbation. We will show that this is the case in a few examples; however, a general method is not yet known to answer this question in whole generality.

Now the question of instabilities and especially the Rheinhardt picture seems to have an analog in this case.

The first order perturbation theory leads to analyze term of the form

$$\langle n|Wn'\rangle = \frac{\langle n|\Delta n'\rangle}{V_n - V_{n'}} \qquad (2)$$

where V_n are the eigenvalues of the unperturbed problem, whereas, Δ is the non-diagonal term of the perturbation and W is the first order term of the unitary diagonalizing our hamiltonian. In order to get a bounded term in (2) we need a compensation of the small divisor $V_n - V_{n'}$ by the numerator. When such a compensation does not occur we get a tunnelling effect.

We will argue that a long range tunneling effect can be the analog of instabilities and the peaks of the corresponding waves function are precisely located at the values of the quantum number n which should correspond in the classical case to the rotation number of the cantorii attracting the orbit for a while. The time of attraction for each cantorus is determined by the distances bewteen the peaks.

This last scheme is not yet well established as a rigorous level, and it could be wrong. However, we shall give an example of solvable hamiltonian with a quasi-periodic interaction exhibiting these properites.

One of the consequences on the spectrum is that we get "Singular continuous spectrum," and strong recurrence properties of the evolution.

This claim requires some explanation: For many physicists the term "recurrence" in quantum mechanics refers to the work of Hogg-Huberman [14], who investigated the time behavior of a quantum system with discrete or point spectrum only. The wave function being quasi-periodic in time is obviously "recurrent" in the sense that it comes back to itself as close as we want, infinitely many times.

However, there is another version of recurrence, as we can see in the paper of Avron and Simon [15]. In most problems, there is in addition a local structure, which tells us where the particle is localized. The point spectrum case of Hogg and Huberman corresponds to the situation where the particle spends all its life in a bounded region. If the hamiltonian has a non-empty continuous spectrum, for any wave vector ϕ in the continuous spectral subspace we get:

$$\lim_{T \to \infty} \frac{1}{2T} \int_{-T}^{T} dt \, \langle\phi | \exp itH \, \phi\rangle = 0 \tag{3}$$

It follows that the mean time spent by the particle in a given bounded region is zero. However, we can compute more precisely this time. It can happen that the total time spent in a bounded region is just finite. This case implies that the spectrum of H is absolutely continuous [16]. It can also happen that the particle comes back infinitely often in the given region, even though it escapes at infinity. This is a recurrent behavior, analogous to the case of a random walk. Then the particle can spend a time $t_A(T)$ in a bounded region A, during the period between -T and +T; we get from (3)

$$\frac{t_A(T)}{T} \to 0$$

but $t_A(T)$ can get bigger and bigger as $T \to \infty$.

This is precisely what occurs for the system we are interested in. When we speak of recurrence we will refer to this case.

Recurrence does not imply that the spectrum is singular continuous as shown by Avron and Simon [15]. There are examples of hamiltonian with absolutely continuous spectrum showing a strong recurrence phenomenon. This is the case when the spectrum as a set is nowhere dense. However, a singular continuous spectrum produces recurrence, as can be shown from the Ruelle results [16].

We shall organize this lecture as follows. In the first section we shall give an elementary equation where small divisors enter. This equation is actually the key point in the K.A.M. theorem.

In the second section we present the A. C. Stark effect, and the results.

In the third section we illustrate the K.A.M. method of this example.

In the fourth section, we argue briefly about the instabilities and the long range tunnelling effect.

In the last section we give an example of a solvable almost periodic hamiltonian, for which a singular continuous spectrum is observed, together with a chaotic behavior of the wave function.

Acknowledgements: The author would like to thank the organizers of the conference for giving him the opportunity of discussing with the expert on the subject of quantum chaos.

I. An elementary equation with small divisors

One considers the partial differential equation:

$$\frac{\partial f}{\partial x} + \alpha \frac{\partial f}{\partial y} = g(x,y) \tag{1}$$

Here, f and g are 2π-periodic continuous function, and α is a real number in $(0,1)$.

A formal solution of (1) can be given via Fourier's transformation:

$$i(m + n\alpha) \hat{f} (m,n) = \hat{g}(m,n) \tag{2}$$

It requires $\hat{g}(m,n) = 0$ each time $m + n\alpha = 0$. If α is irrational, the condition $m + n\alpha = 0$ implies $m = n = 0$.

Therefore:

$$\alpha \notin \mathbb{Q} \ , \qquad \hat{g}(0,0) = \frac{1}{2\pi^2}\int dx\, dy\, g(x,y) = 0 \tag{3}$$

From (2), (3), $\hat{f}$ is defined up to the form $\hat{f}(0,0)$ which is not surprising because adding a constant to f does not change (2).

The question is: When does the formal solution (2) give rise to a solution of (1)?

In other words: Is $\hat{f}(m,n)$ the Fourier transform of a smooth function?

If α is rational, the answer is obviously yes, and f can be chosen as smooth as g .

If α is irrational, the Hurwitz theorem shows that a sequence p_i/q_i of rationals can be found for which

$$\left|\alpha - \frac{p_i}{q_i}\right| \leq \frac{1}{\sqrt{5}}\,\frac{1}{q_i^2} \tag{4}$$

In particular if $(m,n) = \pm(p_i, - q_i)$ one gets:

$$|m + n\alpha| \leq \frac{1}{\sqrt{5}\, n} \tag{5}$$

Thus a subsequence in the Fourier expansion of f , exists exhibiting a divergence, because of this small divisor.

For g a trigonometric polynomial, (5) obviously does not affect the convergence of the series for f . But if g is continuous, a necessary condition for f to be continuous is that

$$\limsup_{i\to\infty} |q_i|\,|\hat{g}(\pm p_i, \mp q_i)| = 0 \tag{6}$$

A convenient way to answer requires the choice of a certain type of number. A diophantine condition on α is that there is $\gamma > 0$, $\sigma > 1$ for which:

$$|q\alpha - p| > \frac{\gamma}{q^\sigma} \qquad \forall\, \frac{p}{q} \in \mathbb{Q} \tag{7}$$

For a given $\sigma > 1$, it is known that almost all α in $(0,1)$ (with respect to the Lebesgue measure) satisfies a condition like (7). In this case, if

$$\sum_{m,n} |\hat{g}(m,n)| \cdot |n|^{\sigma} < +\infty \tag{8}$$

then (1) has a global solution which is continuous.

However f is in general less smooth than g . For instance we get the following results: [17]

1) if g is holomorphic in the strip

$B_r = \{(x,y) \in \mathbb{T}^2 + i\,\mathbb{R}^2 ;\ |\mathrm{Im}x| < r \ ,|\mathrm{Im}y| < r\}$.

and if

$$\| g \|_r = \sup_{(x,y)\in B_r} |g(x,y)| < +\infty \tag{9}$$

then, f is holomorphic in B_r but no longer bounded. Actually:

$$\| f \|_{r-\rho} \le \frac{4\sqrt{7}\,\Gamma(\sigma)}{\gamma\rho^{\sigma}} \|g\|_r \qquad \forall \rho > 0 . \tag{10}$$

2) if $g \in C^r$ with $r > \sigma$ then for each $\varepsilon > 0$, (4) admits a solution f in $C^{r-\varepsilon-\sigma}$.

A Liouville number is an irrational number which satisfies no diophantine inequality. Therefore we can find a sequence $\frac{p_i}{q_i}$ of rational such that

$$|p_i - q_i\alpha| \le \frac{1}{(q_i)^i} \qquad \forall i \ge 1$$

In this case, for a generic g in $C^{\infty}(T^2)$, (1) has no smooth solution [18].

II. The A. C. Stark effect in a perfect crystal [19]

A Bloch electron in a crystal, moving through an alternate external electric field can exhibit dynamical instabilities. The motion is described via a Schrödinger equation:

$$i \hbar \frac{\partial\psi}{\partial t} = -\frac{\hbar^2}{2M} \Delta\psi + \mu V\psi + e(\mathcal{E}(t) \cdot x)\psi \tag{1}$$

where $\psi(x,t)$ belongs at each time t, to $L^2(\mathbb{R}^\nu)$. Here, $\mathcal{E}(t + T) = \mathcal{E}(t)$ and has zero mean. V is a potential with:

$$V(x + a) = V(x) \qquad a \in L \tag{2}$$

where L is a lattice in $\mathbb{R}^\nu$.

To cancel the highly singular term given by the electric field, we can put our wave function in the moving frame of the classical particle. Let $p(t)$, $g(t)$ be the solution of

$$M \frac{dq}{dt} = p \qquad \frac{dp}{dt} = e \cdot \mathcal{E}(t) \tag{3}$$

submitted to the conditions (initial conditions)

$$\int_0^T dt\, q(t) = 0 = \int_0^T dt\, p(t) \tag{4}$$

The moving frame is defined by the unitary operator:

$$\psi(t,x) = e^{-i/\hbar \left[\int_0^t ds \frac{p^2(s)}{2M} + \frac{p(t)x}{M}\right]} \phi(t,x + q(t)) \tag{5}$$

Then we get:

$$i\hbar \frac{\partial\phi}{\partial t} = -\frac{\hbar^2}{2M} \Delta\phi + \mu V(x - q(t))\phi \equiv H(t)\phi \tag{6}$$

If V is smooth enough, the solutions of (6) are described via an evolution operator $U(t,s)$ which has the property:

$$U(t + T, s + T) = U(t,s) \tag{7}$$

because of the periodicity in time of $\mathcal{E}(t)$.

The Floquet operator $U_t = U(t + T, t)$ describes the evolution of ϕ during one period, and, by unitary equivalence, $t = 0$ suffices to describe the time evolution. We will denote it by U.

By the spatial periodicity we can decompose U on Bloch waves namely the Hilbert space $\mathcal{H}_k$ of functions Φ on $\mathbb{R}^\nu$, such that

$$(1) \quad \Phi(x + a) = e^{2i\pi\, k\cdot a}\, \Phi(x) \qquad a \in L$$

$$(2) \quad \|\Phi\|_k^2 = \int_{C^\nu} d^\nu x\, |\Phi(x)|^2 < +\infty \tag{8}$$

where C^ν is the unit cell of the lattice L, i.e. $\mathbb{R}^\nu/_L$.

Obviously k is defined up to elements of the reciprocal lattice L^*, which means that $k \in \mathbb{R}^\nu/_{L^*} = B_\nu$. B_ν is called the Brillouin zone.

Let U_k be the restriction of U on $\mathcal{H}_k$.

If we can find a Borel set $A \subset B_\nu$, such that for $k \in A$, the Floquet operator U_k possesses in $\mathcal{H}_k$ a solution of the eigenvalue problem

$$k \in A \qquad U_k \Phi_k = e^{i\,\omega(k)\cdot T} \phi_k\,,\ \Phi_k \in \mathcal{H}_k \tag{9}$$

then we get a solution of (6) satisfying:

$$\begin{aligned} \phi(t + T, x) &= e^{i\omega(k)T}\, \phi(t,x) \\ \phi(t, x + a) &= e^{2i\pi\, k\cdot a}\, \phi(t,x) \end{aligned} \tag{10}$$

This is the stability problem, for if $\mu = 0$ (which means the crystal is absent), then

$$U = \exp i \frac{\hbar}{2M} T\Delta \tag{11}$$

and the spectrum of U_k on $\mathcal{H}_k$ is just given by eigenvalues

$$\omega_{\underline{m}}(k) = - \frac{\hbar}{2M}\, 4\pi^2 (k + \underline{m})^2 \qquad \underline{m} \in L^* \tag{12}$$

and eigenvectors

$$\phi_{k,\underline{m}}(\underline{x}) = e^{2i\pi(k + \underline{m})\, x} \tag{13}$$

We get, plane waves for (6), and the spectral measure of U is then absolutely continuous.

To simplify notation let $L = a\mathbb{Z}^\nu$ $(a \in \mathbb{R}_+)$, then $L^* = a^{-1}\mathbb{Z}^\nu$ $C_\nu = a[-\frac{1}{2}, \frac{1}{2}]^\nu$, $B_\nu = a^{-}[-\frac{1}{2},\frac{1}{2}]^\nu$. We introduce the dimensionless constant:

$$\alpha = \frac{\pi T \hbar}{M a^2} \tag{14}$$

We redefine k by $\underline{k} = ka$ and $\underline{k}$ is also dimensionless. Then the spectrum of $U_{\underline{k}}$ is given by: (for $\mu = 0$)

$$\Sigma_{\underline{k},\alpha} = \{e^{-2i\pi\alpha(\underline{k} + \underline{m})^2}, \quad \underline{m} \in \mathbb{Z}^\nu\} \tag{15}$$

The <u>key fact is that</u> $\Sigma_{k,\alpha}$ <u>is dense in the unit circle when</u> α <u>is irrational</u>.[20]

Then when $\mu \neq 0$ $\mu \ll 1$, we expect that stability will indeed occur, at least for most of the spectrum, leading to Bloch waves, and absolutely continuous spectrum for U .

Actually we can prove the following:

<u>Theorem 1</u>:[19] Given $\varepsilon > 0$, there is a subset B_ε of the Brillouin zone B_ν, with $|B_\varepsilon| \geq |B_\nu|(1 - \epsilon)$, and a real number $\mu_\varepsilon > 0$ such that if $|\mu| < \mu_\varepsilon$, and $k \in B_\varepsilon$, $U_{\underline{k}}$ has only simple eigenvalues, which are dense in a set of positive Lebesgue measure. Correspondingly, the restriction of U onto the subspace $\mathcal{H}_\varepsilon$ generated in $L^2(\mathbb{R}^\nu)$ by $\{\Phi_{\underline{k},n}; \underline{n} \in \mathbb{Z}^\nu, \underline{k} \in B_\varepsilon\}$ has an absolutely continuous spectrum corresponding to Bloch waves.

In this theorem, we require that $V(x)$ and $E(t)$ are analytic in a strip $|\mathrm{Im}\, x_i| < ra$, $|\mathrm{Im}\, t| < rT$.

The proof of this theorem actually requires the Floquet theory to replace U by a self adjoint operator. This theory goes back to J. S. Howlands [21a] and has been transformed into a technical tool by K. Yajima [21b]. Instead of looking at U we consider the operator:

$$K = -i\hbar\frac{\partial}{\partial t} + H(t) \qquad (\text{resp } K_{\underline{k}} = -i\hbar\frac{\partial}{\partial t} + H_{\underline{k}}(t)) \tag{16}$$

acting on the space K (resp $K_{\underline{k}}$) of function $t \to f(t) \in L^2(\mathbb{R}^\nu)$ (resp $\mathfrak{Z}_{\underline{k}}$) with

$$\text{(i)} \quad f(t+T) = f(t) \tag{17}$$

$$\text{(ii)} \quad \int_0^T dt \, \|f(t)\|^2_{L^2} = \|f\|^2_K \left(\text{resp} \int_0^T dt \, \|f(t)\|^2_{\mathfrak{Z}_{\underline{k}}} \quad \|f\|^2_{K_{\underline{k}}}\right)$$

We get the following equivalence:

<u>Theorem 2</u>: [21b] $e^{-i\frac{TK}{\hbar}}$ (resp $e^{-i\frac{TK_{\underline{k}}}{\hbar}}$ is unitarily equivalent to $\mathbb{1} \otimes U$ (resp $\mathbb{1} \otimes U_{\underline{k}}$) .

We can as well decompose K on Bloch wave, to get the family $K_{\underline{k}}$ of operators on $K_{\underline{k}}$. Then $K_{\underline{k}}$ happens to be given on the canonical basis

$$|n, \underline{m}\rangle = e^{2i\pi n\frac{t}{T} + 2i\pi(\underline{k}+\underline{m})\frac{x}{a}} \qquad \underline{m} \in \mathbb{Z}^\nu, \; n \in \mathbb{Z} \; . \tag{18}$$

by:

$$\langle n, \underline{m}| K_{\underline{k}} |n', \underline{m}'\rangle = \frac{2\pi\hbar}{T}\left(n + \alpha(\underline{k}+\underline{m})^2\right)\delta_{nn'}\delta_{\underline{mm}'} + \mu \hat{V}(n-n', \underline{m}-\underline{m}') \tag{19}$$

Where $\hat{V}$ is the time-space Fourier transform of the function $V(x - q(t))$.

What we shall actually study will be the operator on $\ell^2(\mathbb{Z}^{\nu+1})$ given by the right hand side of (19), proving that if $\underline{k} \in B_\epsilon$ this matrix has only point spectrum.

More precisely:

<u>Theorem 3</u>: [19] If $\hat{K}(\underline{k})$ denotes the operator on $\ell^2(\mathbb{Z}^{\nu+1})$ whose matrix is given by the r.h.s. of (19), and under the assumption of th.1, there is a unitary operator $R(\underline{k},\mu)$ defined for $\underline{k} \in B_\epsilon$, analytic in $\mu, |\mu| < \mu_\epsilon$, Lipshitz continuous in $\underline{k} \in B_\epsilon$, such that:

(i) $R(\underline{k},\mu)\, \hat{K}(\underline{k})\, R(\underline{k},\mu)^{-1}$ is diagonal.

(ii) The eigenvalues of $\hat{K}(\underline{k})$ are given by:
$n + \alpha(\underline{k}+\underline{m})^2 + g(\mu;\underline{k}+\underline{m}) \quad (n,\underline{m}) \in \mathbb{Z}^{\nu+1}$
where g is analytic in μ, Lipshitz continuous in $\underline{k}$.

$$\left|\ \text{(iii)} \quad |\langle n,\underline{m}|R(\underline{k},\mu)|n',\underline{m}'\rangle - \delta_{n,n'}\,\delta_{\underline{m},\underline{m}'}| \leq \text{Const}\,|\mu|\, e^{-\frac{r}{\infty}|n-n'|}\, e^{-\frac{r}{\infty}|\underline{m}-\underline{m}'|}\right.$$

III. The K.A.M. Algorithm

In order to describe properly this algorithm, we start with a slightly more general hamiltonian. Namely:

$$\begin{aligned} K_{\underline{k}} &= (n + \alpha(\underline{k}+\underline{m})^2 + g(\underline{k}+\underline{m}))\delta_{nn'}\,\delta_{\underline{mm}'} + \hat{V}(\underline{k}+\underline{m};n'-n,\underline{m}'-\underline{m}) \\ K_{\underline{k}} &= K_\circ + \hat{V} \end{aligned} \tag{1}$$

where g and $\hat{V}$ are Lipshitz continuous in some region B of B_ν. We assume g, $\hat{V}$ small. (Say of Order ε).

Step 1: First order perturbation theory.

We try to construct a unitary $R = \mathbb{1} + W + O(\varepsilon^2)$ such that $W = O(\varepsilon)$ and

$$RK_{\underline{k}}R^{-1} = \text{diagonal} + O(\varepsilon^2). \tag{2}$$

Clearly, we get:

$$[W, K_\circ] + \hat{V} = \text{diagonal} = \delta K_\circ \tag{3}$$

with

$$\delta_{nn'}\delta_{\underline{mm}'}\delta K_\circ(n\ ;\underline{m}\ ) = \hat{V}(\underline{k}+\underline{m};0,\underline{0})\delta_{nn'}\delta_{\underline{mm}'} \tag{4}$$

This gives the first order correction to the eigenvalues. On the other hand:

$$W(n,n';\underline{m},\underline{m}') = \frac{\hat{V}(\underline{k}+\underline{m};n'-n,\underline{m}'-\underline{m})}{(n-n') + \alpha[(\underline{k}+\underline{m}')^2-(\underline{k}+\underline{m}')^2] + g(\underline{k}+\underline{m}) - g(\underline{k}+\underline{m}')} \tag{5}$$

Thanks to the remark in Section II, the right hand side exhibits a small divisor because α is irrational. In order to get a bounded operator W we need some condition:

Step 2: The small divisor problem.

From the hypothesis on the potential, $\hat{V}$ is expotentially decreasing with the distance $|n'-n|+|\underline{m}'-\underline{m}|$.

This means that the small divisor has to be bounded from below by a slow decreasing function of the distance $|n-n'| + |\underline{m}-\underline{m}'|$.

For this reason we define:

$$B^* = \{k\in B;\ |n-n' + \alpha(\underline{k}+\underline{m})^2 - \alpha(\underline{k}+\underline{m}')^2 + g(\underline{k}+\underline{m})-g(\underline{k}+\underline{m}')| \geq \frac{\gamma}{(|n-n'|+|\underline{m}-\underline{m}'|)^\sigma}\} \tag{6}$$

The problem is now to show that B^* is big enough.

To see this, assume first that $g = 0$. The, B^* is obtained from B by removing a sequence $B^c_{\underline{j},\underline{m},n}$ of strips defined by:

$$B^c_{\underline{j},\underline{m},n} = \{k\in B;\ |n+ \alpha(\underline{k}+\underline{j})^2- \alpha(\underline{k}+\underline{j}+\underline{m})^2| \leq \frac{\gamma}{(|n|+|\underline{m}|)^\sigma}\} \tag{7}$$

It happens that if B is contained in the Brillouin zone $B_\nu=[-\frac{1}{2},\frac{1}{2}]^\nu$, for $\underline{m},n$ given, there is only a finite number $A_{\underline{m},n}$ of $\underline{j}$'s for which $B^c_{\underline{j},\underline{m},n}$ is not empty, and $A_{\underline{m},n} \leq$ const $|\underline{m}|$. On the other hand, the volume of $B^c_{\underline{j},\underline{m},n}$ is of order $\gamma(|n|+|\underline{m}|)^{-\sigma}|\underline{m}|^{-1}$

For each $(\underline{m},n)$, the Lebesgue measure of $\bigcup_{\underline{j}} B^c_{\underline{j},\underline{m},n}$ is therefore bounded by

$$\Big|\bigcup_{\underline{j}} B^c_{\underline{j},\underline{m},n}\Big| \leq \text{Const}\ \frac{\gamma}{(|n|+|\underline{m}|)^\sigma} \tag{8}$$

Then summing on $(\underline{m},n)$, we get: ($|\cdot|$ means Lebesgue measure)

$$|B\backslash B^*| \leq \text{Const}\ \gamma \quad ,\text{provided}\quad \sigma > \nu \tag{9}$$

(Here the constant depends only on ν,α,σ).

If γ is small, we therefore remove only a small amount from B to get B^*.

If $g \neq 0$ is Lipshitz continuous, the same estimate can be performed.

Step 3: Estimates on W and performing the first order.

From (5) we get $\forall\ 0 < \rho < r$

$$|W(\underline{k}+\underline{m},n'-n,\underline{m}'-\underline{m})| \leq K(\sigma,\nu,\alpha)\ \frac{\|\hat{V}\|_{B,r}}{\gamma\,\rho^\sigma}\ e^{-(r-\rho)(|n-n'| + |\underline{m}-\underline{m}'|)} \tag{10}$$

If we want to estimate the Lipshitz norm with respect to k, we have the same kind of estimate. Lipshitz continuity of g is necessary to perform the estimate (8).

Here $||\hat{V}||_{B,r}$ is the smallest constant such that

$$|\hat{V}(\underline{k}+\underline{m};\ n'-n,\ \underline{m}'-\underline{m})| \leq ||\hat{V}||_{B,r}\ e^{-r(|n-n'|+|\underline{m}-\underline{m}'|)} \qquad \begin{array}{c}\underline{k}\in B\\ (\underline{m},n);(\underline{m}',n')\in\mathbb{Z}^{\nu+1}\end{array} \tag{11}$$

Now to perform the first order we pick

$$R = e^{W}\ . \tag{12}$$

and we get:

$$R(K_o + \hat{V})R^{-1} = K_o + \delta K_o + \hat{V}^* \tag{13}$$

with:

$$||\hat{V}^*||_{B^*,r-\rho} \leq \frac{K_2}{\gamma\rho^{\sigma}}\ ||\hat{V}||^2_{B,r} = \frac{0(\varepsilon^2)}{\gamma\rho^{\sigma}} \tag{14}$$

Thus $\hat{V}^* = o(\varepsilon^2)$; the divergent factor $\frac{1}{\gamma\rho^{\sigma}}$ is the result of the small divisors.

Step 4. Recursion and Convergence.

From this algorithm, we construct a sequence of operator:

$$K_o + \hat{V} \to K_1 + \hat{V}_2 \to K_2 + \hat{V}_2 \to \ldots \to K_\ell + \hat{V}_\ell\ .$$

such that:

$$K_\ell \text{ is diagonal of the form } = \delta_{nn'}\delta_{\underline{mm}'}(n + \alpha(\underline{k}+\underline{m})^2 + g_\ell(\underline{k}+\underline{m})) \tag{15}$$

On the other hand:

$$K_\ell = K_{\ell-1} + \delta K_{\ell-1} \qquad\qquad \hat{V}_\ell = \hat{V}^*_{\ell-1} \tag{16}$$

These objects are functions of k on a sequence of subsets

$$B_o = [\tfrac{1}{2},\tfrac{1}{2}]^{\nu} \to B_1 \to \ldots \to B_\ell$$

where (17)

$$B_\ell = B^*_{\ell-1}$$

and in the definition of B_ℓ, we need a constant $\gamma_\ell > 0$, which gives its measure.

At best we get a sequence $r_0 > r_1 > \ldots > r_\ell$ of numbers describing the rate of exponential decreasing of V_ℓ:

If we denote

$$\varepsilon_\ell = \|\hat{V}\|_{B_\ell, r_\ell} . \tag{18}$$

we get:

$$\varepsilon_{\ell+1} \leq K_- \frac{\varepsilon_\ell^{\;2}}{\gamma_\ell (r_\ell - r_{\ell+1})^\sigma} \tag{19}$$

$$|B_\ell \backslash B_{\ell+1}| \leq K_2 \;\; \gamma_{\ell+1} \tag{20}$$

We can choose γ_ℓ , as well as $r_\ell - r_{\ell+1}$ of the form $(\mathrm{Log}\, \frac{1}{\varepsilon_\ell})^{-\beta}$ for some β . Because in this case, from (19) it follows that

$$\varepsilon_{\ell+1} \leq (\mathrm{Ln}\, \frac{1}{\varepsilon_\ell})^{\beta'} \; \varepsilon_\ell^{\;2} \qquad \text{for some } \beta' > 0 \quad .$$

$$\Rightarrow \quad \varepsilon_\ell \leq D(C\varepsilon_0)^{(3/2)^\ell} \tag{21}$$

Therefore ε_ℓ converges extremely rapidly, so rapidly that

$$\sum_\ell \; (\mathrm{Ln}\frac{1}{\varepsilon_\ell})^{-\beta} < + \infty \tag{22}$$

which means that provided <u>ε_0 is small enough</u> we get the result, namely:

(i) $r_\ell \to r_\infty > 0$.

(ii) $\bigcap_{\ell=0}^{\infty} B_\ell = B_\infty$ satisfies $|B_\infty| \geq 1 - \varepsilon$.

(iii) $\|\hat{V}_\ell\|_{B_\ell, r_\ell} \to 0$.

(iv) $\prod R_\ell = \prod e^{W_\ell}$ converges in the norm $\|\cdot\|_{B_\infty, r_\infty}$.

IV. The instability problem: The long range tunnelling effect

We now better understand what could happen when the small divisor problem is so bad that the kind of perturbation theory we previously described does not work. It means that certain eigenvalues of the unperturbed problem are so closed, that they cannot be compensated, in the first order perturbation theory by the local structure of the interaction.

In quantum mechanics we know that such strong resonance effect usually produces a tunnelling effect.

In the models we are looking for, the eigenvalues are attached to a quantum number which corresponds to a localization of the corresponding state in some space. In condensed matter physics, this space is nothing but the usual space in the lattice describing the position of atoms. In other situations, this space is better understood in terms of momentum, or reciprocal lattice.

If one considers a model of this type, for instance on $\mathbb{Z}$:

$$H\psi(n) = 2 \cos 2\pi(x-n\alpha)\, \psi(n) + \mu(\psi(n+1) + \psi(n-1)) \tag{1}$$

which is the Harper equation [22] or also the Almost Mathieu model [23, 24], we can produce examples of values of x or α such that an infinite sequence $(n_k)_{k \in \mathbb{Z}}$ exists for which

$$|\cos 2\pi\,(x - n_k\,\alpha) - \cos 2\pi\,(x - n_{k+1}\alpha)| \le \frac{1}{|k|^{|n_k|}} \quad \forall\, k \tag{2}$$

We thus get a very strong resonance condition between an infinite subset of eigenvalues, spreading everywhere, in a non bounded region.

For such a sequence, we wait for a long range tunneling effect, leading presumably to a special kind of extended states.

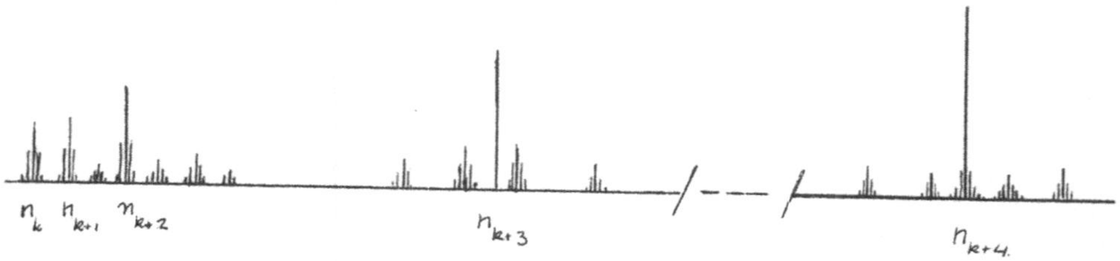

These states have been actually observed in the Maryland model (see Prange's lecture this issue[12b]) where $2 \cos 2\pi x$ is replaced by $\tan \pi x$ provided α is a Liouville number, namely:

there is an infinite sequence $\frac{p_n}{q_n}$ with

$$\left|\alpha - \frac{p_n}{q_n}\right| \le \frac{1}{n^{q_n}} \qquad \forall\, n \tag{3}$$

They exhibit scaling properties, because in this case the q_n's give several scales which are extremely rapidly divergent.

Such extended states have also been observed by Kadanoff et al., and Ostlund et al. [25] using a model close to the Almost Mathieu equation.

The characteristics of these states should be the following:

1) They are localized in means namely

$$\frac{1}{2N+1}\sum_{n=-N}^{n}|\psi(n)| \xrightarrow[N\to\infty]{} 0 \qquad (4)$$

2) The peaks are extremely sparce and very narrow when we go to infinity. It could happen however that

$$\limsup_{n\to\infty} |(\psi(n)| = \infty \qquad (5)$$

3) The location on the amplitude of the peaks is extremely sensitive to the parameters entering into the game. In example 1) and 2) we get extreme sensitivity either with respect to x or to α. In this sense, we can speak of chaotic states rather than exotic [25,26].

4) They are usually associated to a singular continuous spectrum. This means that the spectral measure, defined e.g. by:

$$\int d\mu_\phi(E)\frac{1}{E-z} = \langle\phi|\frac{1}{H_x-z}\phi\rangle \qquad \phi\in \mathcal{H} \qquad (6)$$

has no Dirac mass (the measure of any point is zero) but $\frac{d\mu_\phi(E)}{dE}$ is almost everywhere (in the sense of Lebesgue) either zero or infinite provided ϕ is a vector in the space generated by these states.

5) They produce usually a strong recurrent behavior for the time evolution, in the sense that if ϕ is a normalized vector in the space generated by such states, then the time spent by this vector, after evolution under H_x, in each region is infinite, but the mean time is zero. Everything looks as if the wave packet described by $\phi(t) = \exp(-itH)\phi$ goes far away, is strongly reflected by the potential, and then comes back far to the left, is still reflected, then comes back again farther to the right and so on, mimicking a slowly diffusing random walk.

6) It seems very hard to make a distinction between these states and localized ones, simply because they are extremely sparse and with very little fluctuation they can immediately be transformed into localized ones. On the other hand, the diffusion distance can go as slowly as $(\text{Log } |t|)^{\alpha}$ for some power α, and as we go to numbers, in real physics this means that the particle can at most move from less than a 100, or 1000 lattice spacing during the age of the universe!

Nevertheless, the phenomena of long range tunnelling can sometimes become dominant. In the Almost Mathieu equation, $\mu = 1$ is a critical value between a localized regime, (insulating phase) and a Bloch wave regime (metallic phase).

Numerical calculations of D. Hofstadter [27a] shows that the wave function looks like a Bloch wave at short distance; on the other hand, the Aubry duality argument [27b] shows that this probably is not the case globally. Numerical calculation shows that the spectrum has zero Lebesgue measure at $\mu = 1$, in favor of a singular continuous spectrum. A long range tunneling effect should be a better explanation for these states.

The other interest of this phenomenon is that it can provide perhaps a mechanism in classical mechanics to understand the structure of chaotic orbits in the presence of the invariant torii.

The quantum analog of our invariant torus is just an eigenvalue and the corresponding localized state. We can see the localization region as the value of the rotation number of the corresponding torus. Thus torus is just the spatial aspect of the phenomenon, whereas localization takes place in Fourier analysis (space of rotation number, or action variables).

A long range tunnelling effect means an orbit which is attracted by an infinite set of torii with rotation numbers $\omega_1 < \omega_2 < \ldots < \omega_n$ growing rapidly as $n \to \infty$. By a Fourier analysis this scale of frequencies would produce a motion which is attracted by a sequence of torii. However, if the sequence were fixed, we would have simply a quasi-periodic motion. But a high sensitivity of the sequence of rotation numbers with respect to the initial condition should take place somewhere, as in quantum mechanics, mimicking a chaotic behavior in jumping from one torus to another one.

Obviously these jumps are not allowed for systems with two degrees of freedom, if we have invariant torii. However, the work of S. Aubry [6] and J. Mather [7], together with the analysis of Rheinhardt et al. [8a] show that there are broken torii in these systems, called Mather sets or Cantorii, which allows the orbit to jump.

We don't want to go farther in this direction because it involves too much speculating, and this picture can be wrong.

However, in the last section is an explicit example of a quantum mechanical model producing some of the features described in this section.

III. The Q. M. operator [26]

The quadratic mapping (QM) hamiltonian is an operator acting on $\ell^2(\mathbb{N})$ via:

$$H_{QM}\ \psi(n) = R_{n+1}^{1/2}\ \psi(n+1) + R_n^{1/2}\ \psi(n) \tag{1}$$

The R_n's are hopping coefficients for the diffusion of an electron on a line. To understand the nature of these coefficients, we ask H_{QM} to satisfy an exact renormalization group equation. Namely let D be the distilation operator given by:

$$D\psi(n) = \psi(2n) \tag{2}$$

Clearly D is not a unitary operator, but it is a partial isometry because:

$$D^*D = P \qquad DD^* = \mathbb{1} \tag{3}$$

where P projects on the even numbers $\ell^2(2\mathbb{N})$.

The renormalization group equation is :(for some $\lambda \in \mathbb{R}$) .

$$H_{QM}D = D(H_{QM}^2 - \lambda) \tag{4}$$

If we insist on having H_{QM} self adjoint, we can show that $\lambda \geq 2$. The special case $\lambda = 2$ corresponds to $R_n = 1$ $(n \geq 2)$.

Using (1) and (4) we immediately get:

$$R_o = 0 \qquad R_{2n} + R_{2n+1} = \lambda \qquad R_{2n-1}R_{2n} = R_n \tag{5}$$

This is a recursion relation with a very long memory.

From this recursion relation we can prove that $(R_n)_{n \geq 0}$ is an almost periodic sequence whose frequencies are the rational dyadic numbers. Moreover, one gets:

$$0 < R_{p2^k} < \frac{R_p}{(\lambda-1)^k} \qquad (\lambda > 2) \tag{6}$$

showing that if $\lambda > 2$, the hopping becomes arbitrarily small at points $n = p2^k$.

Using now the renormalization group equation (4) we can prove that the spectral measure with respect to the vector e_o $(e_o(n) = \delta_{n,o})$. defined by:

$$\int d\mu\,(E)\,(E-z)^{-1} = \langle e_o | (H_{QM} - z)^{-1}\ e_o \rangle \tag{7}$$

is <u>both</u> F and F^{-1} invariant, where $F(z) = z^2 - \lambda$.

Since this measure is supported by a compact subset of $\mathbb{R}$, its support K must be F and F^{-1} invariant.

It is known from Julia [28] that such a set is unique; it is <u>a Cantor set</u> of <u>zero Lebesgue measure</u> and it can be described via a coding namely

$$E \in K \Longleftrightarrow \exists\ \underline{\sigma} = (\sigma_o, \sigma_1, \ldots, \sigma_n, \ldots) \qquad \sigma_n = \pm 1 \quad \text{with}$$

$$E = E(\underline{\sigma}) = \lim_{n \to \infty} \sigma_o \sqrt{\lambda + \sigma_1 \sqrt{\lambda + \ldots + \sigma_n \sqrt{\lambda}}} \tag{8}$$

Moreover, we get:

$$F(E(\underline{\sigma})) = E(\underline{\sigma})^2 - \lambda = E(T\underline{\sigma}) \qquad \text{(a)}$$

$$T\underline{\sigma} = (\sigma_1, \sigma_2, \ldots, \sigma_{n+1}, \ldots) \text{ is the one side shift.} \qquad \text{(b)} \tag{9}$$

In addition, thanks to this coding, we can compute explicity the measure μ.

$$\int_K d\mu\, f(E) = \int \underline{d\sigma}\ f(E(\underline{\sigma})) \tag{10}$$

where <u>$d\sigma$</u> affects probability ½ for σ_n to be +1 or -1, and the σ_n's are independent random variables (coin tossing process).

As a conclusion μ is certainly singular continuous because it obviously gives zero measure at each point of K, and it is supported by a set of zero Lebesgue measure.

We can also compute the density of state defined as:

$$k(E) = \lim_{N\to\infty} \frac{\#}{N} \{\text{eigenvalues of } H_{QM} \restriction_{[0,N-1]} \leq E\} \tag{11}$$

It is related to the spectral measure μ via the expression:

$$k(E) = \lim_{N\to\infty} \frac{1}{N} \sum_{n=0}^{N-1} \langle e_n | \chi_{\leq E}(H_{QM}) \; e_n \rangle \tag{12}$$

where $\chi_{\leq E}(x) = 0$ if $x > E$, $\chi_{\leq E}(x) = 1$ if $x \leq E$.

and $\epsilon_n(n') = \delta_{n,n'}$ (13)

Using again the renormalization group equation (4) we can prove that the measure $dk(E)$ defined by k is both F and F^{-1} invariant and leads therefore to

$$k(E) = \int_{-\infty}^{E} d\mu(E') \tag{14}$$

From the Thouless—Herbert-Jones formula [29] , one can compute the Lyapounov exponent:

$$\gamma(E) = \lim_{n\to\infty} \frac{1}{2n} \operatorname{Ln} \left(\frac{|\psi(n)|^2 + |\psi(n+1)|^2}{|\psi(0)|^2 + |\psi(1)|^2} \right) \tag{15}$$

if $H_{QM} \; \psi(n) = E \; \psi(n)$ (16)

Then it happens that $\gamma(E)$ is uniformly bounded in any compact subset of $\mathbb{C}$; moreover, from (4) and (14)

$$2\gamma(E) = \gamma(E^2 - \lambda) \tag{17}$$

In particular, if $E \in K$, $\gamma(E) = 0$, leading to the conclusion that probably the states are extended.

This is actually the case. For if ψ satisfied (16) , then

$$\psi(n) = \frac{P_n(E)}{\sqrt{R_1 \ldots R_n}}$$

where the P_n's are the orthonormal polynomials related to μ :

$$P_o = 1 \ , \quad P_1(E) = E \ , \quad P_{n+1}(E) = EP_n(E) - R_n P_{n-1}(E)$$
$$\int d\mu \ P_n P_{n'} = 0 \quad \text{if} \quad n \neq n' \tag{19}$$

Thanks to (4) we get immediately

$$P_{2n}(E) = P_n(E^2 - \lambda) \tag{20}$$

leading to the relation: (using (5)).

$$\psi(p2^k) = \frac{P_p(F^k(E))}{\sqrt{R_1 \dots R_p}} \tag{21}$$

If $E \in K$, namely in the spectrum, we get:

$$\psi(p2^k) = \frac{P_p(E(T^k\underline{\sigma}))}{(R_1 \dots R_p)^{1/2}} \qquad \text{some } \underline{\sigma} \tag{22}$$

Therefore as $E(\underline{\sigma})$ is choosen randomly in the spectrum with the probability given by μ , the value of ψ at points $p2^k$ are choosen randomly. Choosing $p = 1$, we can prove that $\liminf_{k \to \infty} |\psi(2^k)| > 0$.

Then ψ is likely to have a chaotic behavior and is obviously extended.

However, as remarked by Kohmoto [30] , selecting a subsequence of points like $p2^k$ in the lattice does not allow to conclude that the wave function is chaotic. For if $\psi(n) = \cos 2\pi\alpha n$, for instance, ψ is almost periodic, $\psi(p2^k)$ is a random sequence with respect to k.

In our case, nevertheless, we can prove rigorougly that ψ cannot be a Bloch wave as far as $\lambda > 2$, namely there exists no sequence $\chi(n)$, such that

(1) $\psi(n) = e^{kn} \chi(n)$ for some k and all $n \in N$

(2) $\chi(n)$ has the same almost periodicity as R_n

Thus this model exhibits most of the features described in the previous section. We get a new kind of extended state. The recurrence phenomenon can be seen by looking at the evolution of a wave packet.

If we want to compute the time spent by the function

$$e^{-iH_{QM}t} e_o$$

at the site n, for instance, we need the computation of

$$C_{n,0}(t) = |\langle e_n \, e^{-iH_{QM}t} e_o \rangle|^2 = \left|\int d\mu(E) e^{-itE} P_n(E)\right|^2 \tag{23}$$

For $n = 0$, $P_o(E) = 1$, and we just get the Fourier transform of μ

Since μ is a continuous measure, Wiener's criterion tells us that:

$$\lim_{T \to \infty} \frac{1}{T} \int_0^T C_{n,0}(t)\, dt = 0 \qquad \forall n \geq 0 \tag{24}$$

which means that the <u>mean time</u> spent by the wave packet in each finite region is zero.

On the other hand, since $d\mu$ is singular continuous, the Parseval formula shows that

$$\lim_{T \to \infty} \int_0^T C_{n,0}(t)\, dt = \infty \tag{25}$$

Thus the particle spent an infinite time in any finite region, which means that it comes back to n infinitely often. This is the <u>recurrent behavior</u>.

We can presumably conjecture that $\lim_{t \to \infty} \sup C_{n,0}(t) > 0$, which means that $C_{n,0}(t)$ does not vanish at infinity.

References

[1] a) Kolmogoroff, in "Foundation of Mechanics," by R. Abraham, and J. E. Marsden, Benj. Mass. 1967, p. 263.
b) V. L. Arnold, Usp. Mat. Nauk 18, 5, 13-40(1963), 18, 6, 91-192(1963).
c) J. Moser,"Stable and Random Motion in Dynamical Systems," Princeton University Press (1973).
d) G. Gallavotti, "The elements of Mechanics," Springer-Verlag, New York (1982).

[2] G. D. Birkhoff, "Dynamical Systems," Providence, Rhode Island (1966).

[3] J. Moser, Nachr. Acad. Wiss. Göttingen, Math. Phys. Kl, 11a, no. 1 (1962), 1-20.

[4] a) H. Rüssmann, Preprint (1983).
b) M. Herman, Unpublished Result.

[5] a) J. M. Greene, J. Math. Phys. 20 1183 (1979).
b) F. Doveil, M. Escande, J. Stat. Phys. 26, 257 (1981).

[6] S. Aubry, Y. Le Daeron, "The discrete Frenkel-Kontorora Model and its extensions." To appear in Physica D (1983), and J. Phys. C (1983).

[7] J. Mather, "Existence of quasi-periodic orbits in the twist homeomorphisms of the annulus, " Preprint Princeton (1981). "A criterion for existence of non invariant circles," Preprint Princeton (1982).

[8] a) R. B. Schirts and W. P. Rheinhardt, J. Chem. Phys. 77 (10) (1982) 5204-5217.
b) W. P. Rheinhardt, this issue.

[9] S. Smale, "Differential dynamical Systems," Bull. Amer. Math. Soc. 73 (1967) 747-817.

[10] a) B. Simon, "Almost Periodic Schrödinger Operators," Adv. Appl. Math. (1983).
b) J. Bellissard, in Lect. Notes in Physics no. 153 (1982) Springer.

[11] a) G. Casati, B. V. Chirikov, F. M. Izrailev, J. Ford, In "Stochastic behavior in Classical and quantum hamiltonian systems," Lecture Notes in Physics, no. 93, p. 334 (1979).
b) B. V. Chirikov, " A universal Instability of Many dimensionless Oscillator System," Physics Reports (1979).
c) F. M. Izrailev, D. L. Shepelyanski, Theor. Nat. Fiz. 43 417 (1980).
d) B. V Chirikov, F. M. Izrailev, D. L. Shepelyanski in Soviet Scientific Review, Section C. Vol. 2 (1981).
e) G. Casati, I. Guarneri, Phys. Rev. Lett 50 640 (1983).
f) B. Dorizzi, B. Grammaticos, Y. Pomeau "The periodically kicked Rotator: Recurrence and/or energy growth," Preprint (1983) and this issue.

[12] a) S. Fishman, D. R. Grempel, R. E. Prange, Phys. Rev. Lett. 49,509 (1982).
b) S. Fishman, D. R. Grempel, R. E. Prange, this issue.
[13] B. Souillard, this issue.
[14] T. Hogg, B. A. Huberman, Phys. Rev. Lett. 48, 711 (1982).
[15] J. Avron, B. Simon, J. Funct. Anal. 43, 1-31 (1981).
[16] a) D. Ruelle, Nuovo Cimento, 61A, 655 (1969).
b) G. D. Amrein, V. Georgescu, Helv. Phys, Acta 46, 635 (1973)
[17] H. Rüssmann, in Lecture Notes in Phys., Vol. 38 (1975) pp. 598-624.
[18] M. Hermann, Publication de l'IHES, Paris, #49 (1979).
[19] J. Bellissard, "A-C-Stark effect in a perfect crystal," Preprint Marseille (1983).
[20] See e.g. L. Kuipers, H. Neiderreiter, "Uniform distribution of sequences," (Wiley, New York (1974)).
[21] a) J. S. Howlands, Math. Ann. 207 (1976) 315-335.
b) K. Yajima, J. Kitada, Duke Math. J. 49 (1982) 341-376.
[22] P. G. Harper, Proc. Roy. Soc., London, A68, 874 (1955).
[23] J. Bellissard, R. Lima, D. Testard, Comm. Math. Phys. 88, 207 (1983).
[24] J. Avron, B. Simon, Bull. A.M.S., 6, 81-86 (1982).
[25] a) M. Kohmoto, L. P. Kadanoff, C. Tang, Phys. Rev.Lett. 50, 1870 (1983).
b) S. Ostlund, R. Pandit, D. Rand, H. Schelinhuber, E. D. Siggia, Phys. Rev. Lett. 50, 1873 (1983).
[26] J. Bellissard, D. Bessis, P. Moussa, Phys. Rev. Lett. 49, 701-704 (1982).
[27] a) D. R. Hofstadter, Phys. Rev. B14, 2239 (1976).
b) S. Aubry, G. André, Ann. Israel Phys. Soc. 3, 111 (1980).
[28] See H. Brölin, Ark. Mat. 6, 103 (1965).
[29] a) D. Herbert, R. Jones, J. Phys. C4, 1145-1161 (1971).
[30] I thank Dr. Kohmoto for this remark.

THE RELATION BETWEEN QUANTUM MECHANICS AND CLASSICAL

MECHANICS: A SURVEY OF SOME MATHEMATICAL ASPECTS

Sergio Albeverio
Mathematisches Institut
Ruhr-Universität Bochum
4630 Bochum 1
Fed. Rep. Germany

Teresa Arede *
Centre de Physique Théorique
CNRS, Luminy, Case 907
F-13288 Marseille
Cedex 9

Abstract

We give a survey of some mathematical work on the relation between quantum mechanical quantities like eigenvalues and eigenfunctions of Schrödinger operators and classical mechanical quantities, like classical actions computed along classical paths and lengths of geodesics. In particular we discuss the distribution of eigenvalues for a domain $\mathbb{R}^d$ or a Riemannian manifold. We also single out the manifolds for which the heat kernel and the spectrum of the Laplacian are given entirely by (the lengths of) geodesics, i.e., by classical orbits.

1. INTRODUCTION

The study of the relation between quantum mechanics and classical mechanics, aspects of which have been covered by this Meeting, has been a lively subject since the discovery (1899) of Planck's radiation law. Part of the mathematical results have taken the natural form of basic theorems on the asymptotics of certain partial differential equations. In these asymptotics important relations between symmetry groups on the quantum and classical level arise, giving rise to a fascinating interplay between analysis and geometry ("spectral geometry", see for example [1-5]). In addition, the relation between quantum and classical mechanics is basic to such disparate fields as functional integration (see for example [6-13, 201]), stochastic mechanics (see for example [14-16]), geometrical

* On leave of absence from: Faculdade de Engenharia (DEMEC), Universidade do Porto, R. dos Bragas, 4000 Porto, Portugal.

quantization (see for example [17]), and see also the Proceedings of the preceding Como Meeting [18].

These relations are far from being fully understood and are in a sense at the core of the quantum problem (for those at least that have not given up entirely the hope of understanding our physical world in classical or at least stochastic terms [19].

It is also fascinating that there are systems (like free motion on certain manifolds) where the quantum and classical worlds are intimately connected, and a study of such systems and of their extensions to infinitely many degrees of freedom (quantum fields) appears very interesting also as an alternative to other attempts at constructing quantum field models; for a related point of view see also [20]. Of course a better understanding of the relations between classical and quantum systems will also eventually clarify "quantum chaos" versus "classical chaos". In this short survey we briefly expose some basic facts and techniques giving information on the distribution of eigenvalues and on related quantities of Schrödinger operators, in the case of a domain of the configuration space $\mathbb{R}^d$ or a manifold and about the classical limit of these and related quantum mechanical quantities. We shall mainly be concerned with the case of one particle moving under the influence of a potential (or equivalently of two particles interacting by a two-body potential). We stress somewhat the case of manifolds because it is the natural mathematical framework for the type of results we shall discuss, and also in view of the potentialities mentioned above. Of course, due to the limitations of space and time, this will be little more than a list of topics of interest. We invite the interested reader to look up the references for more information (incidentally, because of the same limitations, the references are far from being exhaustive. Reference X should be understood as "reference X and references given therein". We have not refrained from giving a newer comprehensive reference rather than an older, historically more appropriate but perhaps "isolated" reference).

Two basic observations that will lead us are the following: the asymptotics of the Schrödinger operator in the free case, in the case of the configuration space $\mathbb{R}^d$ (or a manifold), is the study of the operator $e^{-\frac{i}{\hbar}t(-\frac{\hbar^2}{2m}\Delta)}$ as $\hbar \downarrow 0$, (m mass, $\hbar$ Planck's constant divided by 2π, Δ the Laplace-Beltrami (free energy) operator); hence it is equivalent to the study of $e^{\frac{i\hbar t}{2m}\Delta}$ as $t \downarrow 0$, which in turn is closely related to the study of the "heat semigroup" $\exp(ht\Delta/2m)$ as $t \downarrow 0$. Precisely this kind of asymptotics has been studied intensively in the literature. The second observation is that in the case of $\mathbb{R}^d$, the kernel $e^{t\Delta}(x,y)$ of $e^{t\Delta}$ (equal to the one above for $\hbar = 1$, $m = 1/2$) is given for <u>all</u> times by lengths of geodesics (straight lines in this case) between $x,y\in\mathbb{R}^d$, namely $e^{t\Delta}(x,y) = (4\pi t)^{-d/2}\exp((-4t)^{-1}|x-y|^2)$ (and correspondingly for the Schrödinger kernel, i.e., the Green's function of the time dependent Schrödinger equation). By these observations we are induced to link the study of the classical limit of quantum mechanics with the study of the asymptotics for small times t of heat semigroups and, on the other hand, to look for a general class of manifolds on which the heat semigroups is given entirely by length of geodesics (or more generally, by classical orbits).

The contents of the different sections is as follows:

In Section 2 we expose shortly some results on the distribution of eigenvalues for Schrödinger operators on a domain of $\mathbb{R}^d$. In particular we recall some results on the "Weyl-Carleman asymptotics" on the number $N(\lambda)$ eigenvalues $\leqq \lambda$ as $\lambda \to \infty$. We also mention the relation of $N(\lambda)$ with the so-called θ-function associated with the spectrum: $\theta(t) = \Sigma e^{-t\lambda n}$. The behavior of $\theta(t)$ for $t \downarrow 0$ is related to the one of $N(\lambda)$ as $\lambda \to \infty$ (by Karamata-Tauber type theorems). In the case where the λn are the eigenvalues of the negative of the Laplacian, we have, by the fact that $\theta(t) = \text{trace } e^{t\Delta}$, that the study of the limit of $\theta(t)$ for $t \downarrow 0$ is also related to the one of the classical limit $\hbar \downarrow 0$ of the partition function of a quantum system at inverse temperature $\beta = t$. We give some results and references on this circle of problems.

In Section 3 we mention some basic results on the distribution of eigenvalues and eigenfunctions for Schrödinger operators in the case where the configuration space is a Riemannian manifold. In particular the results on the asymptotics of $\theta(t) = \Sigma e^{-t\lambda n}$, as $t \downarrow 0$, then giving information "of Weyl type" on $N(\lambda)$, are mentioned. Also we mention results in the case of compact manifolds, some non compact manifolds and some symmetric spaces.

Section 4 is divided into 4 parts. The two first points 1) and 2) are separated becuase of the different methods used. In part 1) results relating to "Feynman path integrals" type methods are mentioned, whereas in part 2) we discuss results obtained by "Wiener (or Euclidean) path integrals", or more generally by probabilistic methods.

These results give relations between solutions of the Schrödinger equation (and associated eigenvalues) and classical quantities. In part 3) we indicate a class of cases for which "the semi-classical approximation" is exact. We conclude the exposé, part 4) with some remarks about classical versus quantum chaos.

2. DISTRIBUTION OF EIGENVALUES FOR A DOMAIN OF $\mathbb{R}^d$.

We shall mainly be concerned with Schrödinger operators with potentials falling in some suitable sense to zero at infinity or, in the case of N-particle Schrödinger operators, with two-particle potentials going to zero at infinity. This is the case usually considered in traditional scattering theory and we speak in this case for shortness of potentials "of the first type". In the last few years there has been a great interest and many new developments concerning the case of potentials, which we call here for short "the second type", which are periodic, or quasi-periodic or stochastic (ergodic random fields). In all these case, of course, one does not have decrease to zero at infinity and the type of spectral results is very different in this case (typically for $-\Delta + V$ with V of such a form that the spectrum consists of "energy bands", whereas if V is in the above first class so that $V \to 0$ "at infinity" then typically $-\Delta + V$ has essential spectrum at interval (E_0,∞) with eigenvalues accumulating at most thresholds.

The very interesting situation of potentials of the above second type is not considered in this Lecture since there have been at this conference excellent surveys on this topic (by Jean Bellissard, Fabio Martinelli and Bernard Souillard), to which we can refer. See also for example [59,60].

Let us first consider the case of the operator $H = -\Delta + V$ in $L^2(\mathbb{R}^d,dx)$, with V a real-valued function. H is the Hamiltonian for a quantum mechanical particle, in Euclidean d-dimensional space $\mathbb{R}^d$, under the force $-\text{grad } V$. Under suitable assumptions on V, including $V \to 0$ at infinity in a suitable sense, one proves (see for example [21], especially Vol. IV), that the spectrum of $-\Delta + V$ is purely and absolutely continuous in $(0,\infty)$ ("pure scattering states"), with possible negative lower bounded eigenvalues λn. What about the distribution of these eigenvalues, i.e., what about the number $N(\lambda) \equiv \sum_{\lambda_n \leq \lambda} 1$ of eigenvalues $\leq\lambda$ as $\lambda \to 0$?

The ancestor of this problem was given a celebrated solution by H. Weyl in 1911 (the problem was even older, see the historical notes in [22], e.g., p. 18, and also [30]; Weyl's result was conjectured by M. A. Lorentz in 1910). Weyl considered in fact the case of the operator $-\Delta$ in a bounded regular domain Ω of $\mathbb{R}^d$, with a smooth boundary Ω with Dirichlet resp. Neumann boundary conditions. More generally for Ω such a domain in $\mathbb{R}^d$, we have that the spectrum of $-\Delta$ consists of isolated eigenvalues and the interesting asymptotics is of $N(\lambda)$ as $\lambda \to \infty$. Weyl's well-known fundamental result is in this case:

$$N(\lambda) \backsimeq \lambda^{d/2} \frac{2}{d} 2^{-d} \pi^{-d/2} \Gamma(\tfrac{d}{2})^{-1} \text{Vol}(\Omega) = \lambda^{d/2} \frac{\omega_d}{(2\pi)^d} \text{Vol}(\Omega) \text{ as } \lambda \to \infty, \quad (1)$$

with Γ the gamma function, i.e.,

$$\Gamma(\frac{m}{2}) \equiv \begin{cases} (k-1)! & \text{if } m = 2k \\ 2^{-2k}(2k)!\ \pi^{1/2}k! & \text{if } m = 2k+1 \end{cases}$$

, where k is a positive integer, $Vol(\Omega)$ is the Euclidean volume of Ω, and $\omega_d = (2\pi)^d[d2^{d-1}\pi \pi^{d/2-1}(\frac{d}{2})]^{-1}$ is the volume of the unit ball in $\mathbb{R}^d$. It is easy to see and well-known that this can be written as

$$N(\lambda) \asymp (2\pi)^{-d}Vol\{(x,p)\varepsilon\Omega \times \mathbb{R}^d | p^2 \leq \lambda\}, \text{ as } \lambda \to \infty, \quad (2)$$

thus $N(\lambda)$ behaves asymptotically and essentially as the volume of the available region in phase space.

A corresponding result on the eigenfunctions was proven by Caleman (1934), see below.

Let us call $N_o(\lambda)$ the above leading term. For d = 2 it is for example known, for "nice" (Jordan) domains (Courant) that

$$N(\lambda) = N_o(\lambda) + O(\sqrt{\lambda}\ \ln\lambda), \quad (3)$$

with improvements for polygonal domains (also in higher dimension) or domains for which one has separation of variables, see [22] p. 23, e.g., for a square $N(\lambda) = N_o(\lambda) \mp \frac{\sqrt{\lambda}}{4\pi} \ell(\partial\Omega) + O(\sqrt{\lambda})$, ℓ standing for length, - for Dirichlet and + for Neumann conditions. Incidentally, such estimates have nice connections with the number of lattice point problems in number theory, see [22-24]. For $d \geq 3$ one has corresponding results (a classic being Courant mini-max one) concerning Schrödinger operators of the form $-\Delta + \lambda V$ in $\mathbb{R}^d$, $d \geq 3$; a typical one being for example, if $V\varepsilon L^{d/2}(\mathbb{R}^d)$ for $d \geq 3$,

$$N(\lambda V) \asymp \lambda^{d/2} Vol(S^d)(2\pi)^{-d/2} \int_{V(x)\leq 0} [-V(x)]^{d/2}dx \quad (4)$$

as $\lambda \to \infty$, $N(\lambda V)$ being the number of eigenvalues in $(-\infty, 0)$ of $-\Delta + \lambda V$ (counted with multiplicity), see [21] (for results for finite λ see also the papers by Glaser, Grosse, Martin, Lieb and Thirring in [25] and [11], [21,26]; see also Section 3). Results of this type have been extended to general elliptic operators in [27-29].

The next remark, used extensively by Carleman in '34 (and in another related context by Wiener '32), is that the study of $N(\lambda)$ for $\lambda \to \infty$ is closely connected, via a Karamata-Tauber theorem, to the behavior of the Laplace transform of the measure $dN(\lambda)$ as $t \downarrow 0$, e.g.,

$$\theta(t) \equiv \int_0^\infty e^{-\lambda t}dN(\lambda) \asymp Ct^{-\alpha} \text{ for some } \alpha > 0,$$

as $t \downarrow 0$ implies $N(\lambda) \approx \frac{Ct^{\alpha}}{\Gamma(\alpha+1)}$ as $\lambda \to +\infty$.

In such a way one get not only control on $N(\lambda)$ but also on the asymptotic behavior of eigenfunctions. The method has strong connections with analytic number theory and the theory of asymptotic distributions in probability. We recall that one has

$$\theta(t) = \sum_n e^{-\lambda_n t} = \text{Trace } e^{t\Delta} = \int_\Omega K_t(x,x)dx, \tag{5}$$

$K_t(x,y)$ being the fundamental solution (Green function) of the heat equation $\partial_t u = \Delta u$, $t \geqq 0$, i.e., $K_t(x,y) = \Sigma\phi_n(x)e^{-\lambda_n t}\phi_n(y)$, ϕ_n being the orthonormalized system of eigenfunctions to $-\Delta$ in $L^2(\Omega,dx)$. For example, the following heuristic argument works: for large volume Ω and small t, with x,y well inside Ω, one has from "the principle of not feeling the boundary" $K_t(x,y) \approx K_t^o(x,y)$, with $K_t^o(x,y)$ the Green function of heat equation in $\mathbb{R}^d$, and should get Weyl's asymptotics via $\theta(t) \approx \int_\Omega K_t^o(x,x)dx = (\text{Vol } \Omega)(4\pi t)^{-d/2}$. For a proof see [30]. The function $\theta(t)$ is called <u>theta function</u> associated with the operator $-\Delta$ in Ω. The study of the behavior of $\theta(t)$ as $t \downarrow 0$ has been pursued quite extensively by Carleman, Pleijel, Minakshisundaram and others and has become a central part of the so-called "geometrical asymptotics" [1-4]. In particular, for $\Omega \subset \mathbb{R}^2$ it has been proven

$$\theta(t) = \text{Vol}(\Omega)(4\pi t)^{-1} \mp \frac{\ell(\partial\Omega)}{8\sqrt{\pi t}} + \beta + O(\exp(-\delta/t)), \text{ with } \beta = \frac{1}{6} \tag{6}$$

for smooth bounded simple connected Ω, and $\beta = \frac{1-r}{G}$ with r the number of "holes" in Ω, if Ω is not simply connected.

Here we see that $\theta(t)$ and hence the spectrum $\{\lambda_n\}$ is not only related to such geometrical-metrical concepts as domain-volume and boundary-length, but also such topological concepts as "number of holes" (mathematically related to Betti numbers). Such expansions change drastically with the geometry and topology of the domain. These considerations have been strongly influenced by a paper of Kac "Can we hear the shape of a drum?" [30] (see also [31,32,43 and 4]). Of course many other detailed questions can be asked, like, for example, the dependence on boundary conditions. One result is for d = 3 (Grinberg, see [21]), e.g.,

$$\sum_{\lambda_n \leqq \lambda} (\lambda_n^N - \lambda_n^N) \frac{\text{Area}(\partial\Omega)}{16} \lambda^2$$

as $\lambda \to \infty$, for $\Omega \subset \mathbb{R}^3$, with λ_n^D resp. λ_n^N the eigenvalues of $-\Delta$ with Dirichlet resp. Neumann boundary conditions on $\partial\Omega$.

Natural extensions of the above type of results have been, still in the case of domains Ω of $\mathbb{R}^d$, basically in two directions:

a) For more general elliptic and hypo-elliptic [33] operators than $-\Delta$. Typical results here are for m - th order elliptic operators, with a coefficient of the highest order term in derivatives continuous and bounded, and coefficients to lower order derivative terms essentially bounded [34,31]. For a case with degeneracies see also [61]. Also Weyl's asymptotic result holds for such operators in a form corresponding to Equation (2), the role of p^2 being played by the principal part of the symbol of the differential operator.

b) For unbounded domains Ω: recent references for the case $-\Delta$ in certain unbounded domains are for example [35,175] (for older references see also [36]). For a phenomenon of "non Weyl behavior" see [62]. A natural operator to be studied when $\Omega = \mathbb{R}^d$ is of course the Schrödinger operator, for a particle in a potential $-\Delta + V$ (and other Hamiltonians of interest, like, for example, the one for a charged particle in an electromagnetic field).

In this case two different situations arise:

α) V is sufficiently increasing at infinity so that the spectrum of $-\Delta + V$ consists only of eigenvalues and $\theta(t) = \Sigma e^{-\lambda_n t}$ can be defined, at least in the sense of generalized functions. In this case the study of the asymptotics $t \downarrow 0$ (and related ones) has been made by techniques which will be mentioned in Section 4a and b.

β) V is not sufficiently increasing at infinity so that the spectrum of $H = -\Delta + V$ has a continuum part. In this case $\theta(t)$ can not be defined in the above way, a natural object to study here being $\hat{\theta}(t) = \text{trace}\,(e^{-tH} - e^{-tH_0})$, with $H_0 \equiv -\Delta$. Under suitable assumptions on V, one obtains asymptotic expansions of $(4\pi t)^{3/2}\hat{\theta}(t)$ in powers of t, with vanishing constant term, see [37] and references therein.

Let us also mention that the study of the behavior of $\theta(t) = \text{Tr}\, e^{-tH}$ (resp. $\hat{\theta}(t)$) as $t \downarrow 0$ as an instrument to determine the behavior of $N(\lambda)$ as $\lambda \to \infty$ can be extended to the study of other functions depending on the spectrum of the relevant Hamiltonian, like, for example, tr $f(H)$ for a suitable function f, e.g., $f(\lambda) = e^{-it\lambda}$ (see [38] and references therein); $f(\lambda) = (\lambda+z)^{-\alpha}$ for suitable $\alpha > 0$, $z \in \mathbb{C}$ (see [38] and references therein), in order to extract similar information, or more generally information about the location of eigenvalues. For example, an often studied function is $\text{Tr}\, e^{-it(-\Delta)^{1/2}}$, $\text{Tr}\Delta^{-s} = \Sigma_\lambda \frac{1}{{}_n s}$ $= \zeta(s)$ (the ζ-function); one has, that ζ is the Mellin transform of θ:

$$\Gamma(s)^{-1}\int_0^\infty t^{s-1}\theta(t)dt = \zeta(s).$$

For the study of the asymptotics for certain Hamiltonians with singular coefficients, see [72]. For spectral results in the case of special regions (for d = 2: circles, rectangles, ellipses, triangles and certain "spherical domains") or special potentials, see [22,67-71,174,184,150].

Let us mention that the study of $\theta(t)$ as $t \to \infty$ leads to information on the low lying eigenvalues, due to[1])

$$\theta(t) \cong e^{-\lambda_1 t}[c_1 + \sum_{n=2} e^{-(\lambda_n-\lambda_1)t}],$$

see, for example, [63,3,54,64].

This will be discussed further in Section 3. For other results on eigenvalues and eigenfunctions of Schrödinger operators on $\mathbb{R}^d$, see [11,21,65,146,147,196]. For detailed asymptotic expansions for $\theta(t)$ for Hill's operator, see [66]. The study of eigenfunctions of Schrödinger operators has been particularly intensive in recent years, partly by probabilistic methods, which will be mentioned in Section 3. The relation with classical asymptotics goes back to Carleman's study in '34 of spectral projections like $\sum_{\lambda_n \leq \lambda} \phi_n(x)^2$. The behavior is as $(2\sqrt{\pi})^{-d}\Gamma(\frac{d}{2}+1)^{-1}\lambda^{d/2}$, see [73]. For a survey of results see [22]. A basic quantity related to the eigenfunctions is of course the fundamental solution $K_t(x,y)$ of the heat equation (to be discussed further in Section 4) or the fundamental solution $K_t^S(x,y) = e^{-\frac{i}{\hbar}tH}(x,y) = K_{it}(x,y)$ of the Schrödinger equation. One has namely, at least in the sense of generalized functions in t, $K_t^S(x,y) = \Sigma e^{-\frac{i}{\hbar}t\lambda_n}\overline{\phi_n}(x)\phi_n(y)$ and correspondingly for $K_t(x,y)$.

Asymptotic expansions of these quantities in powers of $\hbar$ are available, see Section 4.1, 4.2. A particular interest has been acquired recently, in relation with stochastic mechanics [16] and related ideas applied to classical problems [39-41], the investigation of zeros (nodal sets) of eigenfunctions. For results on this problem, see [4,55].

---

1) See Footnotes.

3. DISTRIBUTION OF EIGENVALUES FOR A MANIFOLD

It is quite natural to extend the study of the Laplacian (and of more general elliptic operators) from a domain of $\mathbb{R}^d$ to the case of Riemannian manifolds M. Physically such operators arise, as is well-known, as Hamiltonians for quantum mechanical particles moving on manifolds, like, for example, in the motion of a quantum particle on unquantized curved space-time, or in the quantization of constrained free classical (geodesic) motion on a manifold or in the case of a particle moving in a crystal with periodic boundary conditions or for a quantum particle with spin in free motion in $\mathbb{R}^3$, the manifold being here $\mathbb{R}^3 \times SO(3)$.

In this Section we shall be concerned with the extension to the case of manifolds and the question discussed in the preceding Section.

3.1 The Case of a Compact Manifold

The definition of the quantities corresponding to θ (case of discrete spectrum of the Laplace-Beltrami operator on the manifold M) is similar to the one for the flat case, with Δ replaced by the Laplace-Beltrami [1] operator on the Riemannian manifold M, supposed to be compact (such that Δ has purely discrete spectrum) [3]. Thus $\theta(t) = \text{Trace } e^{t\Delta} = \sum_n e^{-t\lambda_n} = \int_M K_t(x,x)dx$, with dx the Riemannian volume on M and $K_t(x,y)$ the Green function for the heat equation. In this case there is an extensive literature on the asymptotic expansion as $t \downarrow 0$ of $\theta(t)$, an expansion which extends the one mentioned above in connection with "can you hear the shape of a drum?". A natural problem which arises is whether the spectrum of the Laplacian characterizes completely the Riemannian structure (modulo isometry) of the manifold. This was answered in the negative by Milnor and Kneser, who gave counter-examples of certain tori of dimension $\geqq 12$. There are, however, also results in the positive direction, see [3,5] and Section 4.3. Certainly we expect from what we saw in Section 2 for a plane domain that a great deal of geometrical and topological elements of M will be determined by the spectrum of the Laplacian. The simplest and most studied case is the one of compact Riemannian manifolds. In this case one has that $K_t(x,y)$, and consequently $\theta(t)$, have an asymptotic expansion as $t \downarrow 0$,

$$K_t(x,y) = (4\pi t)^{-d/2} e^{-d(x,y)^2/4t}[u_o(x,y)+u_1(x,y)t+\ldots+u_k(x,y)t^k+0(t^{k+1})] \tag{7}$$

1) See Footnotes.
3) See Footnotes.

with uniquely determined, smooth Riemannian-invariant coefficients u_i related to curvature properties of M, d being the Riemann distance of x,y. Such expansions are called Minakshisundaram-Pleijel expansions, see [1-5,24-31,32,42,150,173,177]. From these one gets

$$\theta(t) = (4\pi t)^{-d/2}[a_0+a_1 t + \ldots + a_k t^k + O(t^{k+1})] \tag{8}$$

with $a_i \equiv \int_M u_i(x,x)dx$. The a_i just depend on the curvature of M and its successive covariant derivatives, in a universal manner. The concrete computation of the a_i is possible and one obtains, for example: a_0 = Vol M, $a_1 = \frac{1}{6}\int_M \tau dx$, τ being the scalar (Gauss) curvature i.e., the trace of the Ricci tensor ρ of M, $a_2 = \frac{1}{360}\{\int 5\tau^2 - 2|\rho|^2 + +2|R|^2 dx\}$ with R standing for the curvature tensor. (The norms $|\rho|^2$, $|R|^2$ are tensor norms induced through the metric.) There are explicit computations also of a_3 and a_4 [43,44]. For other results on $N(\lambda)$ (cf. Section 2), see also [74].

Remark: In general the asymptotic expansion of $\theta(t)$ does not determine the spectrum $\{\lambda_n\}$ (e.g., the 2-torus $T^2 = \mathbb{R}^2/Z^2$ and a "flat Klein bottle" have the same asymptotic expansion, but not the same spectrum of $-\Delta$: see [3], Coroll. B 112). Special results are obtained under further assumptions on M. For example, in the case of compact Riemannian (2-dimensional) surfaces of genus $\geqq 2$, with Poincaré metric (cf. also Section 4.3), with isomorphic fundamental groups, one has that equality of their coefficient a_1 which implies equality of all a_k [45,2]. The above expansion for $\theta(t)$ extends to the case where $-\Delta$, in the definition of $\theta(t)$ as trace of $e^{t\Delta}$, is replaced by an elliptic operator, see for example [42]. A physically interesting simple special case is the one of the rigid rotator in $\mathbb{R}^3$. In this case, in polar coordinates θ,ϕ, the classical Hamiltonian is $H = \frac{1}{2I}(p_\theta^2 + \frac{p^2_\phi}{\sin^2\theta})$, with I the constant momentum of inertia and p_θ, p_ϕ the conjugate momenta of the variables θ,ϕ. In this case, with $I = 1/2$, $H = -\Delta$, with Δ the Laplace-Beltrami operator on S^2, e^{-tH} can be looked upon, with $\beta = t$, as the Gibbs factor for the rotator at inverse temperature β, the heat equation becoming then Bloch's equation.

The $\theta(\beta)$ is then the partition function, which in this case can be computed explicitly to be

$$\theta(\beta) = \sum_{j=0}^{\infty}(2j+1)e^{-\beta j(j+1)}$$

and the above asymptotic expansion reduces to the high temperature expansion $(4\pi\beta)^{-1}(a_0+a_1\beta+a_2\beta^2+\ldots)$ as $\beta \downarrow 0$, with $a_0 = 1$, $a_1 = \frac{1}{3}$,

$a_2 = \frac{1}{15}, \ldots$ (this high temperature expansion goes back to Fowler and Mulholland in 1928). Similarly, for the case of a planar rigid rotator the Hamiltonian can be identified with $-\Delta$, Δ the Laplace-Beltrami operator on S^1, hence

$$\theta(\beta) = \sum_{\ell \in Z} e^{-\beta\pi\ell^2} = (4\pi\beta)^{-1/2}\Big[\sum_{i=0}^{N} a_i \beta^i + O(e^{-1/\beta})\Big],$$

with $a_0 = 1$, $a_i = 0$, $i \neq 0$. $\theta(\beta)$ is here equal to $\theta_3(\beta,0)$, θ_3 being Jacobi's theta function. See [2,46] for further discussions [4]. For results of the type Equation (6) in Section 2 for manifolds with boundary, see [32]. Let us now return to the general case.

As remarked in Section 2 the asymptotic behavior of $\theta(t)$ as $t \downarrow 0$ leads, by Tauber-Karamata's results, to the "Weyl formula" Equation (1) for $N(\lambda) \equiv \sum_{\lambda_n \leq \lambda} 1$, with $\Omega \equiv M$ 73 . Let now V be a measurable function on M and consider the Schrödinger operator $H = -\Delta + V$ on $L^2(M,dx)$. Let $a \leqq 0$ and let $N_a(V)$ be the number of eigenvalues (counted with multiplicity) which are $\leqq a \leqq 0$. Let Ω be an open bounded set on M and let $N_\Omega(\lambda)$ be the number of eigenvalues of $-\Delta$ on $L^2(\Omega,dx)$. By the minimax principle $N_\Omega(\lambda) \leqq N_a(((a-\lambda)\chi_\Omega)$, χ_Ω being the function identically 1 on Ω and zero outside Ω. One can show [47,48] $N_a(V) \leqq L_n \int (V(x)-a)^{n/2} dx$ for all a, at least when V is in a suitable class of potentials and $M = \mathbb{R}^d$ or M is a homogeneous manifold of curvature $\leqq 0$. On L_n one has precise estimates [47,48]. It also follows from these methods that if M is compact, $N(\lambda) \leqq (D_n \lambda^{n/2} + E_n)$ (Vol Ω), with constants D_n, E_n independent of λ, Ω, depending thus only on M.

Let $\theta(t) = \text{trace } e^{-tH}$, with above $H = -\Delta + V$, on a compact manifold M. As remarked above, $-\Delta + V$ being elliptic, one has asymptotic expansions of the form of Equation (8) for $\theta(t)$ as $t \downarrow 0$. One can prove, see [49], $a_k = \int F_k(x,x)dx$, where $F_k(x,x)$ are polynomials in V and its derivatives, e.g., $a_1 = -\int V(x)dx$, $a_2 = \frac{1}{2}\int V((x)^2 dx$, $a_3 = -\frac{1}{6}\int \{V(x)^3 + \frac{1}{2}\nabla V(x)\}^2 dx, \ldots$.

Also $a_k = \frac{(-1)^k}{k!} I_k$, I_k being the k-th conserved integral of the Korteweg-de Vries equation: $U_t = 6UU_x - U_{xxx}$.

Another case where one has asymptotic expansions in powers of t for $t \downarrow 0$ (equivalently high temperature expansions for the corresponding "partition function", using the above interpretation of

---

4) See Footnotes

$\theta(t)$ as a partition function) is the case where M is a manifold which is at the same time a compact symmetric space of rank one 2). Examples of such spaces are $S^n = SO(n+1)/SO(n)$ (the sphere of dimension n, $n \geq 2$). In this case one has for $\theta_{S^n} \equiv \text{trace } e^{t\Delta}$:

$$\theta_{S^{2n}}(t) = \sum_{m=0}^{\infty} \{\frac{2m+2n-1}{2n-1} \prod_{k=1}^{2n-2} \frac{m+k}{k}\} e^{-t[m^2+m(2n-1)]/(4n-2)}, \quad (9)$$

and a similar formula for S^{2n-1}.

For these results and other calculations for such spaces see [51,2]. Also the coefficients a_i in the asymptotic expansion for $\theta(t)$ can be computed explicitly, see the same references. Of course for the case of S^n the spectrum of $-\Delta$ is known explicitly, and given by $\lambda_n = n(n+d-1)$ with multiplicity $m(\lambda_n) = \frac{2n+d-1}{n} \binom{n+d-2}{n-1}$, see for example [52,53].

Formulae of the above type hold however also for $\theta(t) = e^{t\Delta}$ (i.e., for quantum free motion), on general compact symmetric spaces of rank one (hence for the various projective spaces), see [2].

Finally we remark that the question of the generic continuous dependence of the eigenvalues λ_n on the metric on the manifolds has been studied, for a recent reference see [79]. For results locating intervals of $\mathbb{R}$ in which the eigenvalues of $-\Delta + V$ are to be found, see [80]. Concerning the eigenfunctions, see [4,55,203].

A phenomenon of creation of "quasi-modes" around periodic geodesics has been discussed, for example, in [81]. For eigenfunctions invariant under a compact Lie group acting on the manifold, see [82]. For a result characterizing Riemannian manifolds all of whose geodesics are closed of length L and Morse index α as those for which $-\Delta$ has eigenvalues clustering asymptotically for large k to $-\frac{2\pi}{L}(k-\frac{\alpha}{4})$, see [75,76]. For related finer results valid for all eigenvalues, see [77,78].

Let $K_t(x,y)$ be the kernel of the heat equation $\frac{\partial}{\partial t}\psi = L\psi$, with L an elliptic operator, on $\mathbb{R}^d$ or on a complete Riemann manifold. There are both analytic and probabilistic results on the construction of the fundamental solution $K_t(x,y)$ and on the asymptotic behavior as $t \downarrow 0$, see [54,42] and Section 4.2. For the case of the Schrödinger equation, see Section 4.1.

As far as eigenfunctions are concerned, one has, for example, ([55,4]) that if M is a complete Riemannian manifold and $L = -\Delta + V$,

---

2) See Footnotes

V smooth, the zeros of any eigenfunction form a d-1 dimensional C^∞ manifold, except possibly for an exceptional set contained in a d-2 dimensional submanifold.

Concerning the distribution of eigenvalues, in the case of a non compact Riemannian manifold M, e.g., $M = \mathbb{R}^d$, we have that $-\Delta$ no longer has purely discrete spectrum. For this reason one has, as mentioned in Section 2, to modify the definition of the theta function, by subtracting a contribution coming from a "simple part" causing the non discreteness to appear. A typical example is $M = \mathbb{R}^d$, $H = -\Delta + V$, with V going to zero as $|x| \to \infty$, in which case, as mentioned in Section 2, one defines $\hat{\theta}(t) = tr(e^{-tH} - e^{-tH_0})$, with $H_0 = -\Delta$, $t \geqq 0$. The behavior for $t \downarrow 0$ (i.e., the high temperature behavior, when one thinks of t as the inverse temperature and $\theta(t)$ as a modified partition function) is given by the Wigner-Kirkwood asymptotic expansion, for which recent references are [56,83]. For some results for $H = -\Delta$ on unbounded domains Ω of $\mathbb{R}^d$, see [27]. The behavior as $t \downarrow 0$ of $\theta(t)$ (resp. some modification $\theta(t)$ thereof) does, according to Section 1, coincide with the behavior as $\hbar \downarrow 0$ ("classical limit") of $e^{t/\hbar\, \hbar^2\Delta}$.

The study of $\theta(t)$ and its behavior as $t \downarrow 0$ on a non compact manifold has different aspects, which correspond partly to the ones we met in the compact case [5)]

In rather general situations when M is complete one knows at least the leading term of the asymptotics, see [42,84].

For results locating the essential spectrum of $-\Delta$ (the points of the spectrum which are either accumulation points of the spectrum or eigenvalues of infinite multiplicity), see for example [85,86], which also contain references discussing the location of the greatest lower bound of the spectrum $-\Delta$. For such results and results on the nature of the spectrum of the Laplacian on a simply connected complete Riemannian symmetric space of non compact type, see [87]. For complete Riemannian manifolds for which $-\Delta$ has pure point spectrum, see [88]. For a study of the heat kernel and its asymptotics in the case of a connected, not necessarily compact, Riemannian manifold M with a properly discontinuous group of isometries Γ with compact quotient, see [87]. For the case where the manifold M is a (not necessarily compact) symmetric space G/K, see [90]; see also [155-169,185].

---

5) See Footnotes

4. CONNECTION BETWEEN QUANTUM MECHANICAL SPECTRUM AND CLASSICAL ORBITS

4.1 "Hyperbolic Approach" by Oscillatory Integrals: WKB, Feynman Path Integrals and Stationary Phase, Solutions of the Schrödinger Equation

In this Section we shall briefly discuss the more traditional, direct approaches to the study of the relations between quantum mechanics and classical mechanics. Essentially in its various aspects these approaches put in relation the solutions of the Schrödinger equation and associated quantum mechanical quantities (with real time, not as in Section 4.2 with imaginary time) with solutions of the corresponding classical Newton equation and associated classical quantities. The study of the relation, on the other hand, has different aspects, according to which type of problem is considered (propagators, bound states, scattering, ...) and in which direction (from quantum to classical or vice versa) one looks. The methods to be discussed in this Section are called "hyperbolic", in as much as they involve at some stage the consideration of solutions of the time-dependent Schrödinger equation. Mathematically this theory falls into the class of evolution equations of Schrödinger type, the solution theory of which is methodically close to one of the hyperbolic equations of the wave propagation type. In fact both involve essentially the study of the asymptotics as $\hbar \to 0$ of oscillatory integrals of the form

$$\int_{\mathbb{R}^n} e^{\frac{i}{\hbar}\Phi(x)} g_{\hbar}(x)dx \tag{10}$$

for a certain phase function Φ and amplitude function $g_{\hbar}$. Both Φ and $g_{\hbar}$ might depend smoothly on various parameters. The detailed study of integrals of this type belongs to a circle of problems around the method of stationary phase. More properly we have a method of stationary phase when Φ is real-valued, Laplace method when Φ is purely imaginary (in which case there is no oscillation: this case is connected with the problem in Section 4.2, and will not be considered any further in this Section), a steepest descent method when Φ is complex. For obtaining the asymptotics of integrals of above type, with Φ real, the idea is to decompose the integral into contributions coming from the stationary points of the phase, i.e., the point where the derivative of Φ vanishes and the rest. The rest gives a contribution in $\hbar$ which vanishes quicker than any power of $\hbar$, the contribution from the stationary points involving a power of $\hbar$ which depends on "the type of stationary point" (mathematically the type of singularity or degeneracy of Φ at the stationary points). The types of stationary points is studied in the refined version of Morse theory which is also part of the catastrophe theory. For references to detailed studies of oscillatory integrals of above type, see for example [28,75,91-96]. The study of oscillatory integrals of the above type leads to two main developments:

1) Construction of solutions of the Schrödinger equation, by a method of parametrices. For this see [91,95].
2) Asymptotics of solution of the Schrödinger equation in $\hbar$, i.e., finding expressions of the solution of Schrödinger equations $i\hbar \frac{\partial}{\partial t} \psi = H\psi$, $H = -\frac{\hbar^2}{2m} \Delta + V$, for given initial condition $\psi(0,x) = e^{\frac{i}{\hbar}(f(x)} g(x)$ as $\hbar \downarrow 0$, in terms of classical orbits $\gamma_x(t)$ (solutions of Newton's equation of motion, for a particle moving under the force-grad V) starting at time 0 anywhere with velocity $-\nabla f$ and ending at time t in x.

In "generic situations", if the motion is in $\mathbb{R}^d$, to leading order in $\hbar$, one obtains ψ, in the sense of an asymptotic expansion in $\hbar$, as a sum of terms of the form

$$(2\pi t)^{-d/2} D(\gamma_x(t))^{-1/2} e^{\frac{i}{\hbar} S_0(\gamma_x(t))} e^{\frac{i}{\hbar} m(\gamma_x(t))} \tag{11}$$

where $D(\gamma_x(t))$ is the determinant associated with the Gaussian given by the classical action functional

$$S_0(\gamma) \equiv \frac{m}{2} \int_0^t \dot{\gamma}^2 d\tau - \int_0^t V(\gamma(\tau)) d\tau$$

and m is a natural number, called Maslov index associated with the path. This leading term has been obtained by several methods. Among these we mention an approach developed by Maslov, Duistermaat, Hörmander and many others, see for example [97,91] and references therein, as a refined method of characteristics. Other methods involve representing the solution by Feynman path integrals, formally written as

$$\psi(x,t) = \int_{\gamma(t)=x} e^{\frac{i}{\hbar} S_t(\gamma)} \psi(\gamma(0),0) d\gamma,$$

the integral being over "all paths" starting anywhere at time zero, and ending at time t in x. The representation of solutions of the time depending Schrödinger equation by such path integrals goes back to the original work by Kac, see [6-8]. For mathematical justification of formula of above type, see for example [6-8,98,38] and references therein. Dirac (1933) and Feynman (1948) already pointed out that formulae of above type are especially suited for the study of the classical limit by a method of stationary phase in infinite dimensions. This method has been developed on a mathematical level in recent years [6-8,98]. For related methods, which also use basically the Feynman path integrals, see [100-102]. The successive terms in the asymptotic expansion in powers of $\hbar$ have been computed in general situation (on manifolds) in [103], a mathematical proof of the asymptoticity of the expansion is given in [98,198]. Very recently a probabilistic method (related to those to be discussed in Section 4.2) has been adapted to the Feynman path integral case, and has yielded in particular the asymptotic expansion in powers of $\hbar$ under different assymptions for the previously used [104,105].

Similar methods can be used to discuss expansions in powers of $\hbar$ of the Green's function for the time dependent Schrödinger equation (corresponding to initial condition a, delta function), see for example [38].

Another important problem in the study of the relation between quantum mechanics and classical mechanics, in which hyperbolic methods have been used, is the study of the eigenvalue problem. The study of the relation between spectrum of the quantum mechanical energy operator and periodic orbits of the corresponding classical system goes back to the very origin of quantum mechanics ("old quantum theory" of Bohr and Sommerfeld). The simplest relation in this area is that at the basis of the "quantization rules" of Bohr and Sommerfeld. The mathematical and/or practical justification of these rules and more generally of formulae giving eigenvalues and the density of states in terms of quantities of classical mechanics, has involved much work since the advent of quantum mechanics. In particular the JWKB method (which has its actual origins, before quantum mechanics, in methods used in the study of the asymptotics for high frequency of the wave equation) has been developed extensively, especially from the point of view of obtaining good numerical results, particularly for systems of one degree of freedom (or separable systems or systems reducing by the presence of symmetries to the former). For a good review and references, see, for example, [106,107]. More dimensional extensions of the method have been given by Maslov, see [97] (see also refinements, e.g., [144,170,202,94, 108]), under assumptions such as complete integrability (i.e., multiple periodicity) or at least "near integrability" of the classical system (implied, for example, by separability) or presence of Lagrangian submanifolds (classical trajectories lying on invariant tori in phase space, like in cases provided by the KAM theorem).

The quantities studied in this respect are the eigenvalues or the trace of $e^{-i(t/\hbar)H}$, H being the Hamiltonian, or related quantities like the ζ-function, see [38] and references therein. The trace is given asymptotically for $h \downarrow 0$ in terms of contributions coming from classical periodic orbits. The essential difference in behavior between integrable and non-integrable systems, in this respect, has been underlined by Einstein already in 1917 and discussed further by Brillouin, Keller, Maslov, Gutzwiller, Berry,

Percival and others, see [109,110]. The relation between eigenvalue spectrum and classical orbits has been studied in general situations which go beyond the ones (near integrability Lagrangian submanifolds) mentioned above in recent years by Chazarain [93b] and others [27-29, 35] using methods of (hyperbolic) partial differential equations or the mathematical work on the stationary phase for Feynman path integrals in [38,98,8].

In particular, the latter methods yield an asymptotic expansion in powers of $\hbar$ for the theta function $\theta(t) \equiv \sum_n e^{-\frac{i}{\hbar} t\lambda_n} = \text{Trace } e^{-i\frac{t}{\hbar}H}$ associated with $H = -\frac{\hbar^2}{2m}\Delta + \frac{1}{2}(x,A^2x) + V(x)$, A^2 being a symmetric matrix in $\mathbb{R}^d$, $x \in \mathbb{R}^d$, $(,)$ the scalar product in $\mathbb{R}^d$, V a bounded continuous potential in $\mathbb{R}^d$, λ the eigenvalues of H. H is thus the Hamiltonian to an anharmonic oscillator (harmonic oscillator perturbed by a potential V). The expansion is in terms of periodic classical orbits and gives an extension of the classical Poisson formula for the theta function (cf. Section 4.3). Control on related quantities such as the ζ-function (an object which has been studied, e.g., in connection with gauge theories, see for example [111]) $\zeta(s) = \sum \frac{1}{\lambda_n^s}$ has also been obtained, as well as a proof of the Bohr-Sommerfeld quantization formula. For the study of the ζ-function in the case of $d = 1$, $A = 0$, $v(x) = x^4$, see [112,113]. Some remarks concerning the above results are in order. First we remark that these results given an extension towards operators of physical interest of results obtained in the mathematical literature, especially on compact manifolds, in work by Colin de Verdière and Chazarain, refined using Hörmander's theory of Fourier integral operator in work by Duistermaat, Guillemin, Hörmander, Weinstein and others, see for example [75,91-97]. In particular these authors are concerned with generalized functions of the form

$$\text{Tr}(P-\lambda)^{-1}, \ \text{Tr}(e^{-tP}), \ t \in \mathbb{R}_+, \ \text{Tr}(e^{-itP}), \ t \in \mathbb{R}, \ \text{tr } P^z, \ z \in \mathbb{C}$$

(see [1,76]), where P is a pseudo elliptic or hypo-elliptic operator on a compact manifold. These generalized functions of t yield information about the distribution of the eigenvalues of P and their singular support is connected with the periodic orbits of the associated Hamiltonian vector field (see [76] for the singularity at $t = 0$ of $\text{Tr } e^{-itP}$).

The expansion of the Laplace transform $\sum_n \hat{\phi}(\lambda-\lambda_n)$ of $\text{Tr } e^{-\frac{i}{\hbar}tP} \phi(t)$, ϕ a suitable Schwartz function, in powers of λ as $\lambda \to \infty$, λ_n being the eigenvalues of P, has coefficients independent of ϕ which are related to the poles of the zeta function $\zeta(s)$ of P and to the coefficients occurring in the asymptotic expansion of the heat kernel [38]. Such expansions will also be mentioned in Section 4.2. The leading term in such expansions (and related ones, like the ones for the solutions of the Schrödinger equation) in powers of $\hbar$, is called "the semiclassical approximation".

The second remark we want to make concerns the "exactness of the semiclassical approximation", or more generally the question of when $e^{-i\frac{t}{\hbar}H}$ or related quantities like $\theta(t)$ or the eigenvalues λ_n are given entirely in terms of classical orbits. As remarked in the introduction, the simplest instance of this is of course the free motion in $\mathbb{R}^d$, where

$$
\begin{aligned}
e^{-i\frac{t}{\hbar}H}(x,y) &= e^{-i\frac{t}{\hbar}(-\frac{\hbar^2}{2m}\Delta)}(x,y) = e^{i\frac{t\hbar}{2m}\Delta}(x,y) = \\
&= (2\pi i\frac{\hbar}{m}t)^{-d/2}e^{i\frac{|x-y|^2}{2t\hbar/m}} = (2\pi i\frac{\hbar}{m}t)^{-d/2}e^{\frac{i}{\hbar}S_t(\gamma_{x,y})}
\end{aligned}
\tag{12}
$$

with $S_t(\gamma) = \frac{m}{2}\int_0^t \dot{\gamma}^2 d\tau$, the classical action along the path γ, $\gamma_{x,y}$ the classical path in time t between x,y, i.e., the geodesics (here the straight line) between x,y. A large class of manifolds in which the classical dynamics already determines completely the quantum dynamics is discussed in Section 4.3.

Other problems which have been discussed by hyperbolic methods are the relation of quantum and classical scattering, such as control on the scattering quantities in the limit $\hbar \downarrow 0$, see [114,197]. The study of the classical limit of observables on suitable "coherent states" has also been achieved, see [115].

We should also mention various methods of quantization, where the starting point is classical dynamics "deformed" then into quantum dynamics. One such method consists in the study of the algebra of Dirac versus Poisson brackets, see for example [116].

Other methods are the geometric quantization method, which has various forms, see any of the excellent accounts in for example [17] (see also [117,118,179,120]). A stochastic quantization method will be mentioned in Section 4.3.

For a construction of solutions of Schrödinger equation by "Feynman path integrals" on Riemannian manifolds, see [92,121,171] and by methods using "analytic continuation" from Wiener integrals, see [119,188].

4.2 "Parabolic Approach" by Probabilistic Methods: Heat Equation, Stochastic Differential Equation, Stochastic Mechanics

Probabilistic methods in the study of partial differential operators, like the elliptic ones entering in the stationary heat and Schrödinger equations, have their roots in the classical connection between Brownian motion and the Laplacian, discovered independently by Bachelier (1900) (in economics), Einstein (1905) and Smoluchowski (1906) (in physics). These connections have been extensively exploited since Wiener (1923) and Kolmogorov (1933) work. Langevin (1911) introduced stochastic terms in equation of motions, which lead later on, by work of Bernstein, Ito and others to the modern theory of stochastic differential equations. In the late fifties Kac introduced an important new method, on the basis of heuristic ideas of Feynman, which we mentioned in 4.1. Roughly, Kac's approach is to represent the transition semigroup P_t of "the heat equation with killing term V", i.e., $P_t = e^{-t(-\Delta/2 + V)}$, by expectation of a functional of Brownian motion:

$$(P_t f)(x) = E(e^{-\int_0^t V(x+b_s)ds} f(x+b_t)), \quad x \in \mathbb{R}^d, \tag{14}$$

where E means expectation with respect to the standard Brownian motion b_s issued at time 0 from the origin, the formula holding for example for all smooth f of compact support, provided V is continuous and bounded (in fact much less is required on V, see [11]).

The above formula has given an important tool for the study of P_t. In particular, if we introduce the coefficient relevant for the physical discussion we see that we have to study for $\hbar \downarrow 0$ the behavior of

$$(P_t^\hbar f)(x) = E(e^{-\frac{1}{\hbar}\int_0^t V(x+\sqrt{\hbar}\, b_s)ds} f(x+\sqrt{\hbar}\, b_t)). \tag{15}$$

In quantum statistics the partition function can also be expressed by expectations of similar type with respect to Brownian motion, the role of t being there played by the inverse temperature β.

Hence it is interesting to have expansions in $\hbar$ and t of expressions of the type of the righthand side of Equation (15). Such expansions have been done using various probabilistic methods. These methods involve a functional integral version of the Laplace method, see for example [11,57,58,100,102,104,105,122,123,119,99]. Related studies are concerned with the asymptotics for small σ of a stochastic differential equation of the type

$$dX_t = \beta(X_t,t)dt + \sqrt{2\sigma}\,(X_t,t)db_t \tag{16}$$

or a semigroup of the type $P_t = e^{-t(-\sigma\Delta-\beta\nabla)}$ [63,124].

Both kind of problems have been extended to the case of (not necessarily compact) manifolds in which case one arrives at extensions of the results on the distribution of eigenvalues and eigenfunctions mentioned in Section 3. In particular, one gets probabilistic proofs in general situations, for the heat kernel $K_t(x,y)$ of the type mentioned in Section 3.2. The cases where there are discretely many geodesics between x,y, thus x,y are "in general position", i.e., "non conjugate" (i.e., such that the relevant Hessian computed from the action between x and y does not vanish), are covered. In the case where x,y are conjugate, the leading term in the expansion of $K_t(x,y)$ changes from the one given by Equation (7), i.e., $(4\pi t)^{-d/2} e^{-d(x,y)^2/4t} u_0(x,y)$ to the same term multiplied $(4\pi t)^{\alpha} |\text{Log } t|^{\beta}$, for suitable exponents α,β, which depend on the degeneracy (a situation completely corresponding to the one mentioned in Section 4.1).

For the study of diffusion processes (and their approximation) on manifolds, see for example [125,126,127,89,151,172,176,186-188, 192].

We close this Section with a short remark on stochastic mechanical methods. It is well-known in the case of $\mathbb{R}^d$ that Nelson's stochastic mechanics gives an alternative framework for quantum mechanics.

In particular, it gives a natural method of quantization in as much as one can get the Schrödinger equation and all relevant quantities of quantum mechanics starting from a diffusion process X_t satisfying an equation of the form of Equation (16), with $\sigma = \hbar/2m$ and β^t of the gradient type $\beta = 2\sigma\nabla \ln \psi$. By a variational principle or by requiring that Newton's laws hold in the mean (i.e., mass x acceleration = force = $-\nabla V$, where acceleration is understood as mean stochastic acceleration, see [16]) one arrives then at the Schrödinger equation with potential V for ψ.

Vice versa, it associates to any solution of Schrödinger equation, ψ a stochastic process, with constant diffusion coefficient, drift determined by ψ and probability distribution with density given at all times by $|\psi|^2$. For discussions of this connection, see for example [14-16,19,20,39-41]. This relation can be used (as above) to study the limit for $\hbar \downarrow 0$ of several quantities, see [63].

The extension of the relation between quantum mechanics and stochastic mechanics to the case of manifolds has been worked out particularly by D. Dohrn and F. Guerra, see their talk at the preceding Como Conference [18] (see also [16,132,199,189]). Stochastic mechanics on manifolds work completely parallel to stochastic mechanics on $\mathbb{R}^d$, once the correct notion of geodesic transport has been introduced [132], and yields a natural way of deciding upon the natural quantum mechanical Hamiltonian (it chooses $-\frac{1}{2}\Delta + V$, with Δ the Laplace-Beltrami operator: hence no "R/6" or the like terms, here!). One might hold the more general hope that stochastic mechanics will some day be completed by a dynamical classical theory explaining the origin of quantization. We can only refer here to the discussions in the literature, e.g., [16,20,39-41,195].

4.3 Manifolds with Close Relationship between Spectrum of the Laplacian and Geodesics ("When is the Semi-classical Approximation exact?")

We have seen in Section 3 as well as in Section 4.2 that in general situations the kernel $K_t(x,y) = e^{t\Delta}(x,y)$ of the heat equation is given by

$$(4\pi t)^{-d/2} e^{-d^2(x,y)/4t} \, [u_o(x,y) + O(t)].$$

We can ask in which cases the term $u_o(x,y) + O(t)$ can be written as $e^{Ct}\tilde{u}_o(x,y)$ for some constant C, i.e., the leading term in the asymptotic expansion for $K_t(x,y)$ is essentially itself the kernel of a semigroup or, equivalently, when $K_t(x,y)$ is already determined (essentially) by its leading term, i.e., $K_t(x,y)$ is a sum of terms of the from $(4\pi t)^{-d/2} e^{-d(x,y)^2/4t} e^{Ct} \tilde{u}_o(x,y)$. Moreover, we can ask when is $\tilde{u}_o(x,y)$ just expressible in terms of the lengths of geodesics from x to y? In such a case we then have that the whole heat kernel is expressible for all t in terms of lengths of geodesics, the term $e^{-d(x,y)^2/4t}$ giving the contribution to the action coming from the free motion of a classical particle moving along a geodesic from x to y in time t.

We shall say that in this situation "the semi-classical approximation is exact". (More generally, we look for situations in which the quantum mechanical propagator K_t is given entirely, for all times in terms of classical orbits.) This has been discussed in the physical literature in terms of the short time propagator for the time dependent Schrödinger equation, see for example [128,129,130, 131]. We shall briefly discuss following Equation (5) how this problem can be completely handled mathematically in a number of cases.

We shall first present a probabilistic formula which shows explicitly how one can factorize $K_t(x,y)$ in a large class of cases into the term $(4\ t)^{-d/2} e^{-d(x,y)^2/4t}$ times an expectation with respect to Brownian bridge of a term involving essentially a geometric invariant related to the curvature of the manifold (this expectation has to do (to leading order in t (or $\hbar$)) with the curvature term

R/12, R being the constant scalar curvature, much discussed in the literature, see [128-132].

The formula, first proven by Elworthy and Truman in [133], see also [119,126,5], holds in the case where the Riemann manifold belongs to a general class which includes the case of complete manifolds of negative curvature, as well as group manifolds (e.g., the simplest cases being the ones where the group is compact or exponentially solvable).

The condition needed is that there exists a point $y \varepsilon M$, such that the Jacobian $\Theta_y(\cdot)$ of the exponential mapping from the tangent space at y onto M, does not vanish. In such cases we have

$$K_t(x,y) = \sum_i (4\pi t)^{-d/2} \Theta_y(X_i)^{-1/2} e^{-\ell^i(x,y)^2/4t} E\{ \{e^{\frac{1}{2}\int_o^t \Theta_y^{1/2}(X_s^i)\Delta\Theta_y^{-1/2}(X_s^i)ds}\}.$$

The sum is over the points X_i in the tangent space at y which are mapped by the exponential mapping into x; $\ell^i(x,y)$ is the length of the i-th geodesics between x and y. X_s^i is the diffusion process in the tangent space starting at time zero at the origin, ending at time t at the vector X_i, with unit diffusion coefficient and drift $-\frac{X_i}{t-s} - \frac{1}{2}\nabla \log \Theta_y(X_i)$. We note that the term in the exponent under the expectation is just the so-called "quantum potential", evaluated at the path X_s^i. The above problem of knowing "when the semiclassical approximation is exact" reduces here to the question of when the above expectation can be computed in terms of lengths of geodesics. Here are some cases where this is possible, see [5] for more details. One striking case is the one of hyperbolic space forms of constant negative curvature of dimension 3, i.e., the Clifford Klein spaces of dimension 3 (which can be realized by quotienting the manifold H^3 by a discrete subgroup Γ of the Lorentz group O (3,1). In this case the formula becomes simply

$$K_t(x,y) = \sum_{\gamma\varepsilon\Gamma} (4\pi t)^{-3/2} \frac{d(\gamma x,y)}{\mathrm{sh}\, d(\gamma x,y)} e^{-d^2(\gamma x,y)/4t + Rt/6}$$

where the sum is over a discrete set and R is the constant scalar Riemann curvature.

We shall see below that for the trace of K_t one has similar formulae also for hyperbolic spaces of dimension 2 (Selberg trace formula).

For the case where M is the manifold of a compact Lie group we have that the heat kernel is given by

$$K_t(x,y) = K_t(y^{-1}x,e) = \tilde{K}_t(h,e)$$

where h is an element in a Cartan subgroup T (maximal torus) of G, $\tilde{K}_t$ being the heat kernel on T and e the unit in G. $\tilde{K}_t$ is a sum of heat kernels on the Cartan sub-Lie-algebra, thus of the form

$$\tilde{K}_t(h,e) = C \sum_{\gamma\varepsilon\Gamma} (4\pi t)^{-N/2} \Theta_e(H+\gamma)^{-1/2} e^{-\|H+\gamma\|^2/4t} e^{Rt/6} .$$

Γ is a discrete lattice (the dual of the root system), H is such that exp H = h, $\| \ \|$ is the length in the Lie algebra induced by the Riemann structure on M. The "normalizing" factor C depends on the positive roots and is independent of t and h. R is the scalar Riemann curvature of the group [182].

In the case where M is a non compact semi-simple Lie group or a quotient of such a group by a maximal compact subgroup, one gets again formulae of the above type [1]. We also remark that one can use the above results also to handle the cases of manifolds M which are homogeneous spaces M = G/H, G being a compact Lie group and H a closed Lie subgroup (like for example, spheres: $M = S^n = SO(n+1)/ /SO(n)$). One simply uses the fact that the heat kernel K_t of M can be obtained for the one K_t^G of G by integration over H, namely

$$K_t(x,y) = K_t(y^{-1}x,e) = \int_H K_t^G(gh,e)dh$$

for $y^{-1}x = gH$, $g \varepsilon G$.

A classical case of a manifold for which the semiclassical approximation is exact is the one of a torus $M = T^d = \mathbb{R}^d/\Gamma$, Γ being a lattice subgroup with dual Γ'. In this case one has, for example, for $\Theta(t) = \operatorname{Tr} K_t = \int K_t(x,x)dx = \Sigma e^{-t\lambda_n}$ (with λ_n the eigenvalues of $-\Delta$ on T^d:

$$\Theta(t) = \sum_{\gamma'\varepsilon\Gamma'} e^{-4\pi^2 t|\gamma'|^2} = \frac{(\text{Vol } M)}{(4\pi t)^{d/2}} \sum_{\gamma\varepsilon\Gamma} e^{-\frac{|\gamma|^2}{4t}} .$$

We note that $|\gamma|^2$ are the lengths of geodesics and $-4\pi^2|\gamma'|^2$ are the eigenvalues of $-\Delta$ (with eigenfunctions $e^{2\pi i<\cdot,\gamma'>}$). See [3,5] which also discuss other such examples; see also for various extensions of this formula [91-93,38] (and the discussion in 4.1). A quite direct extension, which can also be obtained from the above method, see also [149], is the one to the Selberg trace formulae for a compact Riemann surface (2-dimensional manifold) $M = \Gamma\backslash H$, Γ being a discrete subgroup of SL(2,R) and $H = SL(2,\mathbb{R})/SO(2,\mathbb{R})$ being the upper half plane with Poincaré metric. Thus M can be viewed as a fundamental domain, for Γ, on the upper half plane (or, equivalently

---

1) See Footnotes

as a non Euclidean polygon in the unit circle). The formula for the theta function associated with M is

$$\Theta(t) = c_t \frac{e^{-t/4}}{(4\pi t)^{d/2}} \frac{(\mathrm{Vol}\ M)}{4\pi t} + \sum_{\gamma\varepsilon\tilde{\Gamma}} \ell(\gamma) \sum_{k=1}^{\infty} \frac{e^{-\ell^2(\gamma)/4t}}{2\ \mathrm{Sh}[\frac{1}{2}\ell(\gamma^k)]},$$

the sum being over all conjugacy classes of primitive elements of Γ, $\ell(\gamma)$ being the length of a closed geodesic determined by γ; γ^k is γ run k times; c_t is independent of the $\ell(\gamma)$. For this formula, see [148]. Various extensions of this formula have been given, e.g., to the case of symmetric spaces of rank 1 [135,164]. See also [2,137,152-154,190,191].

4.4 Some Additional Remarks, in particular about Chaos in Quantum and Classical Structures

In this short exposé we have given an overview of various results and techniques used on the mathematical side to study the relations between quantum mechanics and classical mechanics. We have concentrated on problems involving deterministic (non-quasi-periodic) potentials in finite dimensional configuration spaces. We did not treat either stochastic or quasi-periodic potentials since there would be contribution involving those at the same Conference, cf. in particular, the contributions by Bellissard, Martinelli and Souillard. Also we left completely aside problems involving infinitely many degrees of freedom, partly because the techniques used to handle them are in principle related to the ones used in finite dimensions (see for example [6-8,138] on the mathematical side, and for example [9] on the physical side, concerning the hyperbolic resp. Feynman path-integral approach, and for example [11,12,139] concerning the elliptic resp. Euclidean path-integral approach) and partly because the question of principles is already apparent in the finite dimensional situation. Of course there are most interesting specific problems also in the case of infinitely many degrees of freedom. In particular the study of the classical limit of quantum gases, see [140,141], the study of the classical limit of energy eigenvalues for the quantum fields, on which little is known on the mathematical side [142,180,181,188,194] (there is, however, a rich physical literature, see [9]). Many problems of the relation between classical and quantum dynamics which have not been discussed here will receive adequate treatment in other contributions in these Proceedings.

As to the basic question of quantum chaos versus classical chaos, let us make the following comment.

G. Casati has put beautifully in evidence the ergodic classical hierarchy on one hand and asked for a corresponding hierarchy on the quantum mechanical side, see these Proceedings and [143,183,193].

It is possible that the detailed results on the relation of quantum mechanics and classical mechanics, especially those of Section 4.3, might help to clarify these hierarchies: this has, however, not yet been done [2]. Moreover, we like to mention that an ergodic hierarchy for quantum mechanics has been discussed also by G. G. Emch, see [145] and references therein. One simple model to test this hierarchy might be the flow given by the shift operator associated with a Markov semigroup, a model discussed in the Gaussian case in [145], the shift having then the interpretation of reversible dynamics in a microscopic system, the Markov semigroup describing the projected, reduced dynamics on a system in interaction with a heat bath. It is possible that a study of non linear versions of this model will shed some light on above hierarchy.

We like to look at these suggestions as parts in a program of studying quantum mechanics as "stochastic mechanics". Of course the ultimate goal of stochastic mechanics is to understand mechanics in classical terms. This is a deep aim which should not be given up too early.

---

2) See Footnotes

Acknowledgments

We thank Prof. Dr. Raphael Høegh-Krohn and Prof. Dr. Michel Sirugue for many stimulating discussions. The first named author would like to thank Prof. Dr. Giulio Casati and Prof. Dr. Joseph Ford for a very kind invitation to the Como workshop. Both authors would like to take the opportunity to gratefully acknowledge various stays at the Centre de Physique Théorique, CNRS (Marseille), at the Université d'Aix Marseille II and at the center for Interdisciplinary Research (ZiF), University of Bielefeld, in which part of this work was done. The second named author would also like to thank the Mathematics Department of the Ruhr-University, Bochum, for the kind invitations. Both authors are very grateful to Mrs Jegerlehner, Mischke and Richter for their understanding and nice typing.

Footnotes

2. Distribution of Eigenvalues for a Domain of $\mathbb{R}^d$

1) $\theta(t)$ is a special Dirichlet series, built with the eigenvalues of $-\Delta$.

3. Distribution of Eigenvalues for a Manifold

1) If M has Riemannian metric $ds^2 = g_{ik}dx^i dx^k$ then the Laplace-Beltrami operator on M is given in the local coordinates x^i by

$\Delta = \frac{1}{\sqrt{g}} \frac{\partial}{\partial x^i} (\sqrt{g}\, g^{ik} \frac{\partial}{\partial x^k})$, with g the determinant of g_{ik} and $g^{ik} g_{k\ell} = \delta^i_\ell$.

2) Compact symmetric spaces are in general defined as follows: let G be a compact connected Lie group, let σ be an involutive analytic automorphism of G and let K be an open subgroup of the fixed points set of G under σ, i.e., of $G^\sigma = \{h \in G | \sigma(h) = h\}$. The quotient space G/K is then a symmetric space, with Riemannian metric derived from a positive definite invariant bilinear form on the Lie algebra g of G. Rank one means that 1 is equal to the dimension of the maximal abelian subalgebra of the Lie algebra g consisting of all $X \in g$, such that $s(X) = - X$, with s the differential action of σ on g. For these concepts, see [50].

3) In general we suppose the manifold without boundary. In the case of manifolds with boundary we have of course to give the boundary conditions for the differnetial operators involved.

4) For trace formulae, giving $\theta(t)$ for all times t in terms of closed geodesics, see also [66,168].

5) A plane membrane $\Omega \quad \mathbb{R}^2$, held fixed along its boundary, set in motion in the direction perpendicular to the plane, has a displacement $u(x,y,t)$ at its point $x,y \in \Omega$ satisfying the wave equation

$\frac{\partial^2 u}{\partial t^2} = c^2 \Delta u$, with c a constant depending on the physical properties of the membrane and its tension. Periodic solutions of the form $u(x,y,t) = U(x,y)e^{i\omega t}$ represent pure tones and U satisfies $c^2 \Delta U + \omega^2 U = 0$, with the given boundary condition ($U = 0$ on $\partial\Omega$). We see that (U,λ) is an eigenfunction resp. eigenvalue of the (stationary) Schrödinger equation $\Delta U = \lambda U$ with $\lambda = - \omega^2/c^2$.

4. Connection between Quantum Mechanical Spectrum and Classical Orbits

1) It is enough to consider the connected case, since the kernel for the non connected case can be written as a product of the ones for the connected components. Moreover one observes that the kernel of a direct product of groups is the product of the kernel to the components. In this way, for example, the kernel of a semi-simple Lie group can be written as product of those of its simple components.

In the case of a quotient $M = \Gamma\backslash G/K$, G connected simple with finite center, K a maximal compact subgroup of G, Γ a finite subgroup of G, acting freely on G/K and such that $\Gamma\backslash G$ is compact, a trace formula similar to Selberg's trace formula discussed above has been given in [164b]. It gives essentially the theta function $\theta(t)$ in terms of lengths of closed geodesics. It extends previous related formulae, see [2].

2) For the problem of asymptotic expansions in the case of certain operators with singular coefficients arising in quantum field theory see for example [72,111].

REFERENCES

1. V. Guillemin and Sh. Sternberg, Geometric asymptotics, Math. Surv., 14, AMS., Providence (1977).
2. K. E. Hurt, Geometric quantization in action, D. Reidel, Dordrecht (1983).
3. M. Berger, P. Gauduchon and E. Mazet, Le spectre d'une varieté Riemannienne, Lecture Notes Maths., 194, Springer, Berlin (1971).
4. U. Simon and H. Wissner, Geometrische Aspekte des Laplace-Operators, pp. 73-92, in: Jahrb. Überlicke Math. 1982, Ed. S. D. Chatterji et al., Bibl. Inst., Mannheim (1982).
5. T. Arede, Géometrie du noyau de la chaleur sur les varietés, Thèse de 3.èmecycle, Université d'Aix-Marseille II (1983).
6. S. Albeverio and R. Høegh-Krohn, Mathematical Theory of Feynman Path Integrals, Lecture Notes in Mathematics, 523, Springer, Berlin (1979).
7. S. Albeverio et al., Edts., Feynman Path Integrals, Proc. Marseille 1978, Lecture Notes in Physics, Vol. 106, Springer, Berlin (1979).
8. S. Albeverio, Ph. Blanchard and R. Høegh-Krohn, book in preparation for Encycl. Maths., Ed. G. C. Rota, Addison-Wesley.
9. V. N. Popov, "Functional Integrals in Quantum Field Theory and Statistical Physics", D. Reidel, Dordrecht (1983).
10. L. S. Schulman, "Techniques and applications of path integration", Wiley, New York (1981).
11. B. Simon, "Functional Integration and Quantum Physics", Academic Press, New York (1979).
12. A. Jaffe and J. Glimm, "Quantum physics, a functional integral point of view", Springer, Berlin (1982).
13. F. Langouche, D. Roekarts and E. Tirapegui, "Functional Integation and Semiclassical Expansions", D. Reidel, Dordrecht (1982).
14. E. Nelson, "Dynamical Theories of Brownian Motion", Princeton Univ. Press, Princeton, New Jersey (1967).
15. F. Guerra, "Structural aspects of stochastic mechanics and stochastic field theory", Phys. Repts., 77:263-312 (1981).
16. E. Nelson, "Quantum Fluctuations, Cours de III Cycle", Lausanne (1983) and book in preparation.

17a J. Sniatycki, "Geometric quantization and quantum mechanics", Springer, Berlin (1980).

17b D. J. Simms and N. M. J. Woodhouse, Lectures on geometric quantization, Lect. Notes Phys., 53, Springer, Berlin (1976).

18. G. Casati and J. Ford, Stochastic Behavior in Classical and Quantum Hamiltonian Systems, Lect. Notes Phys., 93, Springer, Berlin (1979).

19. E. Nelson, Connection between Brownian motion and quantum mechanics, in: Einstein Symp. Berlin, Lect. Notes Phys., Springer, Berlin (1979).
20. F. Guerra, Contribution to the IV Bielefeld Encounters in Mathematics and Physics, "Trends and developments in the eighties", Ed. S. Albeverio and Ph. Blanchard, World Publ., Singapore, in preparation.
21. M. Reed and B. Simon, "Methods of Modern Mathematical Physics", I-IV, Acad. Press (1978).
22. H. P. Baltes and E. Hilf, "Spectra of finite systems", Bibl. Inst., Mannheim (1976).
23. F. Fricker, Einführung in die Gitterpunktlehre, Birkhäuser, Basel (1982).
24. D. Heijal, The Selberg trace formula and the Riemann zeta function, Duke Mathematical Journal, 43:441-482 (1976).
25. E. Lieb, B. Simon and A. S. Wightman, Edts., "Studies in Mathematical Physics", Essays in Honor of Valentine Bargmann, Princeton Univ. Press, Princeton, New Jersey (1976).
26a V. Glaser and A. Martin, "Comment on the paper ...", CERN Preprint 3510 (1983).
26b H. Grosse, "Quasi classical estimates on moments of the energy levels", Preprint.
27. J. Fleckinger, Estimation des valeurs propes d'opérateurs elliptiques sur des ouverts non bornés, Ann. Fac. Sci., Toulouse, 2:157-180 (1980).
28. D. Robert, Calcul fonctionnel sur les opérateurs admissibles et applications, J. Funct. Anal., 45:pp. 74-94 (1982).
29a B. Helffer and D. Robert, Comportement semi-classique du spectre des Hamiltoniens quantiques elliptiques, Ann. Inst. Fourier, Grenoble, 31, 3:pp. 169-223 (1981).
29b B. Helffer and D. Robert, Comportement semi-classique du spectre des Hamiltoniens quantiques hypoelliptiques, CRAS., t. 292: ser. I, p. 47 (5 Janvier 1981).
30. M. Kac, "Can we hear the shape of a drum", Amer. Math. Monthly, 73:1-23 (1967).
31. C. Clark, The asymptotic distribution problems, SIAM Rev., 9:627-646 (1967).
32. H. P. McKean and I. M. Singer, Curvature and the eignevalues of the Laplacian, J. Differential Geometry, 1:43-69 (1967).
33. K. Yosida, "Functional Analysis", Springer, Berlin (1968).
34. R. Beals, A general calculus of pseudo differential operators, Duke Math. J., 42:1-42 (1975).
35. J. Fleckinger, Distribution of the eigenvalues of operators of Schrödinger type, Spectral Theory Differential Operator, Proc. Conf. Birmingham/USA., 1981, North-Holland Math. Stud., 55:pp. 173-180 (1981).
36. I. M. Glazman, Direct Methods of Qualitative Spectral Analysis of Singular Differential Operators, Israel Progr. Scient. Transl. (1965).

37a Y. Colin de Verdière, Une formule de traces pur l'opérateur de Schrödinger dans $\mathbb{R}^3$, Ann. Sci. Ee. Norm. Supér., IV Sér. 14:pp. 27-39 (1981).

37b O'Brien, A trace formula for Schrödinger operators with step potentials, Adelaide Preprint (1982).

37c D. Bollé and S. F. J. Wilk, Low energy sum rules for two-particle scattering, J. Math. Phys., 24:1555-1563 (1983).

38. S. Albeverio, Ph. Blanchard and R. Høegh-Krohn, Feynman Path Integrals and the trace formula for Schrödinger operators, Commun. Math. Phys., 83:49-76 (1982).

39. S. Albeverio, Ph. Blanchard and R. Høegh-Krohn, A stochastic model for the orbits of planets and satellites: an interpretation of Titius-Bode law, Exp. Math., 1 (1983).

40. S. Albeverio, Ph. Blanchard and R. Høegh-Krohn, Diffusions sur une varieté Riemannienne: barrières infrachissables et applications, Coll. L. Schwartz, Astérisque, Soc. M. Fr. (1984).

41. S. Albeverio, Ph. Blanchard and R. Høegh-Krohn, Newtonian diffusions and planets, with a remark on non standard Dirichlet forms and polymers, to appear Proc. LMS Conference Stochastic Analysis, Swansea, ed. D. A. Truman and D. Williams, Lect. Notes Maths., Springer, Berlin (1983).

42. R. Azencott et al., Géodesiques et diffusion en temps petits. Séminaire de probabilités de Paris VII Astérisque, Soc. Math. de France (1981).

43. T. Sakai, On Eigenvalues of Laplacian and Curvature on Riemannian Manifolds, Tohoku Math. J., 23:589-603 (1971).

44 P. B. Gilkey, The Index Theorem and the Heat Equation, Publish on Perish Press (1974).

45. R. S. Cahn, P. Gilkey and J. A. Wolf, Heat equation, proportionality principle and volume of fundamental domains, pp. 43-54 in: "Differential Geometry and Relativity", Ed. M. Cahen, M. Flato, D. Reidel, Dordrecht (1976).

46. N. Hurt and R. Hermann, Quantum Statistical Mechanics and Lie Group, Harmonic Analysis, Part A, Math. Sci. Press (1980).

47. E. Lieb, The classical limit of quantum spin systems, Comm. Math. Phys., 31:327-340 (1973).

48a E. Lieb, Estimates for the eigenvalues of the Laplacian and Schrödinger operators, Bull. Am. Math. Soc., 82:751-753 (1976).

48b E. Lieb, On the lowest eigenvalue of the Laplacian for the intersection of two domains, Princeton Univ., Preprint (1983).

49. A. M. Perelomov, Schrödinger equation spectrum and KdV type invariants, Ann. Inst. H. Poincaré, A24:161-164 (1976).

50 S. Helgason, "Differential Geometry, Lie Groups and Symmetric Spaces", Acad. Press (1978).

51 P. S. Cahn and J. A. Wolf, Zeta functions and their asymptotic expansion for compact locally symmetric spaces of negative curvature, Comm. Math. Hel., 51:1-21 (1976).

52. A. Weinstein, Quasi-classical mechanics on spheres, Istituto Nazionale di Alta Matematics Symposia Matematics Vol. XIV (1974).
53. Ii. Kiyotaka, On the multiplicites of the spectrum for quasi-classical mechanics on spheres, Tohoku Math. J., 30:517-524 (1978).
54. S. Molchanov, Diffusion process and Riemannian geometry, Russ. Math. Surveys, 30:1, 1-63 (1975).
55a S. Y. Cheng, Eigenfunctions and nodal sets, Comment. Math. Helv., 51:43-55 (1976).
55b K. Uhlenbeck, Generic properties of eigenfunctions, Amer. J. Math., 98:1059-1078 (1976).
56a T. A. Osborn and R. Wong, Schrödinger spectral kernels: high order asymptotic expansions, J. Math. Phys., 24:1487-1501 (1983).
56b Y. Fujiwara, T. A. Osborn and S. F. J. Wilk, Wigner-Kirkwood expansions, Phys. Rev., A25:14-25 (1982).
56c T. A. Osborn and Y. Fujiwara, Time evolution kernels: uniform asymptotic estimates, J. M. Phys., 24:1093-1103 (1983).
57. S. R. S. Varadhan, Diffusion processes in a small time interval, Comm. Pure Appl. Math., 20:659-685 (1962).
58. M. Donsker and S. R. S. Varadhan, Asymptotic evaluation of certain Markov expectations for large time I, II, III, Comm. Pure Appl. Math., 28:1-47 (1975); 29:279-301 (1976); 29:389-461 (1976).
59. W. Kirsch, Random Schrödinger operators and the density of states, to appear in Proc. Marseille Conf. (1983).
60. W. Kirsch and F. Martinelli, Random Schrödinger operators: recent results and open problems, to appear in Proc. Bielefeld. Encounters "Trends and Developments in the Eighties", Ed. S. Albeverio and Ph. Blanchard, World Publ., Singapore (1984).
61. M. Sablé-Tougeron, Comportement asymptotique des valeurs propres pour une classe d'opérateurs elliptiques dégénérés, Tohoku Math. J., 30:583-605 (1978).
62a B. Simon, Non classical eigenvalue asymptotics, Pasadena Preprint 1983, to appear in J. Funct. Anal., (1983).
62b B. Simon, Schrödinger semigroups, Bull. AMS., 7:447-526 (1982).
62c B. Simon, Semiclassical analysis of low-lying eigenvalues I, II, Ann. Inst. H. Poincaré.
63. G. Jona-Lasinio, F. Martinelli and E. Scoppola, Description of the semiclassical limit of quantum mechanics in terms of diffusion processes, pp. 132-133 in: Math. Publ. in Theor. Phys., Proc. Berlin, 1981, Ed. R. Schrader et al., Lect. Notes Phys., 153, Springer, Berlin (1982).
64. B. Gaveau, Theorèmes de comparaison pour les opérateurs elliptiques en mécanique et en géometrie, Paris VI, Preprint.
65. A. M. Berthier, Spectral theory and wave operators for the Schrödinger equations, Res. Notes Maths., 71, Pitman, Boston (1982).

66. T. Sunada, Trace formula for Hill's operators, Duke Math. J., 47:529-546 (1980).
67. C. Agostinelli, Sulla determinazione degli autovalori nel problema delle vibrazioni diuna membrana con contorno epicicloidale fisso, Rend. Acc. Lincei Roma, 7:316-320 (1949).
68. P. H. Bérard, Spectres et groupes cristallographiques I, II, Inv. Math., 58:179-199 (1980).
69. H. D. Fegan, Special function potentials for the Laplacian, Can. J. Math., 34:1183-1194 (1982).
70. M. Serra, Opérateur de Laplace-Beltrami sur une varieté quasi-périodique, Thèse de III Cycle, Marseille (1983).
71. C. Baird, On a higher dimensional analogue of Neuman motion and harmonic maps into spheres, MIT-Preprint (1983).
72. C. J. Callias, The heat equation with singular coefficients, Commun. Math. Phys., 88:337-385 (1983).
73. S. Minakshisundaram and A. Plejel, Some properties of the eigenfunctions of the Laplace-operator on Riemannian manifolds, Can. J. Math., 1:242-256 (1949).
74. P. J. Bérard, On the wave equation on a compact Riemannian manifold without conjugate points, Math. Z., 155:249-276 (1977).
75. J. J. Duistermaat and V. Guillemin, The spectrum of positive elliptic operators and periodic geodesics, Proc. AMS Summer Inst., Diff. Geom., Stanford, AMS (1973).
76. I. M. Singer, Eigenvalues of the Laplacian and invariant of manifolds, pp. 187-199, Proc. Int. Congr. Math., Vancouver (1974).
77. A. Weinstein, Asymptotics of the eigenvalues clusters for the Laplacian plus a potential, Duke Math. J., 44:883-892 (1972).
78. Y. Colin de Verdière, Quasi-modes sur les varietés Riemanniennes, Inv. Math., 43:15-52 (1977).
79. S. Bando and H. Urakawa, Generic properties of the eigenvalues of the Laplacian for compact Riemannian manifolds, Tohoku Math. J., 35:155-172 (1983).
80. D. Barthel and R. Kümritz, Laplacian with a potential, pp. 19-29 in: "Global Differential Geometry", Lect. Notes Maths., 838, Springer, Berlin (1979).
81a J. V. Ralston, Constructive of approximate eigenfunctions of the Laplacian concentrated near stable closed geodesics, J. Diff. Geom., 12:87-100 (1977).
81b A. Weinstein, Fourier integral operators, quantization and the spectra of Riemannian manifolds, Coll. Int. CNRS., 237 (1974).
81c H. Widom, Eigenvalue distribution theorems for certain homogeneous spaces, J. F. A., 32:139-147 (1979).
81d W. Lichtenstein, Qualitative behavior of special functions on compact symmetric spaces, J. F. A., 34:433-455 (1979).
82. J. Brüning, Invarious eigenfunctions of the Laplacian and their asymptotic distribution, pp. 69-81 in: "Global Differential Geometry", Lect. Notes Math., 838, Springer, Berlin (1979).

83. Ph. Combe, R. Rodriguez, M. Sirugue and M. Sirugue-Collin, High temperature behavior of quantum mechanical thermal functionals, RIMS., 19-355 (1983).
84. J. Vauthier, Théorèmes d'annulation et de finitude d'espaces des 1-formes harmoniques sur une varieté de Riemann ouverte, Bull. Sc. Math., 103:129-177 (1979).
85a R. Brooks, On the spectrum of non compact manifolds with finite volume, Maryland, Preprint (1983).
85b R. Brooks, A relation between growth and the spectrum of the Laplacian, Math. Z., 178:501-508 (1981).
86. J. M. Bismut, The Atiyah-Singer theorems for classical elliptic operators: a probabilistic approach, Orsay Prepr.
87a H. Donnelly, Spectral geometry for certain non compact Riemannian manifolds, Math. Z., 169:63-76 (1979).
87b H. Donnelly, Asymptotic expansions for the compact quotient of property discontinuous group actions, Ill. J. Maths., 25:485-496 (1979).
88. H. Donnelly and P. Li, Pure point spectrum and negative curvature for non compact manifolds, Duke Math. J., 46:497-503 (1979).
89. Ya. I. Belopolskaya and Yu. L. Daletskii, Ito equations and differential geometry, Russ. Math. Surv., 37:109-163 (1982).
90. N. Wallach, The asymptotic formula of Gelfand and Gangolli for the spectrum of G, J. Diff. Geom., 11:91-101 (1976).
91. J. J. Duistermaat, Oscillatory integrals, Lagrange immersions and unfolding of singularities, Comm. Pure Appl. Math., 27:207-281 (1974).
92a Y. Colin de Verdière, Spectre du Laplacien et Longueurs des géodésiques periodiques I, II, Comp. Math., 27:83-106 (1973); 27:159-184 (1973).
92b Y. Colin de Verdière, in: Sémin. Goulaouic-Schwartz, Paris (1973/74).
93a J. Chazarain, Formule de Poisoon pour les varietés Riemanniennes, Inv. Math., 24:65-82 (1974).
93b J. Chazarain, Spectre d'un Hamiltonian quantique et mécanique, classique, Comm. P.D.E., 5:595-644 (1980).
94. J. Leray, Solutions asymptotiques et groupe symplectique, pp. 73-97 in: Fourier Integral Operators and Partial Differential Equation, Ed. J. Chazarain, Lect. Notes Maths., Springer, Berlin (1975).
95. L. Hörmander, Fourier integral operators I, Acta Math., 127: 79-183 (1971); II (with J. J. Duistermaat), 128:183-269 (1972).
96. B. Malgrange, Méthode de la phase stationnaire et sommation de Borel, Complex analysis, microlocal calculus and relativistic quantum theory, Proc. Colloq., Les Houches 1979, Lect. Notes Phys., Springer, Berlin, 126:170-177 (1980).
97a V. P. Maslov, Théorie des perturbations et méthodes asymptotiques, Dunod, Paris (1972).
97b V. P. Maslov and M. V. Fedoriuk, Semiclassical approximation in Quantum Mechanics, D. Reidel, Dordrecht (1981).

98. S. Albeverio and R. Høegh-Krohn, Oscillatory integrals and the method of stationary phase in infinitely many dimensions with applications to the classical limit of quantum mechanics I, Inv. Math., 40:59-106 (1977).
99. K. Watling, A note on [119], work in preparation, Univ. Warwick.
100a I. Davies and A. Truman, On the Laplace asymptotic expansion of conditional Wiener integrals and the Bender-Wu formula for the X^{2N}-anharmonic oscillator, J. Math. Phys., 24:255-266 (1983).
100b I. Davies and A. Truman, Laplace asymptotic expansions of conditional Wiener integrals and generalized Mehler kernel formulas, J. Math. Phys., 23:2059-2070 (1982).
100c I. Davies and A. Truman, Laplace expansions of conditional Wiener integrals and applications to quantum physics, 40-55 in [134].
101a D. Fujiwara, Fundamental solution of partial differential operator of Schrödinger type II, Proc. Jap. Acad., 50:699-701 (1974).
101b A. Takershita, An approach to quasi classical approximation for the Schrödinger equation, Proc. Jap. Acad., 48:365-369 (1972).
102. D. Elworthy and A. Truman, Feynman maps, Cameron-Martin formulae and anharmonic oscillators, Warwick/Swansea Preprint (1983).
103. C. De Witt-Morette, A. Maheshwari and B. Nelson, Path integration in non relativistic quantum mechanics, Phys. Repts., 50:255-372 (1979).
104. R. Azencott and H. Doss, L'equation de Schrödinger quand h tends vers zero: une approche probabiliste, to appear in Proc. Marseille Conf. March 1983, Ed. S. Albeverio et al., Springer, Berlin.
105. J. M. Bismut, Large deviations and the Malliavin calculus, Paris-Sud Preprint (1983).
106. M. V. Berry and K. E. Mount, Semiclassical approximation in wave mechanics, Rep. Prop. Phys., 35:315-397 (1972).
107. I. C. Percival, Semiclassical theory of bound states, Adv. Chem. Phys., 36:1-61 (1977).
108. J. P. Eckman and R. Sénéor, The Maslov-WKB method for the (an)-harmonic oscillator, Arch. Rat. Mech. An., 61:153-173 (1976).
109a I. C. Percival, Regular and irregular spectra, J. Phys., B6: L 220-232 (1973).
109b I. C. Percival, Volta Memorial Conference, Como 1977, Eds. G. Casati and J. Ford, Stochastic behavior in classical and quantum Hamiltonian systems, Lect. Notes in Physics., 93, Springer (1979).
110. M. V. Berry, Waves and Thom's theorem, Adv. in Phys., 25:1-26 (1976).
111. C. Callias and C. H. Taubes, Functional determinants in Euclidean Yang-Mills theory, Comm. Math. Phys., 77:229-250 (1980).

112. A. Voros, The zeta function of the quartic (and homogeneous anharmonic) oscillator, the Riemann problem, complete integrability and arithmetic applications, Proc. Sem. France and USA 1979-1980, *Lect. Notes Math.*, 925:184-208 (1982).
113. G. Parisi, Trace identities for the Schrödinger operator and the WKB method, Nucl. Phys. B.
114. K. Jajima, The quasi-classical limit of quantum scattering theory I, II, *Comm. Math. Phys.*, 69:101-129 (1979); *Duke Math. J.*, 48:1-22 (1981).
115a K. Hepp, The classical limit for quantum mechanical correlation functions, *Comm. Math. Phys.*, 35:265-277 (1974).
115b G. A. Hagedorn, Semiclassical quantum mechanics, *Comm. Math. Phys.*, 71:77-93 (1980).
116. G. G. Emch, Prequantization and KMS structures, *Int. J. Th. Phys.*, 20:891-904 (1981).
117. H. B. Doebner and J. Tolar, Quantum mechanics on homogeneous spaces, *J. Math. Phys.*, 16:975-984 (1975).
118. K. K. Wan and C. Viaminsky, Quantization in spaces of constant curvature, *Progr. Theor. Phys.*, 58:1030-1044 (1977).
119. D. Elworthy and A. Truman, Classical mechanics, the diffusion (heat) equation, and the Schrödinger equation on a Riemannian manifold, *J. Math. Phys.*, 22:2144-2166 (1981).
120. F. A. Berezin, Quantization in complex symmetric spaces, *Math. USSR.*, Izv. 9:341-379 (1975).
121. A. Inoue and Y. Maeda, On integral transformations associated with a certain Riemannian metric I - the path integral in the curved space, Preprint.
122. M. Schilder, Some asymptotic formulas for Wiener integrals, *Trans. Am. Math. Soc.*, 125:63-85 (1966).
123. R. S. Ellis and J. S. Rosen, Laplace's method for Gaussian integrals with an application to statistical mechanics, *Ann. of Prob.*, 10, 1:47-66 (1982).
124. D. Ventcell and M. I. Freidlin, On small random perturbation of dynamical system, *Russ. Math. Surv.*, 25:1-55 (1970).
125. N. Ikeda and Sh. Watanabe, "Stochastic differential equations and diffusion processes", North-Holland Publ., Amsterdam.
126. D. Elworthy, Stochastic Differential Equations on Manifolds, LMS Lect. Notes, Series 70, Cambridge Univ. Press.
127. E. Jørgensens, The central limit problem for geodesic random walk, Z. Wahrsch.-Theorie verw. Geb., 32:1-64 (1975).
128. L. Schulman, A path integral for spin, Ph. Thesis, Princeton (1967).
129. J. S. Dowker, Covariant Schrödinger equation, *in* Functional Integration and its Applications, Ed. A. M. Arthurs, Oxford (1975).
130a M. S. Marinov and M. V. Terentyev, Dynamics on the groups manifold and path integral, *Fortschr. d. Phys.*, 27:511-545 (1979).

130b M. S. Marinov, Path integrals in quantum theory: an outlook of basic concepts, Phys. Repts., 60:2-57 (1980).
130c G. C. Gerry and S. Silverman, Path integral for coherent states of the dynamical group SU (1,1), J. Math. Phys., 23:1935-2003 (1982).
131. B. De Witt, Dynamical theory in curved spaces, Rev. Mod. Phys., 29:377-397 (1957).
132. D. Dohrn and F. Guerra, Nelson's stochastic mechanics on Riemannian manifolds, Lett. Nuovo Cim., 22:121-127 (1978).
133. D. Elworthy and A. Truman, The diffusion equation and classical mechanics: an elementary formula. Stochastic Processes in Quantum Theory and Phsyics: Proc. Marseille 1981, Lect. Notes in Phys., A73, Springer Verlag.
134. S. Albeverio, Ph. Combe and M. Sirugue-Collin, Eds., Stochastic Processes in Quantum Theory and Statisitical Physics, Proc. Marseille 1981, Lect. Notes Phys., 173, Springer, Berlin (1982).
135a A. B. Venkov, Selberg's trace formula, Math. USSR., Izv. 12: 448-462 (1978).
135b A. B. Venkov, Expansion in automorphic eigenfunctions of the Laplace-Beltrami operator in classical symmetric spaces of rank one and the Selberg trace formula, Proc. Stekl. Inst. Math., 125:1-48 (1973).
136. R. Gangolli and G. Warner, Zeta functions of Selberg's type for some non compact quotients of symmetric spaces of rank one
137. G. Warner, Selberg trace formula for non uniform lattices, the R-rank one case, Adv. Math., 6:1-142 (1979).
138. S. Albeverio, Ph. Blanchard, Ph. Combe, R. Høegh-Krohn and M. Sirugue, Local relativistic invariant flows for quantum fields, Comm. Math. Phys., 90:329-351 (1983).
139. S. Albeverio and R. Høegh-Krohn, Dirichlet forms and diffusion processes on rigged Hilbert spaces, Z. Wahrscheinlichk. th. verw. Geb., 40:1-57 (1977).
140. J. Ginibre, Some applications of functional integration in statistical mechanics, pp. 327-427, in: "Statistical Mechanics and Quantum Field Theory", Ed. De Witt, Gordon and Breach (1971).
141. S. Albeverio and R. Høegh-Krohn, Homogeneous random fields and statistical mechanics, J. Funct. Anal., 19:242-272 (1975).
142. J. P. Eckmann, Remarks on the classical limit of quantum field theories, Lect. Math. Phys., 1:387-394 (1977).
143. G. Casati, Irreversibility and chaos in quantum systems, to appear in Lect. Notes in Maths., Springer, Berlin.
144. A. Voros, An algebra of pseudo-differential operators and the asymptotics of quantum mechanics, J. Funct. Anal., 29:104-132 (1978).
145. G. G. Emch, S. Albeverio and J. P. Eckmann, Quasi-free generalized K-flows, Repts. Math. Phys., 13:73-85 (1978).

146. M. Requardt, Some new estimates on eigenfunctions, eigenvalues, expactation values, number of bound states in Schrödinger theory, Göttingen, Preprint 1983, to appear Helv. Phys. Acta.
147. J. M. Combes, P. Duclos and R. Seiler, The Born Oppenheimer approximation, in: Rigorous Atomic and Molecular Physics, Ed. G. Velo and A. Wightman, Plenum (1981).
148. H. Huber, Über die Eigenwerte des Laplace-Operators auf kompakter Riemannscher Fläche, Comment. Math. Helv., 51:215-231 (1976).
149. H. P. McKean, Selberg's trace formula as applied to a compact Riemann surface, Comm. Pure App. Math., 25:225-246 (1972).
150. H. P. W. Gottlieb, Hearing the shape of an annular drum, J. Austr. Math. Soc., B24:435-438 (1983).
151. S. T. Yau, On the heat kernel of a complete Riemannian manifold, J. Math. Pures et Appl., 57:191-201 (1928).
152. B. Randol, Small eigenvalues of the Laplace operator on compact Riemann surfaces, Bull. AMS., 80:996-1000 (1974).
153a H. Huber, Zur analytischen Theorie Hyperbolischer Raumformen Bewegungsgruppen II, Math. Ann., 142:385-398 (1960/61).
153b W. Müller, Spectral theory for Riemannian manifolds with cusps and a related trace formula, Math. Nachr., 111:197-288 (1983).
154. A. Selberg, Harmonic analysis and discontinuous groups in weakly symmetric Riemannian spaces with applications to Dirichlet series, J. Ind. Math. Soc., 20:47-87 (1956).
155. M. Pinsky, The spectrum of the Laplacian on a manifold of negative curvature, J. Diff. Geom., 13:87-91 (1978).
156a G. G. Emch, Quantum and classical mechanics on homogeneous Riemannian manifolds, Rochester Preprint (1982).
156b G. G. Emch, A geometric dequantization program for the solution of the Dirac problem, Rochester Preprint (1982).
157. K. Ichihara, Curvature, geodesics and the Brownian motion on a Riemann manifold II, Nagoya Math., 87:115-125 (1982); I, Nagoya, 87:101-114 (1982).
158. J. J. Duistermaat, J. A. C. Kolk and V. S. Varadayan, Spectra of compact locally symmetric manifolds of negative curvature, Inv. Math., 52:27-93 (1979).
159. B. Gilkey, Lefshetz fixed point formulas and the heat equation, pp. 91-147, in: Partial Differential Equations and Geometry, Ed. C. I. Byrnes and M. Dekker, New York (1979).
160. M. A. Shubin and V. N. Tulowskii, An asymptotic distribution of eigenvalues and pseudo-differential operators on $\mathbb{R}^n$, Math. USSR., Sb. 21:565-583 (1973).
161. V. Guillemin and S. Sternberg, The metaplectic representation, Weyl operator and spectral theory, Differ. Geometric Methods in Math. Phys., Proc. Aix-en-Provence and Slamanca, Lect. Notes in Math., 836:420, Springer Verlag, Berlin (1979).
162. H. D. Fegan, The heat equation on a compact Lie group, TAMS., 246:339-357 (1978).

163. E. Cartan, Notice historique sur la notion de parallelisme absolu, Math. Ann., 102:698-706 (1930).
164a R. Gangolli, On the length spectra of some compact manifolds of negative curvature, J. Diff. Geom., 12:403-424 (1977).
164b R. Gengolli, Zeta functions of Selberg's type for compact space forms of symmetric spaces of rank one, Ill. J. Maths., 21:1-41 (1977).
164c R. Gangolli and G. Warner, On Selberg's trace formula, J. Math. Soc. Jap., 27:328-343 (1975).
164d N. Wallach, On the Selberg trace formula in the case of compact quotient, Bull. AMS., 82:171-196 (1976).
165. R. Brooks, The spectral geometry of solutions, Am. J. Math. (1982-83).
166. A. Uhlman, On a Riemannian method in quantum theory about curved space-time, Cz. J. Phys. (1981).
167. R. Brooks, The fundamental group and the spectrum of the Laplacian, 56:581-598 (1981).
168a T. Sunada, Trace formula and heat equation asymptotics for a non compact manifold, An. J. M., 104:795-812 (1982).
168b N. Subia, Formule de Selberg et formes d'espaces hyperboliques compactes, Lect. Notes Maths., 497:674-700, Springer (1975).
168c A. M. Berthier, On the point spectrum of Schrödinger operators, Ann. Scient. Ec. Norm. Sup., 15:1-15 (1982).
168d Vauthier, Fonction de Green et surfaces nodales des fonctions de l'Hamiltonian $-\frac{1}{2}\Delta + V$, Preprint.
169a J. M. Combes, R. Schrader and R. Seiler, Classical bounds and limits for energy distributions of Hamiltonian operators in electromagnetic fields, Ann. Phys., 111:1-18 (1978).
169b H. Hess, R. Schrader and D. B. Uhlenbrock, Kato's inequality and the spectral distribution of Laplacians on compact Riemannian manifolds, J. Diff. Geom., 15:27-37 (1980).
169c H. Hogreve, J. Potthoff and R. Schrader, Classical limit of quantum particles in external Yang-Mills potentials, in preparation.
170a H. Kitada, A calculus of Fourier integral operator and the global fundamental solution for a Schrödinger equation
170b H. Kitada and H. Kumano-Go, A family of Fourier integral operators and the fundamental solution for a Schrödinger equation
170c D. Fujiwara, Fundamental solution of partial differential operator of Schrödinger's type I, Proc. Jap. Acad., 50: 566-569 (1974).
170d D. Fujiwara, A Construction of the fundamental solution for the Schrödinger equations, Proc. Jap. Acad., 55, Ser. A: 10-14 (1979).
171. J. Tarski, Path integrals over manifolds, Preprint.

172a S. Kusuoka, The Malliavin calculus and the hypo-ellipticity of 2. order degenerate elliptic differential operators, in preparation.
172b S. Kusuoka and D. Stroock, in preparation.
173. P. B. Gilkey, Recursion relations and the asymptotic behavior of the eigenvalues of the Laplacian, Comp. Math., 38:201-240 (1979).
174. B. Grammaticos and A. Voros, Semi-classical approximation of nuclear Hamiltonians, Ann of Phys., 123:359-380 (1979).
175. A. Mayda and J. Ralston, An analogue of Weyl's theorem for unbounded domains, III, Duke Math. J., 46:725-731 (1979).
176a M. A. Pinsky, Brownian motion on a small geodesic ball, to appear in Proc. Schwartz Coll., Astérisque (1983).
176b M. A. Pinsky, Stochastic Taylor formulas and Riemannian geometry, Northwestern Univ., Preprint.
177. H. Tamura, Asymptotic formulas for eigenvalues of elliptic operators of second order, Duke Math. J., 49:87-119 (1982).
178. J. Ginibre and G. Velo, The classical field limit of non-relativistic many-body systems, I Comm. Math Phys., 66:37-76 (1979); II Asymptotic expansions for general potentials, Ann. Inst. H. Poincaré, Nouv. Ser. A33:363-394 (1980).
179a M. V. Berry, Quantization of mappings and other simple classical models, Proc. N.Y. Ac. Sci., Nonlinear Dynamics, Ed. R. Helleman (1979).
179b M. V. Berry, N. L. Balasz and M. Tabor, Quantum maps, Ann. of Phys., 122:26-63 (1979).
180. D. Williams, Instanton gas parameters in the double well model, pp. 295-297 in Proc. Ref. [19].
181. S. Albeverio, Ph. Blanchard and R. Høegh-Krohn, Some application of functional integration, pp. 265-275 in Proc. Ref. [19].
182a L. D. Eskin, The heat equation and the Weierstrass transform on certain symmetric Riemannian spaces, Am. Math. Soc. Transl., 75:239-254 (1968).
182b A. Benabdallah, Noyau de diffusion sur les espaces homogènes compacts, Bull. Soc. Math. Fr., 101:263-265 (1973).
183. G. Casati, Chaos in quantum mechanics, Milano, Preprint.
184a M. C. Gutzwiller, Path integrals and the relation between classical and quantum mechanics, pp. 163-198, in: "Path Integrals", Ed. G. J. Papadopoulos and J. T. Devreese, Plenum (1978).
184b M. C. Gutzwiller, Classical quantization conditions for a dynamical system with stochastic behavior, pp. 316-326 in Proc. Como Meeting 1977, Ref. [18].
185. J. Dodziuk, Eigenvalues of the Laplacian and the heat equation, Am. Math. Monthly, 88:686-695 (1981).
186. M. Pinsky, Stochastic Riemannian geometry, pp. 198-236, in: "Probabilistic analysis and related topics", Ed. A. T. Bharucha-Reid, Acad. Press, New York (1978).

187a M. A. Pinsky, Can you feel the shape of a manifold with Brownian motion, to appear in Exp. Math. (1984).
187b A. Gray and M. Pinsky, The mean exit time for a geodesic ball in a Riemannian manifold, Bull. Sci. Math., 107:1-26 (1983).
187c M. A. Pinsky, Isotropic transport process on a Riemannian manifold, Tras. AMS., 218:353-360 (1976).
188. D. Elworthy and A. Truman, The classical limit of quantum mechanics in a curved space background, in: Proc. Int. Conf. Funct. Integr. Louvain, Ed. J. P. Antoine, Plenum (1979).
189. F. Guerra and L. M. Morato, Quantization of dynamical systems and stochastic control theory, CNRS Marseille Preprint (1982).
190. L. Bernard-Bergery, Laplacian et geodésiques sur les formes d'espace hyperbolique compactes, pp. 107-122, in: Lect. Notes Maths., 317, Springer, Berlin (1973).
191. A. L. Besse, Manifolds all of whose geodesics are closed, Springer, Berlin (1978).
192. K. D. Elworthy, Stochastic methods and differential geometry, Sem. Bourbaki 33e année, (1980/81) (Feb. (1981).
193. G. Casati, I. Guarnieri and F. Valz-Gus, Preliminaries to the ergodic theory of infinite dimensional system: a model of radiant cavity, Preprint.
194. O. Steinmann, Soliton quantization in gauge theories, Z. Phys., C6:139-154 (1980).
195. D. Dürr, Diffusions and central limit theorem, to appear in Proc. Ref. [104].
196. A. Peres, Regular and irregular energy spectra, these Proc.
197a J. Harthong, Formula de Poisson pour les varietés à bord. Une nouvelle méthods inspirée par G. D. Birkhoff, Publ. Rech. Math., Strasbourg.
197b J. Harthong, La propagation des ondes, Publ. Rech. Math., Strasbourg.
198. J. Rezende, "Remark on the solution of the Schrödinger equation for anharmonic oscillators via Feynman path integral", BI TP 82/22, Bielefeld preprint, August 1982 and work in preparation.
199. P. A. Meyer, Géométrie différentielle stochastique (bis), Sém. Prob. XVI, 1980/81, Lect. Notes Maths., 921, Springer, Berlin (1982).
200. S. Levit, Semi-classical approximation to path integrals-phases and catastrophes, pp. 481-491 in Ref. [184a].
201. J. P. Antoine and E. Tirapegui, Functional Integration, Plenum Press, New York (1980).
202. D. Fujiwara, Remarks on convergence of Feynman path integrals, pp. 40-48, in: "Theory and Application of Random Fields", Proc. Bangalore, 1982, Ed. G. Kallianpur, Lect. Notes Control and Inf. Sci., Springer, Berlin (1983).

203a R. Carmona, Pointwise bounds for Schrödinger eigenstates, Comm. Math. Phys., 62:97-106 (1978).
203b R. Carmona and B. Simon, Pointwise bounds on eigenfunctions and wave packets in N-body quantum systems, V, Comm. Math. Phys., 80:59-98 (1981).

LIFSHITZ SINGULARITY OF THE INTEGRATED DENSITY OF STATES AND ABSENCE OF DIFFUSION NEAR THE BOTTOM OF THE SPECTRUM FOR A RANDOM HAMILTONIAN

Fabio Martinelli

Istituto Matematico
Università di Trento
Povo Trento, Italy

Helge Holden

Matematisk Institutt
Universitet i Oslo
Blindern, Oslo 3, Norway

Let us consider the following simple model of a stochastic Hamiltonian which describes the motion of a quantum particle interacting with a random potential:

$$H_\omega = -\Delta + \Sigma_i q_i(\omega)\chi_{C_i}(x) \text{ on } L^2(R^\nu). \tag{1}$$

Here C_i is a covering of R^ν by unit cubes around the sites of Z^ν and $q_i(\omega)$ are independent identically distributed random variables with common distribution $dP(q_0 < \lambda)$. For simplicity we assume $\langle \exp(t|q_0|)\rangle < +\infty$ for any $t > 0$, where $\langle ...\rangle$ denotes expectation with respect to $dP(q_0 < \lambda)$. An important role in the physics of the above model is played by the distribution of the eigenvalues of H_ω $N(E)$ defined below. We will refer in the following to $N(E)$ as the integrated density of states. In this note we will report on some old and recent results concerning $N(E)$ and on their application to the analysis of the diffusive behavior of the model. For general background concerning stochastic Hamiltonians we refer to [1,2] (see also Souillard in this book).

Let now Λ be a bounded cube and let $H^D_\Lambda(\omega)$, $H^N_\Lambda(\omega)$ be the restriction of H_ω to $L^2(\Lambda)$ with Dirichlet and Neumann boundary conditions respectively and let $\lambda_k(H^D$, $\lambda_k(H^N)$ be their eigenvalues counting multiplicity. We define the finite volume integrated density of states by:

$$\begin{aligned} N^D_\Lambda(E,\omega) &= |\Lambda|^{-1}\ (k;\ \lambda_k(H^D) < E) \\ N^N_\Lambda(E,\omega) &= |\Lambda|^{-1}\ (k;\ \lambda_k(H^N) < E) \end{aligned} \tag{2}$$

It is easy to see that $N_\Lambda^N(E,\omega) > N_\Lambda^D(E,\omega)$ for any ω, E, Λ. Let us now divide the cube Λ into smaller cubes Λ_i in such a way that $\Lambda = \cup\Lambda_i$ and $\Lambda_i \cap \Lambda_j = \emptyset$ for $i \neq j$. Then the Dirichlet-Neumann bracketing gives:

$$N_\Lambda^D(E,\omega) > \Sigma_i N_\Lambda^D(E,\omega)|\Lambda|^{-1}|\Lambda_i|$$
$$N_\Lambda^N(E,\omega) < \Sigma_i N_\Lambda^N(E,\omega)|\Lambda|^{-1}|\Lambda_i|. \tag{3}$$

Using Equation (3), the subadditive ergodic theorem (see [3]) and the assumption $\langle\exp(t|q_0|)\rangle < \infty$ we can prove [4]:

Theorem 1 $\qquad \lim_{\Lambda\to R^\nu} N_\Lambda^D(E,\omega) = \lim_{\Lambda\to R^\nu} N_\Lambda^N(E,\omega) = N(E)$ a.s.

where N(E) is a non random non decreasing function of E constant on the complement of spec(H_ω) given by:

$$\inf_\Lambda \langle N_\Lambda^N(E,\omega)\rangle = N(E) = \sup_\Lambda \langle(N_\Lambda^D(E,\omega)\rangle$$

This theorem is particularly useful since we can read off properties of the integrated density of states N(E) from the properties of its finite volume analogue.

For example it is possible to obtain rather easily [4] the asymptotic behavior of N(E) as $E \to \pm\infty$:

Theorem 2 $\qquad$ a) $\lim_{E\to+\infty} N(E)E^{-\nu/2} = (2\pi)^{-\nu}\tau_\nu$

where τ_ν is the volume of the unit sphere in R^ν.

b) $\int_{-\infty}^{-1}|E|^\gamma dN(E) < K_{\gamma,\nu}\langle|q_0|^{\gamma+\nu/2}\rangle$.

Statement a) is just the Weyl's result (see, for example, S. Albeverio in this book) while b) suggests that for $E \to -\infty$, N(E) behaves as $|E|^{\nu/2}P(q_0 < E)$. In the case of Gaussian q_0 one can actually prove that N(E) is asymptotically Gaussian for large negative energies E.

An alternative way to prove theorem 2 is to compute the Laplace transform of N(E):

$$L(t) = \int\exp(-tE)dN(E). \tag{4}$$

Using subsequently Tauberian type of theorems, one obtains the asymptotic behavior of N(E) as $E \to \pm\infty$ by analysing the short (long) time behavior respectively of L(t). The ergodic theorem and the Feynmann-Kac formula give:

Theorem 3 $L(t) = (2\pi t)^{-\nu/2}\int_{C_o} E_{x,x}^{o,t} < \exp(-\int_0^t ds V_\omega(X_s)) > dx$

where $V_\omega(x) = \Sigma q_i(\omega)\chi_{C_i}(x)$ and $E_{x,x}^{o,t}$ denotes the expectation over the pinned Brownian motion X(s) starting at x and ending at x for $s = t$.

We remark here that estimate b) of the above theorem gives a very bad estimate of N(E) near the left edge of the spectrum of H_ω in the case of bounded q_i's. The reason for this is that in this situation (e.g., $q_i \varepsilon (0,1)$ a.s.), contrary to the case of, for example, Gaussian q_i's, the low lying states are no longer substantiated by local (i.e., in few cubes C_i) abnormal fluctuations of the potential V_ω but by the collective behavior of the q_i's in a large region Λ, how large depending on E. It is then natural to expect some kind of universal behavior for N(E) near the left edge of the spectrum. This in fact is the case and one proves:

Theorem 4 Let inf support of $P(q_o < \lambda) = 0$. Then there exist two constants k_1 and k_2 greater than zero such that:

$$\exp(-k_2 E^{-\nu/2}) < N(E) < \exp(-k_1 E^{-\nu/2})$$

as $E \to 0^+$.

We remark here that the above theorem has been proved for a very large class of models (see [5]). For a special model different from ours, it has been shown (see [6]) using theorem 3 and Donsker-Varadhan type of techniques that N(E) is asymptotically equal to $\exp(-k\, E^{-\nu/2})$ for an explicit k.

A simple proof of theorem 4 is based on theorem 1 and on the following two estimates valid for a cube $\Lambda(E)$ of size $L(E) \sim E^{-1/2}$:

$$P(\lambda_1(H^D_{\Lambda(E)}) < E) > \exp(-k_2 E^{-\nu/2})$$
$$P(\lambda_1(H^N_{\Lambda(E)}) < E) < \exp(-k_1 E^{-\nu/2}). \qquad (5)$$

The reason for the first one to hold is that on $\Lambda(E)$ one has: $-\Delta_{\Lambda(E)} > \text{const}.E$; thus in order to have $\lambda_1(H^D_{\Lambda(E)}) < E$ we need: $V_\omega(x) \sim E$ for all $x \varepsilon \Lambda$ and this clearly happens with a probability of order of the righthand side of Equation (5). A slightly more sophisticated argument is required for the second estimate since in this case one needs a lower bound for $\lambda_1(H^N)$. We refer the reader to [1] for a more precise but still heuristic version of the above argument. Completely open is the question whether N(E) admits a density $\rho(E) = \frac{dN(E)}{dE}$. Although we have no results in this direction we can prove, using essentially a result due to Wegner for the Anderson model [7], the following estimate on the finite volume integrated density of states:

Theorem 5 $$\frac{\langle dN_\Lambda(E)\rangle}{dE} < \frac{|dP_0(\lambda)|_\infty |\Lambda| \text{const.} E^{\nu/2}}{d}.$$

This estimate, although non uniform in the volume Λ, plays an important role in the Fröhlich-Spencer approach to Anderson localization.

Absence of Diffusion near the Bottom of the Spectrum

In the second part of this note we explain briefly the role played by theorem 4 in the analysis of the diffusive behavior of the model in consideration for small energies. For what follows we assume:

1) $q_0 \varepsilon (0,1)$ a.s. ; $P(q_0 < 1/2) = \alpha < 1$

ii) $\delta^{-1} = \left|\frac{dP}{d\lambda}(q_0 < \lambda)\right|_\infty < +\infty$.

Let Ψ_E now be a function well localized in space and belonging to the spectral subspace of H_ω with energy within $E \pm \sigma$, $0 < \sigma <<$, and let us consider the following quantity:

$$r_E^2(t) = \langle \int dx |x|^2 |(\exp(itH_\omega)\Psi_E)(x)|^2 \rangle. \tag{6}$$

The asymptotic behavior of $r_E^2(t)$ for large t gives a measure of the diffusion of the particle. More precisely in the case of finite diffusion one expects that $r^2(t)$ behaves as:

$$r_E^2(t) \sim D(E)t \text{ as } t \to +\infty$$

where $0 < D(E) < +\infty$ is the diffusion constant. It is not difficult to show that the study of $r^2(t)$ as $t \to +\infty$ can be reduced to the analysis of:

$$\lim_{\varepsilon \to 0} \varepsilon^2 \langle \int dx |x|^2 |G(\omega, E + i\varepsilon, 0, x)|^2 \rangle \tag{7}$$

with $E \varepsilon \operatorname{spec}(H_\omega)$ and $G(E + i\varepsilon, \omega, 0, x) = (H_\omega - E - i\varepsilon)^{-1}(0,x)$.

For such a quantity in the Anderson model (i.e., the lattice version of our model), Fröhlich and Spencer recently developed [8, 9] a very powerful technique to prove that for energies in a suitable range, the Green's function $G(E + i\varepsilon, \omega, 0, x)$ decays exponentially in $|x|$ with probability one with "mass" m(E) bounded away from zero uniformly in ε. This in turn implies that for energies E in the same range, the corresponding $r_E^2(t)$ has an ergodic mean which vanishes. Their proof is based on quantum mechanical perturbation theory about an infinite sequence of block Hamiltonians. The blocks correspond to regions where the potential V_ω is singular in the sense that the eigenvalues of the corresponding Hamiltonian with Dirichlet boundary conditions are close to the given energy E. As the size of the blocks increases the eigenvalues of the corresponding Hamiltonian are permitted to get closer to E. The divergent terms appearing in $(H_{block} - E - i\varepsilon)^{-1}$ are killed by the exponential decay of $(H_{block} - E - i\varepsilon)^{-1}$ on a scale of order of the size of the block. In order to remove perturbatively the Dirichlet conditions at the boundaries of the blocks it is necessary to control the tunneling between singular regions and for this the blocks have to be very well separated one from the other. Let $S_E(\omega)$ now be the classically allowed region, i.e.,

$$S_E(\omega) = \{j\varepsilon Z^\nu;\ q_j(\omega) < E\}. \tag{8}$$

Then as the set of singular sites we could choose the set $S_{2E}(\omega)$ since if $\Lambda \cap S_{2E}(\omega) = \emptyset$ then $H^D(\omega) > 2E$ and the corresponding Green's function $(H^D - E - i\varepsilon)^{-1}$ decays exponentially with mass of order $E^{1/2}$. With this choice of the set of singular sites one could perform Fröhlich-Spencer perturbation arguments for energies E so small that the corresponding S_E would consist only of isolated lakes well separated one from the other. However, it is possible to widen the range of energies for which one has exponential decay of the Green's function if we keep into account also the tunneling effects in momentum space by changing our choice of the set of singular sites. A suggestion in this sense comes from theorem 4 and is the following:

Let $Z^\nu(E) = L(E)Z^\nu$ where $L(E) \sim E^{-1/2}$ and let $\{C_E(j)\}\ j\varepsilon Z(E)$ be a covering of R^ν by cubes of size L(E) around the sites of $Z^\nu(E)$; then we can say that $j \varepsilon Z^\nu(E)$ is singular if

$$\lambda_1(H^N_{C_E(j)}) < 2E. \tag{9}$$

Using Dirichlet-Neumann bracketing it is easy to see that if $\Lambda \subset R^\nu$ is such that any $j \varepsilon \Lambda$, $j \varepsilon Z^\nu(E)$ is non-singular in the above sense then $H^D_\Lambda > 2E$ so that one has again exponential decay of the Green's function of H^D_Λ. Furthermore, the probability of 0 to be singular has been estimated in theorem 4 and the estimate shows that even if the classically allowed region S_E contains a sea (= infinite cluster of nearest neighbor cubes C_i, $i\varepsilon Z$), it never contains an infinite cluster of nearest neighbor cubes $C_E(j)$ for E sufficiently small. By performing Fröhlich-Spencer arguments on clusters of cubes $C_E(j)$ one gets exponential decay of the Green's function of H_ω for energies less or equal to a threshold $E^0(\alpha,\delta)$ of the form:

$$E^0(\alpha,\delta) = \min\{E^0(\alpha),\ \ln\left(\frac{1}{\delta_E{}^1(\alpha)}\right)^{-2/\nu}\} \tag{10}$$

where $E^0(\alpha)$ and $E^1(\alpha)$ are small constants independent of δ. For a rigorous proof of this last result we refer the reader to [10].

REFERENCES

1. I. M. Lifshitz, Sov. Phys., Usp. 7, no. 4:549-73 (1965).
2. H. Kunz and B. Souillard, Comm. Math. Phys., 78:201 (1980).
3. M. Akcoglu and U. J. Krengel, Reine Angew. Math., 323:53-67 (1981).
4. W. Kirsch and F. Martinelli, J. Phys. A.:Math. Gen., 15:2139-2156 (1982).
5. W. Kirsch and F. Martinelli, Comm. Math. Phys., 89:27-40 (1983).
6. S. Nakao, Jap. J. Math., 3:111-139 (1977).
7. F. Wegner, Z. Physik, B44:9-15 (1981).
8. J. Fröhlich and T. Spencer, Comm. Math. Phys., 88:151-184 (1983).
9. J. Fröhlich and T. Spencer, A rigorous approach to Anderson localization, Preprint ETH (1983).
10. H. Holden and F. Martinelli, Absence of diffusion near the bottom of the spectrum for a random Schrödinger operator on $L^2(R^\nu)$, Preprint Bochum (1983), (to appear in Comm. Math. Phys.).

FLUCTUATIONS IN NUCLEAR SPECTRA*

S.S.M. Wong
Department of Physics
University of Toronto
Toronto, Canada

Introduction

The nucleus is a many-body system consisting of anywhere between 2 to around 250 nucleons interacting with each other through a two-body force. Since there are roughly 1600 different possible nuclei, each capable of existing in a large number of excited states accessible to experimental measurements, nuclear physics becomes one of the ideal grounds for statistical studies. We shall be concerned only with the fluctuations of energy level positions here. Space limitation prevents us from getting into the vast and interesting subject of the statistics of nuclear reaction and transition strengths. A more comprehensive review can be found in ref.[1,2]

For our purpose, we shall think of the nucleus as a system of interacting nucleons and ignore the small differences between a proton and a neutron. The possible existence of three and higher particle rank forces has neither being established nor completely discarded in nuclear physics. Either way, their effects, if any, are sufficiently negligible for us to disregard. The same is not true of one-body terms, which may arise from the kinetic energy of individual nucleon motion or Hartree-Fock single particle effects. However, they do not affect the fluctuation properties of level positions in a direct way: we can therefore forget

*Work supported in part by the Natural Science and Engineering Research Council of Canada.

about them for most purposes. For these reasons, we can describe our interest simply as the fluctuations of energy levels due to two-body interaction between fermions.

The study must be based on measured quantities or else we will lose sight of the physics. Experimental data are, however, subjected to technical limitations. As a result, numerical simulation of level positions becomes an important supplement to the study. For nuclei, the shell model is the primary tool for such work. In addition, random matrix ensembles are introduced so as to avoid the danger of drawing conclusions based on one or two assumed Hamiltonians. For an ensemble to be useful, it must have as much as possible the known information of the nuclear system built into it. The two-body random ensemble (TBRE) is, perhaps, the one which comes the closest to meet this criterion. On the other hand, for mathematical tractability, the Gaussian random ensemble (GOE) is historically used instead. The basis of taking the ensemble approach comes from the statistical mechanics notion that, if the ensemble distribution of a quantity is sharply peaked, the ensemble average provides a good representation of the value that can be obtained with the true system.

In this talk, we shall first, for the benefit of people outside nuclear physics, briefly review the experimental situation. A short desccription of the numerical simulation of data as well as the two most commonly used random matrix ensembles are given next. The nuclear-table ensemble, formed out of experimental data near the ground state, is then shown to provide certain interesting and important results. From these studies, we shall show that, although TBRE and GOE are quite different in their physical origin and low order moments, they have identical fluctuation properties. Furthermore, as far as the available measures are concerned, these properties are the same for experimental data as well as numerical simulation results and extend over well separated energy domains. The fact that the same fluctuation patterns are obtained from widely different assumptions as well as from experimental data implies that the results must be common to quantum mechanical systems of interacting fermions under a fairly general set of conditions.

Experimental situation

In order to study the distribtuion of energy levels, we need a large number of states within a narrow energy span. The small energy interval is to ensure that the average level spacing D, as well as other possible underlying scaling parameters, remains unchanged for all the levels. In addition, the nuclear Hamiltonian has certain exact symmetries, such as spin (J), parity (π) and sometimes isospin (T). Consequently, each state is characterized by a set of good quantum numbers (J^{π}, T). Mainly because of the lack of level repulsion, relationship between states with different good quantum numbers are quite different from those between states of the same ones. Hence, in addition to energy, complete (J^{π}, T) identification of a sequence of levels is also necessary for fluctuation studies.

Experimentally there are mainly two energy domains that adequate data are avialable. The first is the low-lying region starting from the ground state up to energies where the level density becomes too high and too many competitive decay channels avialable for clear spin and parity assignments to all the levels. For light nuclei, say $A \simeq 40$, the upper limit is around 7 MeV containing about 20 levels with 10 different (J^{π}, T) combinations. As a result, only one to two spacings between levels of the same (J^{π}, T) in each nucleus are available for fluctuation studies. The situation is not improved with heavier nuclei.

A second region lies just above the neutron separation threshold, around 15 MeV in light nuclei and 7 MeV in heavy ones. The absorption of slow neutrons, up to few keV in energy, in many nuclei is dominated by s-wave ($l = 0$) resonaces. If the target nucleus, such as ^{166}Er, has $J^{\pi} = 0^+$ ground state, then the final states formed by capturing a s-wave neutron have $J^{\pi} = 1/2^+$ (since a neutron has intrinsic spin 1/2). As a result, series up to ≈100 levels within an energy span of a few keV have been identified.[3] The purity of the sequence depends in part on the strength of p-wave ($l = 1$) resonances, some of which may be strong enough to be detected and mistaken to be weak s-wave ones, and in part on the instrumental resolution which determines the minimum strength of a resonance before it can be registered. In the ^{166}Er case, it is believed that out of a sequence of 109 levels there are four possible p-wave resonances included and five s-wave ones undetected. Unambiguous J^{π} identification for a level normally requires an angular distribution study. This is, however, too tedious to be carried out in a region of such high level density for neutrons.

Proton resonance data are also available at somewhat lower energies.[4] The level density in this region is much smaller and, as a result, a larger energy interval is needed. It is therefore necessary to "unfold" the spectrum so as to take away the variation of D as a function of energy. This is in general fairly reliable thing to do except when there is the complication of *doorway* states. Because of the lower level density and the fact that protons are involved, better identification of orbital angular momentum l can be achieved with proton resonances.

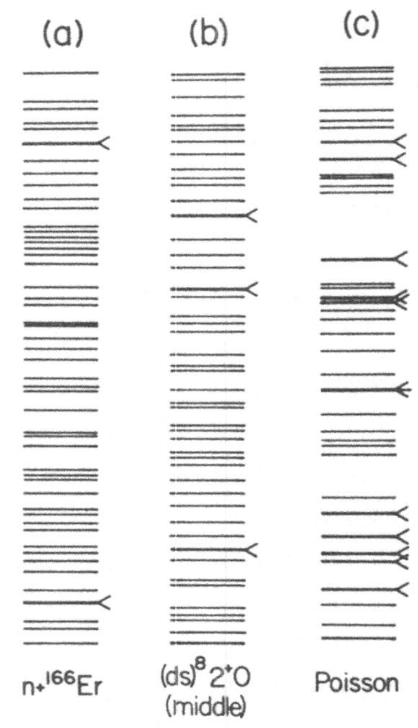

Fig. 1. A section of 50 levels taken from (a) neutron resonance on ^{166}Er target, (b) shell model eigenvalues for $(ds)^8(2^+,0)$ and (c) Poisson spectrum. Arrowheads mark spacings less than $D/4$.

Most of the experimental measurements have gone through certain amount of screening using statistical tests so as to have some idea of the quality of the data. The amount of *adjustments* made to the data is not insignicant even in case of ^{166}Er, one of the best experimental sequences available. As we shall see later, these "tests" are by no means unrelated to the statistical measures used to study the fluctuation properties of the spectrum. Caution is therefore called for in many cases. Much more rigorous statistical analyses are possible[5] but have not been used in most of the experimental data reductions.

A 50 level section of the neutron resonance on ^{166}Er data is shown in Fig. 1a. The density of states is constant for the entire 109 levels: the spectrum, however, departs noticeably from a smooth, *i.e.*, evenly spaced, one. To see the *fluctuation*, we can plot the distribution of nearest neighbour spacing

$$S_i = (E_{i+1} - E_i), \tag{1}$$

where E_i is the energy of the ith level, in the form of a histogram. The result for the entire 108 spacings is shown in Fig. 2a. The smooth curve

$$P(S) = \frac{\pi S}{2D^2} \exp\{-\frac{\pi S^2}{4D^2}\} \qquad S \geq 0 \qquad (2)$$

is called the Wigner surmise[6,7] derived using level repulsion argument. The form was shown by Mehta and Gaudin[8] to be extremely good approximation of the exact results.

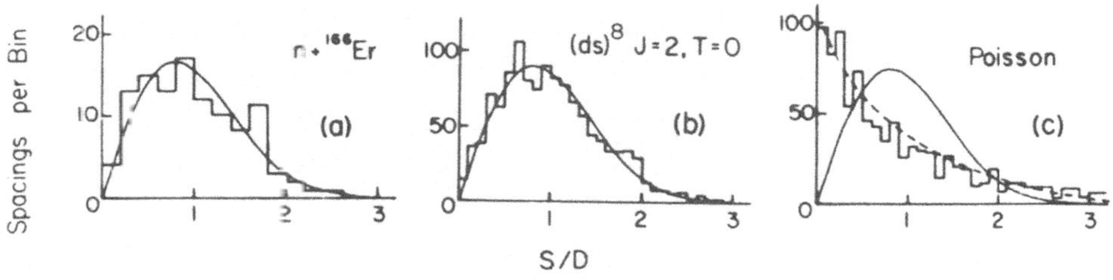

Fig. 2. Nearest neighbour spacing distribution for the three cases of Fig. 1. The smooth curves are based on Wigner surmise and the dashed curve in (c) is $\sim \exp -S/D$

In contrast, it is interesting to compare with the case of replacing the energy levels, E_i, by a set of random numbers arranged in ascending order. The levels in such a ***Poisson*** spectrum are completely uncorrelated: therefore there is no repulsion between them. The distribution of S, instead of (2), changes to $P(S) = \frac{1}{D}\exp\{-\frac{S}{D}\}$. More quantitative measures of fluctuation are described later.

Numerical simulation of data using the nuclear shell model

Partly because of the scarcity of good quality data, numerical modelling becomes an integral part of fluctuation studies. If the nuclear Hamiltonian is assumed to be known, the energy levels can be calculated, say, using the nuclear shell model. The need of a model comes from the following reasons. The Hilbert space for a nucleus is in principle an infinite one. By choosing the basis representation carefully, it is possible to solve the nuclear eigenvalue problem, at least for the low-lying states, in a finite space of dimension d.

The basic assumption of the nuclear shell model is that the many-particle states are made of antisymmetrized (required by Pauli exclusion principle) and normalized products of single particle states which are solutions of an A-particle Hartree-Fock equation. This is equivalent to a rotation of the basis representation so that the dominant part of the nuclear Hamiltonian is contained in the single particle energies. In this way, the ground state of a nucleus can be considered, in a first approximation, as made up by filling the lowest A single particle states. The highest filled level is given the name Fermi level. Excited states are formed by promoting particles below the Fermi level (holes) to orbits above (particles). The excitation energy is then given by the difference between those of particles and holes. Such a one-body Hamiltonian approach is quite successful in understanding certain gross features of nuclei such as level density. However, for fluctuations studies, we need the details in the energy level positions given by the two-bdoy residual interaction, the part of nuclear interaction that cannot be accounted for by single particle energies alone.

In the A-particle basis representation we have chosen here, the one-body Hamiltonian can only contribute to the diagonal matrix elements. The residual interaction, on the other hand, has non-zero off-diagonal matrix elements and can therefore admix different basis states. However, if the Hartree-Fock assumption is a valid one, the residual interaction will be ineffective to mix state that are well separated by the one-body part. As a result, we can ignore the high-lying single particle orbits if our interest is in the low-lying part of the spectrum. Thus, the shell model space is a finite one defined by a set of single particle states.

The eigenvalues obtained from diagonalizing the matrix are the energy levels of the A-nucleon system for the given Hamiltonian and active space. It is therefore not difficult to see that, to a large extext, the one-body Hamiltonian determines the level density in a given energy span while the two-body part fixes the detailed position where a particular level is to be located.

So long as the number of active fermions $m \gg 2$, the distribution of shell model eigenvalues seldom departs from Gaussian unless the Hamiltonian used has very limited number of degrees of freedom. This phenomenon is now well understood.[9,10] Let the maximum particle rank of the interaction be represented by p ($p = 2$ for two-body interaction), so long as $m \gg p$, the eigenvalue distribution is Gaussian. As p is increased to be close to m, the level density distribution becomes semi-circular.

This is the convenient place to point out that the off-diagonal matrix elements between states of different exact quantum numbers (J^π, T) vanish becase of the exact symmetries of the Hamiltonian. As a result, states with different (J^π, T) are not mixed by the Hamiltonian. In fact, they can be handled as separate matrices. The relationships between eigenvalues obtained for different (J^π, T) come only from the fact that the same Hamiltonian and active space are used. In contrast, states of the same (J^π, T) are much more strongly correlated by virtue of the fact they are the eigenvalues of the same matrix. Unless there are other symmetries in the Hamiltonian used, it is highly unlikely that two of the eigenvalues are degenerate, a phenomenon sometimes known as ***level repulsion.***

Since the number of active nucleon is usually greater than 2, the particle rank of the residual interaction, many of the off-diagonal matrix elements remian to be zero. Thus the shell model matrix is in general quite sparse and has the form of a band matrix. However, not much use has been made of these properties. More important to us, it is fairly certain than none of the conclusions we shall be drawing depends in a direct way to this particular mathematical form of the matrix.

Shell model studies are quite successful in explaining many of the observed nuclear properties at low energies. The major reason for discrepancies can usually be attributed to the small size of the space and/or unphysical Hamiltonian used. Our interest in the nuclear shell model here lies solely as a tool for numerical simulation of "data" since it seems to be fairly realistic in representing various nuclei. With shell model eigenvalues, there is no longer any question as to the purity of the sequence. However, since the eigenvalue problem must be solve numerically, there is a very severe restriction on the maximum size of the matrix one can handle. As a result, it is not possible to obtain a long sequence within a region of constant level density. This difficulty is usually circumvented by unfolding, *i.e.,* map the eigenvalue distribution to a constant density one by using some adopted form of the secular variation with energy of level density parametrized in terms of low order moments of the distribution. So long as there is a clear separation of the role of low and high order moments, the removal of low order ones by unfolding cannot affect the fluctuation behaviour of the eigenvalue distribution. As we can see from Figs. 1 and 2, the shell model spectrum and its nearest neighbour spacing distribution are very similar to those observed in experimental data. This is also true for the other fluctuation measures described ahead.

Two-body random ensemble

Since the nuclear interaction is not completely known, there is the danger that conclusions drawn from shell model calculations are merely artifices of the particular Hamiltonian used. It is therefore worthwhile to try many different *reasonable* Hamiltonians. Let us ignore the one-body part for the moment; the nuclear Hamiltonian is then completely defined by giving all the matrix elements in the space of two particles. The most unbiased two-body Hamiltonian is, then, one with all the two-body matrix elements given by random numbers. Symmetry considerations require that the matrix to be real and symmetric and all the off-diagonal elements between basis states with different good quantum numbers vanish: no other selection goes into the choice of the Hamiltonian. Propagation of the two-body matrix to the m-particle space of interest can be carried out using the same shell model technique described earlier. The eigenvalues in the m-particle space then give the spectrum for an "arbitrary" two-body Hamiltonian. A collection of many such spectra calculated for the same (J^{π}, T) case in the m-particle space but starting with different uncorrelated sets of random numbers as two-body matrix elements constitutes the two-body random ensemble (TBRE).

For the most part, TBRE must be worked out numerically. Analytical studies can be made for some of the aspects if one is willing to give up (J^{π}, T) symmetries and assume a dilute system, *i.e.*, the number of active nucleons is much less than the number of available single particle states so that the effects of Pauli exclusion principle may be ignored.[1,10] The amount of numerical effort can also greatly reduced if only the projection of spin $\vec{J}$ operator is retained as the good quantum number.[11] None of the conclusions to be drawn later is known to be aff ected by these approxiations.

Gaussian orthogonal ensemble

GOE has been used from the beginning of random matrix studies for nuclear spectra.[7,12] Instead of starting from two-particle space, GOE works directly with the m-particle space.* All the elements of

*Since no single particle structure is used explicitly, the number of active nucleons does not appear in an obvious way.

the m-particle matrix are assumed to be random. In addition to be real and symmetric, the only other constraint is that the distribution of the values of the matrix elements is invariant under a similarity transformation of the basis. Although this requirement does not seem to affect any of the results, it is important for the following physical argument. Since the representation we have chosen to realize the matrix is an arbitrary one, the result must be invariant under a rotation of the basis. In order that the distribution of the matrix elements is not changed by such a rotation, the off-diagonal one should be Gaussian with zero centroid and unit variance (to set the scale) and the diagonal ones with a variance of two. GOE is very attractive also from the point of view that many elegant analytical treatments can be carried out.[13)]

It is well known that the eigenvalue distribution from such a matrix is semi-circular with very small probability for departure. It was the concern in the early days that the distribution has the wrong form to match the low-energy behaviour of experimental density which has an exponential rise roughly of the form $\sim \exp(a\sqrt{E})$. The reason for the difference from TBRE and shell model results is quite interesting from a physical point of view. Since a GOE matrix starts from the m-particle space, it has the m-particle space as the defining space. For such a matrix, the underlying Hamiltonian is dominated by m-particle interaction. This can be understood from the following dimensional arguments alone. For

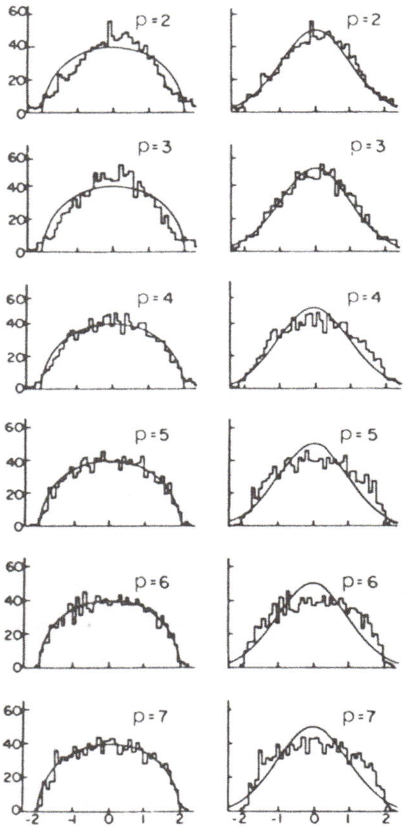

Fig. 3. Level density distribution for a seven-particle 50-dimensional ensemble showing that as p, the particle rank of the interaction, is increased from 2 to 7, the density distribution changes from GAUssian to semi-circular (taken from ref.[9)]).

a space with a total of N single particle states, the dimension of two-particle space is given by $d_2 = \binom{N}{2}$ and the m-particle space by $d_m = \binom{N}{m}$. For a symmetric matrix of dimension d, the number of independent matrix elements is $d(d+1)/2$ if we ignore the constraints imposed by (J^π, T). Hence, for $m \gg 2$ (and $N > m/2$), the number of high particle rank ($p \lesssim m$) matrix element greatly overwhelms those of low particle ranks ($p \approx 2$). The propagation of Hamiltonian matrix elements from two to m particle space does not change the argument very much since on the average the propagator has the value $\binom{m}{2}$.[14]

However, the level density is only a function of the low order moments of the eigenvalue distribution. Fluctuations, on the other hand, happen on much shorter energy scales and are therefore mainly sensitive to the extremely high order moments, perhaps those comparable to the matrix dimension d. As we shall see ahead, in spite of the fundamental differences in the ***physics*** of the ensembles, their fluctuation behaviours are essentially identical.

Nuclear-table ensemble

As mentioned earlier, there are a few spacings between states of the same (J^π, T) in the ground state region known for many nuclei. Together they constitute a large sample of experimental spacing data with well identified energy, spin, parity and isospin. The only trouble is that the underlying scaling parameter D varies from nucleus to nucleus. As a result, we cannot make statistical studies with them unless we have a model for the variation of D as a function of nucleon number A. This is a kind of "unfolding" except, instead of energy, we are trying to remove the A dependence of D.

In general average spacing in nuclei decreases with increasing nucleon number A. Local departures from this simple rule exist for the very light ones, those near closed shells and "collective" nuclei. If data from these "special" cases are excluded from the sampling, a $D \sim A^{-1}$ rule is found to be adequate for scaling all the remaining spacings. Such a collection of experimental data is called the nuclear-table ensemble.[15] The nearest neighbour spacing distribution of the ensemble is shown in Fig. 4a. The agreement with the Wigner surmise is quite good. In contrast, if the ground state spacing, *i.e.*, the spacing between the ground state and the first excited state regardless of the (J^π, T) values (they are different from each other in most of the cases), the distribution, after scaled by the same $D \sim A^{-1}$ rule, is close to a Poisson case as shown in Fig. 4b.

The result is significant for several reasons. The first is that nearest neighbour spacings between states of different (J^{π}, T) are uncorrelated. This is not a surprise but experimentally this is one of the very few places that such observations can be made. Numerically it is confirmed by shell model simulations.

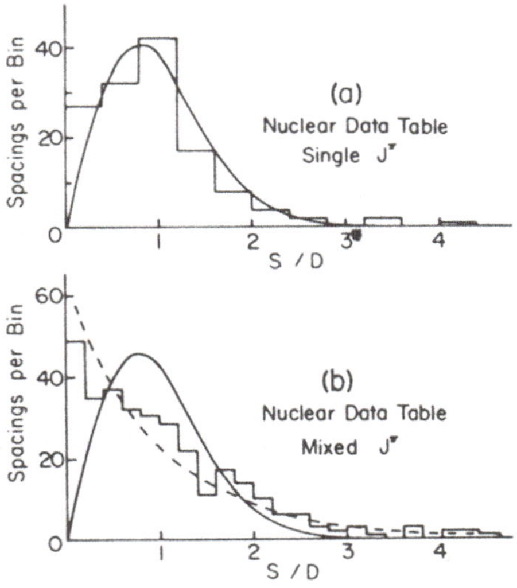

Fig. 4. Nearest neighbour spacing distribution in the ground state domain: (a) for spacings between states of the same (J^{π}, T) and (b) for spacings irregardless of (J^{π}, T). See also Fig. 2.

Secondly, in so far as nearest neighbour spacing distribution can determine, the fluctuation of energy level positions is the same in the ground state domain as at higher energies. This is not necessarily an expected result since there may well be good arguments that the ground state of a quantum mechanical system is special by virtue of being the lower bound of an otherwise infinite spectrum. The result of the nuclear-table ensemble, however, says that the ground state is just another member of the spectrum sharing the same fluctuation properties as the rest.

Unfortunately there is not adequate data for applying other statistical meaures. Random matrix ensembles and shell model results are not of great help since the energy variation of D is the fastest at both ends of the spectrum. Furthermore, the accuracy of unfolding procedures used are least reliable here. As a result, it is not easy to make any meaningful statement for the ground state region.

Some support for the conclusion can be obtained in the following way. One of the characteristics of energy levels is their rigidity. That is, in an ensemble, a particular level does not move very much from one member to another. Numerical calculations, confirmed by analytical studies, says that it is unlikely for a level to move more than a distance D away from its average position in the ensemble. For example, using the centroid of each spectrum as the reference point, the ensemble average position of the r-th level will be at position rD. In any one member of the ensemble, the level is most likely to be bound in the interval $(r \pm 1)D$.* One way to see whether the same fluctuation pattern extends to the ground state region is to check whether spectrum rigidity also extends to the ground state.

In Fig. 5, the distributions of ground state location as measured from the centroid of each member are plotted for 294-dimensional GOE and TBRE. In order to avoid uncertainties in unfolding, the eigenvalue distribution width is used instead of D as the energy unit. For GOE, the distribution is quite sharply peaked with a width less than one average local spacing. For TBRE the spread is much larger but still characterised by a standard deviation of around one level spacing in this region (width $\approx$ 1.3D). The larger spread in ensemble distribution for TBRE comparing with GOE is mainly due to the far smaller numbers of degrees of freedom in the defining matrix elements of the system and is observed in the distributions of other measures as well.

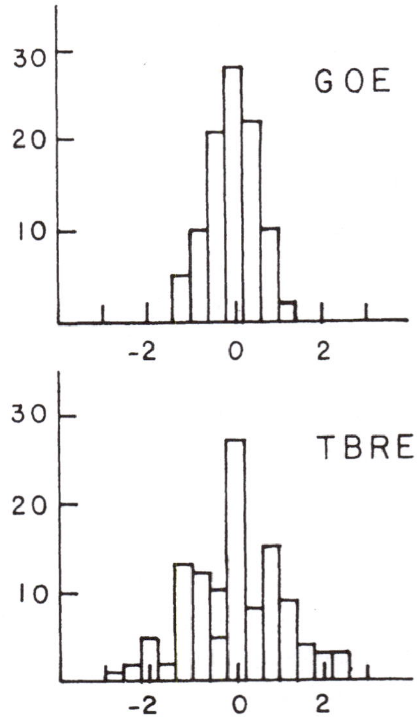

Fig. 5. Distribution of ground state position for members of an ensemble. The horizontal axis is in units of local spacing.

Fluctuation measures

We have seen that nearest neighbour spacing distribution is a way to display the fluctuation of energy level positions and eigenvalues. In analogy with eq. (1), we can define the k-th order spacing as

$$S_i^k = (E_{i+k+1} - E_i), \tag{3}$$

*It is necessary to use a value of D that is calculated for each member as the unit for that member in order to avoid any finite size effects.[14]

i.e., , the spacing distance between two levels with k levels in between. As k increases, the distribution of S^k in units of D, the average of S, approaches a Gaussian distribution centered at $k+1$ for both GOE and TBRE. This means that levels far away from each other are essentially uncorrelated. In fact, the differences from Gaussian become indistinguishable for $k \gtrsim 8$ as shown in Fig. 6. Experimental data also seem to follow the same trend however there is usually not enough levels for a meaningful plot of the higher k cases. Similarly it is not significant to display the higher order spacing distribution for shell model results as it becomes more more susceptible to errors in unfolding.

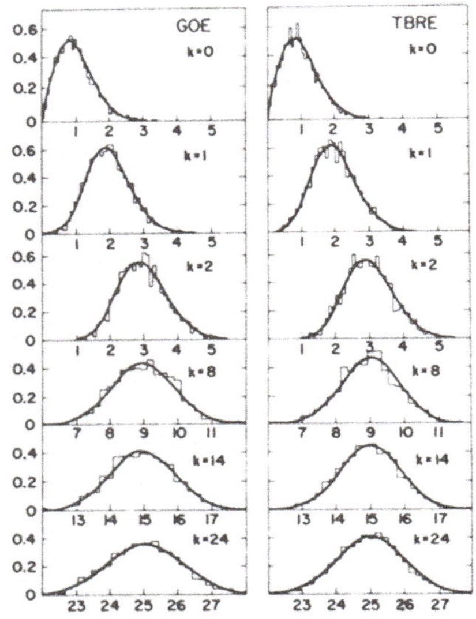

Fig. 6. Histogram of k-th order spacing for different k values. The smooth curves are theorectical GOE results for $k = 0, 2$ and Gaussian for higher k values.

In order to be more quantitative, it is necessary to characterize fluctuation in terms of parameters or statistical measures. In order for a statistic to be useful, it must have a narrow ensemble distribution, preferably with known variance. In this way, any departure from the ensemble average value can be compared with (the square root of) the variance so as to determine whether the deviation is a significant one. In addition, the measure must be easy to evaluate and as insensitive to impurities in the sequence as possible.

Departures from the Wigner surmise for nearest neighbour spacing distribution can be characterize in terms of the level repulsion parameter ω.[16] The statistic is useful in many circumstances; however, since its ensemble behaviour is not well known, we shall not deal with it here. The same goes for the Λ statistic[17] which is also based on the nearest neighbour spacing distribution.

The distribution of the S^k can be characterized in term of the width σ_k, square root of the variance σ_k^2. For $k = 0$, a fairly accurate value of σ_0 can be worked out from eq. (2). More precise results for GOE up to $k = 8$ have been obtained by Mehta[17] and are listed in Table

I. The ensemble widths of the distribution of σ_k have been calculated from 294-dimensional GOE and TBRE and they are of the order of 0.1 for low k and rising to around 0.2 for $k \approx 10$.

Instead of spacing distributions, the same information can also be casted in terms of the correlation coefficient $C(r)$ between two nearest neighbour spacings S separated by r spacings in between

$$C(r) = \frac{n}{n-r-1} \frac{\sum_i (S_i - D)(S_{i+r+1} - D)}{\sum_i (S_i - D)^2}. \tag{4}$$

It is related to σ_k by[19,20]

$$C(r) = \frac{1}{2\sigma_0^2} \left\{ \sigma_{r+1}^2 - 2\sigma_r^2 + \sigma_{r-1}^2 \right\}. \tag{5}$$

For $r = 0$, $C(0) = -0.271$. The negative sign means that a small spacing is likely to be followed by a large one, a reflection of the level repulsion and spectrum rigidity phenomena. Since $C(r)$ and σ_k give the same set of information, only the value of $C(0)$ is listed in Table I in conformity with practice in the literature.

Another way to measure fluctuation is to find, in an interval rD, how much does the observed number of levels differs from r. The value of r should be small enough so that for a sequence of n levels, as many independent spans of rD as possible can be measured. On the other hand, r must be large enough comparing with unity so that the interval will have enough levels for the statistic to be meaningful. Since D is fixed by the entire span of n levles, the average deviation from expected number is zero. The variance, average of the squares of the deviation, is positive with ensemble average given by

$$\Sigma^2(r) = \frac{2}{\pi^2} \left\{ \ln(2\pi r) + 0.5772 \cdots + 1 - \frac{\pi^2}{8} \right\} + O\left(\frac{1}{\pi^2 r} \right). \tag{6}$$

In Table I, the square root of $\Sigma^2(r)$ for $r = 10$ is given for various ensembles and "data". A more exact value than (6) obtained for GOE is 0.95.[1)]

The F-statistics was introduced by Dyson[21)] for the purpose of testing the purity of an experimental sequence. It is based on the expected correlation between spacings in an interval of length L. The correlation function used is

$$f(x) = \frac{1}{2}\ln\left\{\frac{1+(1-x^2)^{1/2}}{1-(1-x^2)^{1/2}}\right\} \qquad \text{for } |x| < 1$$

and equals to 0 otherwise. For the ith level, the value of F is

$$F_i = \sum_{j\neq i} f\left(\frac{E_j - E_i}{L}\right). \tag{7}$$

For $L = rD$, the expectation value of F_i is

$$\langle F_i \rangle = r\pi - \ln r\pi - 0.656$$

with a variance of $\sigma^2(F) = \ln r\pi$.

At a spurious level, $\langle F_i \rangle = r\pi$. Thus, for $r \approx 10$, the presence of impurity can be sported. For a pure sequence, the distribution of F_i is Gaussian: departures therefore serve also as indications of impurity. However, Monte Carlo tests have shown that the distribution of F_i will return to be Gaussian, albeit with larger variance, once the impurity is more than 10 per cent or so. Hence, a Gaussian distribution of F_i alone is not adequate as the evidence for a pure sequence.

Two other commonly used statistics, Δ_3 and Q were introduced by Dyson and Mehta in 1963.[22)] Δ_3 measures the fluctuation by evaluating the deviation of levels from a smooth spectrum in terms of a staircase function $F(x)$ with energy as the horizontal axis. At the position of an energy level, the value of $F(x)$ goes up by unity.

$$\Delta_3(n) = \frac{1}{2L}\min_{A,B}\int_{x-L}^{x+L}\{F(x') - Ax' - B\}^2 dx', \tag{8}$$

where $2L = nD$ is the interval length of the sequence. For GOE, the ensemble average value is

$$\langle \Delta_3 \rangle = \frac{1}{\pi^2}\{\ln(n) - 0.0687\} \tag{9}$$

whereas for a completely random sequence, the value is $n/15$ instead. The major advantage of Δ_3 is that the ensemble variance is fairly constant independent of n. Hence it is a good measure for long sequences. Recent improvements on this measure makes it even more attractive as a statistic for fluctuation studies.[23)]

The Q statistic measures the short range correlations between levels involving a weighted sum of $\ln\{2\pi|E_i - E_j|/D\}$ with counter terms added to moderate the strong variations of Q with the motion of individual level across members of the ensemble. The form of the measure is too long to be given here. For GOE the ensemble average and variance for an interval of n levels are

$$Q(n) = n\left\{0.365 - \frac{1}{r\pi^2}\right\}$$
$$\sigma^2(n) = n\left\{0.266 - O\left(\frac{1}{r\pi^2}\right)\right\} \tag{10}$$

where r should be taken to be around 2 to 4.

Results and conclusions

We have used shell model results and ensembles of random matrices to supplement experimental data for the study of energy level fluctuations in nuclei. Although our concerns are mainly with TBRE and GOE, many other ensembles are available.[1,24)] However, in order for an ensemble to be useful to us, it must be physically relevant as well as mathematically convenient. As we learn more and more about the nuclear system, it may be possible to construct an ensemble that is more restrictive than TBRE so that we may be able to average over an even smaller set of ***reasonable*** Hamiltonians. Unforturnately a numerical approach may be the only way to study such ensembles.

There are also two other questions the anwsers to which are not easily forthcoming. The first one is ergodicity.[25)] Although we do not have time-development here as in regular statistical mechanics systems, nevertheless, it is necessary to know that reasonable Hamiltonians do span all the members of an ensemble with equal probability. The second question is whether the ***truely reasonable*** Hamiltonian has a significant weight in the ensemble. If this is not case the ensemble average may not be related to the average of the subset of interest regardless of the narrowness of the ensemble distribution. An example of this is found in the the case of level density distribution. Since GOE is a much more general ensemble, it contains TBRE as a small part of it. Even though the low order energy moments have very narrow ensemble distribution in GOE, it does not represent those for TBRE and experimental data.

The measures described in the previous section, as well as a few others, have been applied to a variety of ensemble results, shell model eigenvalues and experimental spectra. A sample of the results is given in Table I. It is seen that as far as the measures can detect, the fluctuations are the same in all the cases. Except for those indicated by an asterisk, all the values fall within one standard deviation of the results expected from GOE. The number outside the one standard deviation criterion is within that expected for a "random" sample. In a more complete survey,[1] greater numbers of deviations are found. These are inevitably associated with particular pieces of data. In other words, if a spectrum does not *fit* the GOE value for one of the measures, there is a high probability that it also fails to meet the GOE expectations for other measures as well. One of the possible conclusion is that the datum itself is faulty and this is often corroborated by experimental evidences, such as the existence of large numbers of possible missing and spurious levels.

Table I. Fluctuation statistics for different case†

Case	n	C(0)	σ_0	σ_3	σ_6	$\sigma(F)$	$\Sigma(10)$	Δ_3	Q
GOE(∞)	100	−0.27	0.53	0.75	0.82	1.9	0.95	0.46	34.0
GOE(294)	100	-0.27 ± 0.08	0.54	0.75	0.84	1.9 ± 0.3	0.88	0.46	34.2
TBRE(294)	100	-0.26 ± 0.08	0.52	0.73	0.80	2.0 ± 0.4	0.96	0.48	34.5
$(ds)^8 2^+0$(1206)	120	−0.30	0.50	0.67	0.68	1.3*	0.81	0.33*(0.48)	30*(41)
$(ds)^{12} 0^+0$(839)	84	−0.24	0.52	0.87	0.97	1.9	0.95	0.51(0.44)	35*(29)
^{166}Er	109	−0.22	0.53	0.76	0.81	1.8	0.96	0.46(0.47)	30(36)
^{168}Er	50	−0.29	0.50	0.73	0.65	1.9	0.78	0.29(0.39)	15(16)
^{152}Sm	70	−0.26	0.56	0.77	0.81	1.9	0.91	0.40(0.42)	25(23)
^{232}Th	178	−0.19	0.54	0.83	0.86	2.2	1.05	0.39*(0.51)	69*(60)
^{238}U	146	−0.24	0.50	0.60	0.62*	1.7	0.80	0.42(0.49)	32*(49)
^{172}Yb	55	−0.24	0.55	0.88*	1.00*	1.7	1.04	0.41(0.40)	18(18)

†The results for GOE(∞) are analytical; all others are obtained numerically. For Δ_3 and Q, the theoretical values depend on n and are given in brackets; for all others, the theoretical values are given by the GOE(∞) case. Asterisk indicates that the value deviates from ensemble average value by more than one standard deviation.

On the other hand, one should not be surprised to see the high degrees of correlations among the results for different measures. In addition to exact relations between them, such as that given by eq. (5), all the known measures are essentially given by a two-point or covariance function[1)]

$$S^{\rho}(x,y) = \overline{\rho(x)\rho(y)} - \bar{\rho}(x)\bar{\rho}(y), \tag{11}$$

where $\rho(x)$ is the density and the overbar indicates ensemble averaging. In other words, all the available fluctuation measures depend essentially on the correlations between two points in the spectrum. In principle, higher point measures, *i.e.*, those depending on the correlations between several levels, can be designed but none is known to have been applied.

As far as the available measures can tell, fluctuations are the same for both GOE and TBRE in spite of the important difference in their physical basis and in their low order moments. The fact that they are also the same for experimental data gives us the confidence that we are studying something that is physical, not merely a construct of our ensembles. Very few assumptions are made in defining these ensembles. The main one is that it is a time-reversal invariant quantum mechanical system of fermions interacting with a p-body force. By the similarity between GOE and TBRE, we have eliminated any dependences of fluctuation on the particle rank of the interaction. We are therefore led to the inevitable conclusion that fluctuations found in nuclear spectra are general properties of a quantum mechanical system.

One time, there seems to be some hope of using fluctuation measures to distinguish between different systems, *e.g.*, two-body *versus* many-body interaction. Certainly this is impossible unless a set of measures independent of the presently available ones can be found and capable of making the distinction between different ensembles. However, this seems to be unlikely.

The failure is not necessarily a disappointment. If fluctuations in nuclear spectra is indeed a general quantum mechanical property, it is of interest to find out how wide the universality is. Besides extending the statistical analyses to other systems such as atomic data,[26)] quite different systems, such as small metallic particles* should be examined. The occasion of such a interdisciplinary gathering as the present one is certainly a good occasion to make prgress in this direction.

---

*For a short review, see ref.[1)]

References

1. T.A. Brody, J. Flores, J.B. French, P.A. Mello, A. Pandey, S.S.M. Wong, Rev. Mod. Phys. **53** (1981) 385.
2. C.E. Porter, editor, *Statistical Theories of Spectra: Fluctuations* (Academic, New York, 1965).
3. H.I. Liou, H.S. Camarda, S. Wynchank, M. Slagowitz, G. Hacken, F. Rahn and J. Rainwater, Phys. Rev. **C5** (1972) 974.
4. E.G. Bilpuch, A.M. Lane, G.E. Mitchell and J.D. Moses, Phys. Rep. C **28** (1976) 145.
5. C. Coceva and M. Stefanon, Nuclear. Phys. **A315** (1979) 1.
6. J. von Neumann and E.P. Wigner, Phys. Z. **30** (1929) 467.
7. E.P. Wigner, SIAM Rev. **9** (1967) 1.
8. M.L. Mehta, Nucl. Phys. **18** (1960) 395.
9. J.B. French and S.S.M. Wong, Phys. Lett. **35B** (1971) 5.
10. K.K. Mon and J.B. French, Ann. Phys. (N.Y.) **95** (1975) 90.
11. E. Yépes, doctoral dissertation, 1975 (Universidad Nacional Autonoma de México, unpubished).
12. E.P. Wigner, Ann. Math. **62** (1955) 548.
13. M.L. Metha, *Random Matrices and the Statistical Theories of Energy Levels* (Academic, New York, 1967).
14. S.S.M. Wong and J.B. French, Nucl. Phys. **A198** (1972) 188.
15. T.A. Brody, E. Cota, J. Flores and P.A. Mello, Nucl. Phys. **A259** (1976) 87.
16. T.A. Brody, Lett. Nuovo Cimento **7** (1973) 482.
17. J.E. Monahan and N. Rosenzweig, Phy. Rev. **C5** (1972) 1078.
18. M.L. Mehta, as reported in ref.[19], unpublished.
19. O. Bohigas and M.J. Giannoni, Ann. Phys. (N.Y.) **89** (1975) 393.
20. J.D. Garrison, Ann. Phys. (N.Y.) **30** (1964) 269.
21. F.J. Dyson, as reported in H.I. Liou, H.S. Camarda and F. Rahn, Phys. Rev. **C5** (1972) 1002.
22. F.J. Dyson and M.L. Mehta, J. Math. Phys. **4** (1963) 701.
23. R.U. Haq, A. Pandey and O. Bohigas, Phys. Rev. Lett. **48** (1982) 1086.
24. E.P. Wigner, in *Statistical Properties of Nuclei*, edited by J.B. Garg (Plenum Press, 1972).
25. A. Pandey, Ann. Phys. (N.Y.) **119** (1979) 170.
26. H.S. Camarda and P.D. Georgopulos, Phys. Rev. Lett. **50** (1983) 492.

CHARACTERIZATION OF FLUCTUATIONS OF CHAOTIC QUANTUM SPECTRA

Oriol Bohigas, Marie-Joya Giannoni and Charles Schmit

Division de Physique Théorique*, Institut de Physique Nucléaire, 91406 Orsay Cedex, France

INTRODUCTION

The problem of finding criteria to characterize quantum chaos has become of interest in recent years, once the essential features of classical chaotic motion have been rather well settled. As will probably appear in this Conference, the tentative definitions of quantum chaos are still sparse and conjectural, reflecting our poor understanding of the question. Our purpose here is to propose signatures of quantum chaos by means of statistical properties of the energy levels of the discrete spectrum. In what follows, we shall consider dynamical systems which are conservative and time-reversal invariant, and their classical analogue strongly chaotic (K-systems).

From different numerical experiments and theoretical studies of such systems performed so far, the main feature that seems to emerge is the presence of the phenomenon of repulsion of levels, i.e. the vanishing of the distribution of the nearest-level spacings p(x) as x goes to zero. This property is well known to occur in the spectra of resonances of the compound nucleus and was first mentioned by Berry and Tabor[1] as being possibly a general feature of levels of irregular spectra, in contrast with the clustering of levels (Poisson distribution $p(x)=e^{-x}$) found by these authors for integrable systems. The first numerical indication (see Fig.1a) of level repulsion for chaotic systems with few degrees of freedom was obtained by McDonald and Kaufman[2] (and confirmed by Casati, Valz-Gris and Guarneri[3]) for the stadium. A calculation by Berry[4] for the Sinai's billiard led to the same conclusion (Fig.1b). Concerning

*Laboratoire Associé au CNRS.

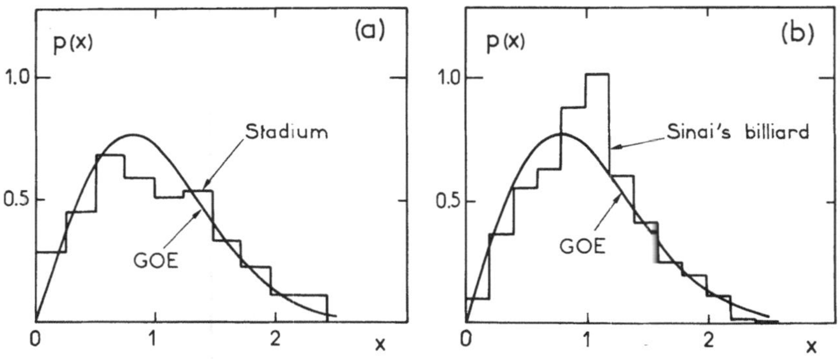

Fig. 1. Distribution of the nearest-neighbour spacings for : a) desymmetrized stadium (histogram taken from Ref.2) ; b) desymmetrized Sinai's billiard (histogram taken from Ref.4). The curve corresponds to the prediction of the Gaussian Orthogonal Ensemble of random matrices, as explained in the text.

theoretical predictions, Zaslavsky guessed[5] that for classical chaotic systems the behaviour of p(x) for small values of x is x^{γ}, where γ depends on the rate of exponential separation of trajectories. However, this conjecture is contradicted by Berry, whose arguments lead to a linear vanishing of p(x) as $x \to o$ for a class of systems ("generic systems") which also include non chaotic systems (see refs. 6,7 and these proceedings). Therefore, it seems that the knowledge of p(x) as $x \to o$ does not provide an unambiguous identification of quantum chaos. On the other hand, it should be noticed that the information carried by p(x=o), although interesting, is quite limited. Many other quantities which are useful in studying level fluctuations have been extensively investigated in the context of Random Matrix Theory (RMT), as for instance, the degree of rigidity of the spectrum, which, in contrast with the level repulsion, does contain information on the correlations between spacings. In Section 2 we give a brief review of the ideas involved in the study of fluctuation properties of quantal spectra[8-9], and of the success of RMT when applied to real physical systems[10,11]. The material presented in Section 2 will be used in Section 3 to study the spectral properties of a classically chaotic system.

2. FLUCTUATION PROPERTIES OF SPECTRA AND RANDOM MATRIX THEORY

a) Parameters Characterizing Fluctuations

In the statistical study of a quantum discrete spectrum, one has to distinguish two kinds of properties : i) global properties, such as the smoothed level density $\bar{\rho}(E)$, and ii) local properties, i.e. fluctuations of levels around $\bar{\rho}(E)$. These two types of properties are very different in nature and completely disconnected. To study the chaotic behaviour of spectra the average density of states is uninteresting and in what follows we shall concentrate on measures of fluctuations. We first recall that, to get rid of spurious effects on the local properties due to variations of the density, one has to work at constant density on the average. For this purpose it is essential to have a good method providing the smoothed cumulative density $\bar{N}(E)$

$$\bar{N}(E) = \int_0^E \bar{\rho}(E')dE' \tag{1}$$

One can then "unfold" the original spectrum, i.e. map the spectrum of eigenvalues $\{E_i\}$ onto the spectrum $\{\varepsilon_i\}$ through

$$\varepsilon_i = \bar{N}(E_i) \tag{2}$$

We will take as energy unit the average spacing $\bar{x}$ between two adjacent levels of the unfolded spectrum

$$\bar{x} = \bar{s}_i = \overline{(\varepsilon_{i+1} - \varepsilon_i)} \tag{3}$$

The spacing distribution p(x) satisfies then

$$\int p(x)dx = \int xp(x)dx = 1 \tag{4}$$

It should be emphasized that, when studying local properties, we are only interested in results which are translational invariant over the spectrum, i.e. the results in the interval $[\varepsilon, \varepsilon+L]$ should be independent of ε. This may be hard to check when performing numerical studies, especially for systems which reach their asymptotic fluctuation properties at very high excitation energies.

Let us now come to the characterization of fluctuation properties. The first interesting (and most popular) quantity is the spacing distribution p(x) already mentioned and which carries information on the correlation between two adjacent levels. But p(x) tells nothing about the correlations between two adjacent spacings s_i and s_{i+1}, as can be easily realized by constructing a spectrum "level by level" with the prescription that ε_{i+1} has the probability p(x)dx of lying in an interval dx at a distance x of ε_i ; in this way, one gets a spectrum with no correlations between spacings (by taking $p(x)=e^{-x}$ one obtains the Poisson case). The most obvious parameter that can be introduced is the correlation factor C between s_i and s_{i+1}. More interesting is the statistic Δ_3 of Dyson and Mehta[12] :

$$\Delta_3(L;x) = \frac{1}{L} \underset{A,B}{\mathrm{Min}} \int_x^{x+L} [n(\varepsilon) - A\varepsilon - B]^2 d\varepsilon , \tag{5}$$

which measures the least-square deviation of the staircase representing the cumulative density $n(\varepsilon)$ from the best straight line fitting it in any interval $[x,x+L]$. If the spectrum is translational invariant, averages of Δ_3 will be independent of the position x of the interval $[x,x+L]$. Δ_3 describes the so-called "degree of rigidity" of the spectrum : the most perfectly rigid spectrum is the picket fence with all spacings equal (for instance, the one-dimensional harmonic oscillator spectrum), therefore maximally correlated, for which $\Delta_3(L) = 1/12$, whereas, at the opposite, the Poisson spectrum ($p(x) = e^{-x}$ and no correlations between spacings) has a very large average value of Δ_3 ($\bar{\Delta}_3(L) = L/15$), reflecting strong fluctuations around the mean level density. Notice that by studying, for instance, the average value of Δ_3 as a function of L, one can choose the range L (in units of the mean spacing) over which fluctuations are investigated. A detailed insight on the information contained in Δ_3 can be found in Refs. 8-10 ; for other fluctuation measures, see e.g. the contribution of S.S.M. Wong to this Conference.

b) Random Matrix Theory : the Gaussian Orthogonal Ensemble (GOE) and its Applications

The idea of representing the Hamiltonian of a complex quantum system by a random matrix was initiated by Wigner, and developed by several authors [8,9] ; the physical purpose of such a theory was to find an appropriate frame to describe the fluctuation properties of slow neutron resonances in heavy nuclei. The ensemble of random matrices which has been the most extensively used in the study of nuclear (and atomic) spectra is the Gaussian Orthogonal Ensemble (GOE). The GOE of NxN real symmetric matrices can be defined in several equivalent ways. Let us remind the main assumptions made in its derivation : the system is time-reversal and rotational invariant (physically justified assumptions), and all matrix elements H_{ij} ($i \leq j$) are independent random variables (for reasons of mathematical tractability but physically unjustified). In the limiting case when $N \to \infty$, it is possible to get analytical results for several properties. In particular, for the quantities considered in 2.a) one has :
i) p(x) is very accurately approximated by the so-called Wigner surmise :

$$p(x) \simeq \frac{\pi}{2} x \exp\left(-\frac{\pi}{4} x^2\right) ; \qquad (6)$$

ii) the correlation factor C between adjacent spacings has the average value

$$\bar{C} = -0.271 \qquad (7)$$

iii) the average value of Δ_3 bahaves asymptotically (large L) as :

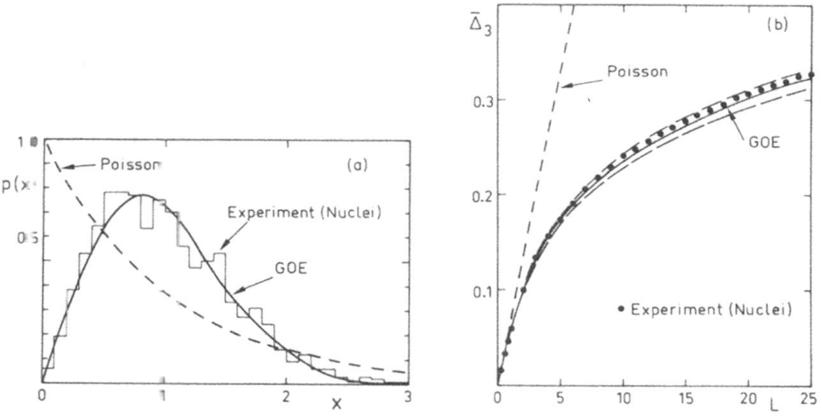

Fig. 2. Fluctuation properties of nuclear levels (taken from Ref. 10). 1762 resonance energies have been included in the analysis ; a) nearest-neighbour spacing histogram, b) average value of Δ_3 as a function of L. The curves corresponding to the Poisson (uncorrelated) and GOE cases are plotted for comparison. Dashed lines close to the GOE values of $\overline{\Delta}_3$ correspond to one standard deviation when finite sampling effects, provided by GOE, are taken into account.

$$\overline{\Delta}_3(L) \simeq \frac{1}{\pi^2} \ln L - 0.007 \qquad (8)$$

One can obtain the exact value[10] of $\overline{\Delta}_3(L)$ by a numerical integration, using the known expression of the two point correlation function[13]. For $L = 5$ the value obtained from Eq.(8) differs from the exact one by 10% and for $L = 15$ by 2.5%. The most salient feature of GOE fluctuation properties is a very strong spectral rigidity, which shows up in the logarithmic behaviour of $\overline{\Delta}_3$ (compare to the linear behaviour for the Poisson case).

What is, now, the predicting power of GOE when applied to level fluctuations of complex nuclei? Recently, the whole body of accurate nuclear data (~ 1750 resonances coming from ~ 30 different nuclei) has been analyzed simultaneously and several new spectral measures have been used[10]. The results for p(x) and $\overline{\Delta}_3$ ($0 < L < 25$) are reproduced in Fig. 2. Other fluctuation measures have been considered and the comparison between GOE predictions and experimental data has yielded a remarkable agreement, even for very sensitive quantities like the variance of Δ_3. These results provide a conclusive evidence for the validity of GOE fluctuations, raising again the old question : why is this parameter-free theory, in which no information about the specific features of the system is included (as, for instance, the interaction between nucleons), so efficient? Does the success of GOE denote that fluctuation properties of nuclear levels

Table 1. Fluctuations of Atomic Energy Levels.

		Nd	Nd	Nd^+	Nd^+	Nd^+	Sm^+	Sm^+	Tb
J^π		4^-	6^-	$7/2^-$	$13/2^-$	$15/2^-$	$3/2^-$	$9/2^-$	$9/2^-$
L		35	38	34	28	32	26	31	45
$\bar{\Delta}_3(L)$	exp.	0.39	0.45	0.30	0.37	0.39	0.37	0.40	0.31
	GOE	0.35 ±0.11	0.36 ±0.11	0.35 ±0.11	0.33 ±0.11	0.34 ±0.11	0.32 ±0.11	0.34 ±0.11	0.38 ±0.11

Results obtained by Camarda and Georgopulos[11] for Δ_3 for different series of levels of neutral and singly ionized atoms. Each series is identified by the angular momentum and parity (J^π). For GOE the value of $\bar{\Delta}_3$ is followed by the square root of the ensemble average of the variance of Δ_3 (asymptotic value =0.11).

result from a general law of nature? That this is probably the case is corroborated by a recent analysis[11] of 269 atomic energy levels corresponding to 8 different atoms. We reproduce in Table 1 some of the results for Δ_3. One can see that the agreement between theory and experiment is good.

3. LEVEL FLUCTUATIONS OF QUANTUM SINAI'S BILLIARD

We present now numerical results for the level fluctuations of the desymmetrized quantum Sinai's billiard (see Fig. 3a). Use will be made of the methods outlined before. We proceed as follows : We determine the eigenvalues $E_i = k_i^2/2m$ of the Schrödinger equation

$$(\Delta + k_i^2)\,\psi_i = 0 \tag{9}$$

with Dirichlet boundary conditions by using the method of Korringa-Kohn-Rostoker as described in Ref.4. We compute several sets of eigenvalues $\{E_i(R)\}$ for different values of R. To unfold the spectrum we use the Weyl-type formula which gives the average number of levels up to energy E[14]

$$\bar{N}(E) = \frac{1}{4\pi}\,(SE - L\sqrt{E} + K)\,, \tag{10}$$

where S and L are respectively the surface and the perimeter of the billiard and K is a constant of the order of unity. To insure

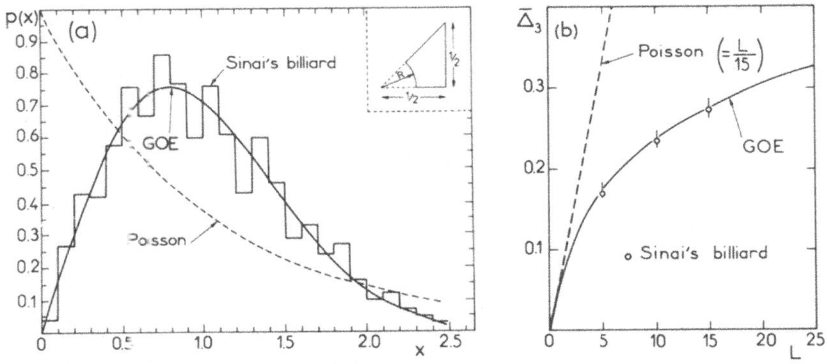

Fig. 3. Results of energy level fluctuations for desymmetrized Sinai's billiards as specified on the upper right corner of Fig. 3a. 740 levels have been included in the analysis corresponding to the 51-th to 268-th level for $R = 0.1$, 21-th to 241-th level for $R = 0.2$, 16-th to 194-th level for $R = 0.3$, 11-th to 132-th level for $R = 0.4$. Fig. 3a : results for the spacing distribution p(x). Fig. 3b : results for the average value of the $\Delta_3(L)$ statistic of Dyson and Mehta for $L = 5$, 10 and 15. Curves corresponding to the Poisson case (stretch of uncorrelated levels) and to the random matrix theory predictions (GOE) are drawn for comparison. The error bars on Fig. 3b (one standard deviation) correspond to finite sampling effects as predicted by GOE.

translational invariance we have excluded from the final analysis the first levels of every spectrum. This is natural if one observes the spectra corresponding to different values of R : one realizes that the smaller the value of R the more the number of levels, starting from the ground state, that can be obtained by perturbing the spectrum $\{\varepsilon_i(R=0)\}$ (triangular billiard). And, as it will come out from our analysis, the fluctuation properties of Sinai's billiards are essentially different from the triangular billiard and are characteristic of levels that cannot be attained by perturbation, i.e., of levels that are at relatively high energy. To increase the statistical significance of the results several spectra $\{\varepsilon_i(R)\}$ corresponding to four different values of R have been analyzed as corresponding to a single stretch of 740 levels. In doing so, care has been taken that one is working with "independent information" : the different values of R should not be chosen to be too close to one another. Otherwise two different spectra corresponding to R and $R+\delta R$ would be almost deducible one from the other and one would just be dealing with redundant information.

In Fig. 3a is shown the spacing distribution p(x) which compares beautifully with the GOE prediction not only for small spacings (level repulsion) but over the entire range of spacings. As can be seen in Fig. 1b, the results obtained in Ref.4 do show departures

Table 2. Fluctuation measures of Sinai's billiard

	Sinai	GOE	Poisson
C	-0.30	-0.27 ± 0.04	0.0
σ^2	0.273	0.286 ± 0.015	1.0

Correlation factor C between adjacent spacings and variance σ^2 of the spacing distribution p(x). See caption of Fig. 3 for further explanation.

from GOE predictions : this is due to lack of translational invariance in the spectra considered (only the lowest ~ 20 first levels of each spectrum were included) and probably also to redundancy in the spectra (too close values of R were considered). In Table 2 the variance σ^2 of p(x) is reproduced for Sinai's billiards and shown to be consistent with the GOE prediction. We consider next spacing correlations. In Fig. 3b are shown the results for the average value of $\Delta_3(L)$ for $L = 5$, 10 and 15 which again compare remarkably well with the corresponding GOE predictions. In Table 2 is shown the correlation factor between two neighbour spacings which again agrees with GOE. We therefore conclude that all level fluctuation measures of Sinai's billiard investigated so far are consistent with GOE predictions.

4. SUMMARY AND CONCLUSIONS

The present work gives a short review of the remarkable success of the GOE in predicting nuclear and atomic energy level fluctuations. Some of the more characteristic properties of GOE fluctuations are described. It is emphasized that the repulsion of levels does not exhaust, by any means, the richness of GOE fluctuations. Indeed, properties like the rigidity of the spectrum are of most relevance. In contrast, in the previous studies of spectral behaviour of regular and irregular systems, attention has been exclusively paid to the presence or absence of level repulsion (which is unable to distinguish between chaotic and non chaotic systems). We propose to use the tools developed in random matrix theory for searching signatures of chaotic spectra.

We then proceed to study numerically the level fluctuations of Sinai's quantum billiard (a chaotic system). We find that all fluctuation measures investigated are consistent with the corresponding GOE predictions. One can then conjecture that chaotic systems (which ones should be made precise in the future) show GOE fluctuations. The correctness of the conjecture would bring a new perspective in the understanding of the origin and domain of validity of GOE fluctuations.

REFERENCES

1. M.V. Berry and M. Tabor, Proc. Roy. Soc. London 356 (1977) 375
2. S.W. McDonald and A.N. Kaufman, Phys. Rev. Lett. 42 (1979) 1189
3. G. Casati, F. Valz-Gris and I. Guarneri, Lett. Nuov. Cim. 28 (1980) 279
4. M.V. Berry, Ann. Phys. (N.Y.) 131 (1981) 163
5. G.M. Zaslavsky, Phys. Rep. 80 (1981) 157
6. P.J. Richens and M.V. Berry, Physica 2D (1981) 495
7. M.V. Berry, Lectures given at the Les Houches Summer School 1981, to be published by North Holland
8. C.E. Porter, editor, Statistical Theories of Spectra : Fluctuations (Academic, New York 1965) ; M.L. Mehta, Random Matrices and the Statistical Theory of Energy Levels (Academic, New York 1967)
9. T.A. Brody, J. Flores, J.B. French, P.A. Mello, A. Pandey and S.S.M. Wong, Rev. Mod. Phys. 53 (1981) 385
10. R.U. Haq, A. Pandey and O. Bohigas, Phys. Rev. Lett. 48 (1982) 1086 ; O. Bohigas, R.U. Haq and A. Pandey in Nuclear Data for Science and Technology (ed. by K.H. Böckhoff), pages 809-813, (Reidel Publishing Company, Dordrecht 1983)
11. H.S. Camarda and P.D. Georgopulos, Phys. Rev. Lett. 50 (1983) 492
12. F.J. Dyson and M.L. Mehta, J. Math. Phys. 4 (1963) 701
13. A. Pandey, Ann. Phys. (N.Y.) 119 (1979) 170
14. H.P. Baltes and E.R. Hilf, Spectra of Finite Systems (Wissenschaftsverlag, Mannheim 1976).

SPECTRAL FLUCTUATIONS AND CHAOS IN QUANTUM SYSTEMS

Giulio Casati and Giorgio Mantica
Dip. di Fisica dell'Università di Milano
Via Celoria 16, 20133 Milano, Italy

Italo Guarneri
Dip. di Fisica Nucleare e Teorica dell'Università
di Pavia, Via Bassi, 27100 Pavia, Italy

1. INTRODUCTION

A whole session in this Conference has been devoted to fluctuations in Nuclear Spectra and related topics. In the organizers' mind, this fact was meant as an acknowledgment of the close relationship existing between two originally very different problems.

The first problem is Quantum Chaos itself. Even though the very existence of something like Quantum Chaos is a controversial point, we think that the following, very neutral, statement of the problem, might be generally accepted: do quantum systems which in the classical limit are chaotic (QCS) exhibit any peculiarity that can be stated intrinsically, i.e., without making reference to the classical limit itself? If so, are these peculiarities relevant to quantum dynamics, and how?

The second problem is the description of spectra of heavy nuclei. As it is well-known, because of their complicated structure, one gives up a precise description of such spectra and attempts instead a statistical description, which is gained by introducing suitable "natural" ensembles of matrices [1,2]. Even though this approach has proved successful in the description of nuclear, and even atomic, spectra [3,4] its foundational basis is still to be cleared. One usually holds that heavy nuclei are "complicated" systems with many components, so that even the precise Hamiltonian is not known; however, what should be meant by "complicated", or, in other words, what characteristics of the (unknown) Hamiltonian justify the adoption of ensembles distinguished by so strong, and even physically questionable, statistical assumptions, is still an open question.

The two problems sketched above look indeed very different in nature. Yet there are now many reasons to think that they have much in common. Many authors working in Quantum Chaos have conjectured that Hamiltonians of QCS should be "random" in some sense of the word. Berry [5] suggested that some spectral features of QCS might be remindful of random matrix theory; and it had already been stated by Zaslavski [6] (and it was later stressed by Casati et al. [7]) that the vague notion of "complexity" which was being used as an intensive support to the use of Random Matrix Theory (RMT) in Nuclear Physics, instead of being related to the large number of degrees of freedom, might be related more closely to the classical notion of complicacy, which is by now a much more refined concept, usually referred to as "chaos" [8].

Should the fundamental similarity of the problem of Quantum Chaos with the problem of spectral fluctuations be established, a methodological progress, at least, would be accomplished: the methods of RMT, or some variant of them, might be used in Quantum Chaos in order to get predictions on the statistics of the spectra; "vice versa" the study of statistical properties of "chaotic" Hamiltonians might give some foundational justification to the use of RMT. The purpose of this paper is to indicate some directions along which this relationship between different fields might be fruitfully developed.

2. THE USE OF ENSEMBLE OF HAMILTONIANS TO DESCRIBE QUANTUM CHAOS

Even accepting, for argument's sake, that a link between Quantum Chaos and RMT exists - so that suitable ensembles of operators have something to do with "chaotic" systems, two problems, at least, immediately arise:

1) RMT is concerned with ensembles of operators. Instead, in Quantum Chaos one is interested in properties of <u>one</u> operator. Why and how to use an ensemble of Hamiltonians in order to get information about one Hamiltonian?

2) How can an ensemble of matrices be held representative of an ensemble of operators in ∞ - dimensional Hilbert space?

The second question, albeit more technical, is important in that the assumptions (such as rotational invariance, etc.) which make Gaussian Orthogonal Ensemble (GOE) a "natural" ensemble <u>cannot</u> be carried over to ensembles of operators. This is a further indication that these assumptions are excessive, and perhaps that there are definite limits of applicability of results of RMT to realistic ensembles.

As to the first problem, we remark that the procedure of building an ensemble of Hamiltonians in order to model one Hamiltonian is already familiar in physics - think, for example, of Anderson's model [9]. In this Conference we had a review [10] on exact mathematical results, obtained by perturbing a free-particle Hamiltonian with a random potential. What one has in mind in that case is to describe the motion of a particle in a "disordered" potential; in order to do that, one takes into account an ensemble of potentials and then looks for those properties that are "typical" in the ensemble. In doing this a technical property, known as "self-averaging" is useful and, in fact, the method of ensembles is justified by this very property. This is an ergodic-like property, according to which, in an appropriate limit (e.g., as the dimension of the box containing the particle $\rightarrow \infty$) the dispersion of relevant spectral quantities over the ensembles $\rightarrow 0$. (It is important to mention here that also matrix ensembles, e.g., GOE, GUE, have this property in the limit $N \rightarrow \infty$, where N is the rank of matrices.) Therefore, "typical" values of these quantities differ very little from average values, which are much more easily computed: as a matter of fact, this is a major advantage of the method of statistical ensembles. Yet, it may seem that we are forcing very different objects under the same artificial heading. For, in ensembles with "disordered" potentials it is immediately apparent what the elements of randomness are. Instead, with one QCS, such as Sinai's billiard, there is no apparent disorder in the Hamiltonian - what is chaotic though, is Classical Dynamics. Still, we suggest that the Hamiltonian itself here is disordered: either because, as J. Ford suggests, diagonalizing it make up a problem of positive algorithmic comlexity, or, in the following sense:

Suppose that one attempts a numerical solution of the spectral problem for Sinai's billiard. One will get a matrix, and possibly a "random" one: for example, by considering an ensemble of different billiards - as done by Bohigas et al. [4] in their numerical study - one makes up an ensemble of matrices, and it may well happen that this ensemble has "good" statistics, more or less remindful of GOE.

In short our contention is that "chaotic" quantum systems are distinguished by properties that are "typical" in certain ensembles of operators. The problem is then to understand what ensembles are "good". By this we mean: what statistical properties an ensemble should have, in order that "typical" members have a chaotic limit?

A remark is necessary here: one may argue that if anything like chaotic quantum motion exists, the best way to find it is to place a particle in a random potential. So, since in typical 1-dimensional cases localization occurs, this will be another indication that no chaos is to be found in Quantum Mechanics comparable to the classical chaos. This would be a misleading attitude. First of all, one dimensional motion is not chaotic at all, even in classical mechanics.

Moreover, in one, or higher-dimensional ensembles, there is the possibility of occurrence of "singular continuous spectrum" which seems to correspond to some truly chaotic behavior [11,12]. In other words, localization, which is a very strange phenomenon itself, may not be the most chaotic behavior that one can expect in a random potential.

Also, we would like to recall Shnirelman's result [13] according to which $u_n^2(x)$, the squared n-th eigenfunction of the ergodic billiard, tends (weakly) to the uniform distribution as $n \to \infty$.

In other words, eigenstates of QCS are delocalized. Therefore, ensembles with disordered potentials in which localization occur are not "good". So, ensembles hitherto studied in some detail are of two types. First we have the "realistic" ones, that are obtained by introducing random parameters in physically meaningful Hamiltonians: e.g., $H = -\Delta + V$, V random. In this case they do not seem adequate to model QCS except perhaps in cases when a singular continuous spectrum occurs (further analysis is however necessary here). On the other hand, we have ensembles of matrices, which are highly unrealistic, because it is certainly hard to give a meaning to the classical limit in this case. Despite this difficulty, ensembles of the latter type may be used with remarkable success to make predictions about the statistics of the spectra of QCS; and also, as we are going to explain, they possess some "good" properties in the sense explained at the beginning of this section.

3. THE BEHAVIOR OF AUTOCORRELATION FUNCTIONS

It is generally accepted in classical statistical mechanics that some understanding of the approach to equilibrium can be gained by the method of ensembles. The problem there is to understand what statistical properties an ensemble of trajectories (of one dynamical system) should have, in order that some sort of thermodynamic behavior ensues. The most proper way to answer this question would certainly be to get as much information as possible about the trajectories themselves; nevertheless, in practise it was found more fruitful to identify a number of "abstract" ensemble properties, such as mixing, positive metrical entropy, and so on, leaving to a later (and nowadays actual) stage the task of verifying whether, and to what extent, actual systems possess such properties.

In a quite different context, we have the problem of understanding what properties an ensemble of quantum Hamiltonians should have in order that typical members of it are classically stochastic. Here also, a proper way of solution would be to work out the ensemble structure from detailed information about individual "chaotic" Hamiltonians; and here, also, we suggest instead as a rewarding task, the identification of "abstract" statistical properties which may ensure that the ensemble is made up with quantum chaotic systems. By chaotic quantum systems we simply mean systems that are classically chaotic; and, in order to give a non-ambiguous meaning to the latter, we consider the autocorrelation function (ACF) of a given dynamical variable.

On the classical level, it is known [14] that for generic smooth integrable Hamiltonian systems with N degrees of freedom the microcanonically averaged ACF decays to its limit value as $t \to \infty$. For mixing, or K-systems, in general an algebraic decay is to be expected as well [15]. However, for more strongly chaotic systems (Anosov flows) ACF decay exponentially.

Coming to Quantum Mechanics, the first problem is the very definition of ACF. Here we will refer to the formalism of Wigner symbols, as the most suited to deal with classical limits.

Consider, for simplicity, a system with configuration space $\mathbb{R}^2$. Let $\hat{f}$ be the operator describing a dynamical variable f. Its Wigner symbol is a function on the classical phase space $\mathbb{R}^4$:

$$f_W(\underline{x},\underline{p}) = 2\hbar^{-1}\sqrt{\pi} \int_{\mathbb{R}^2} d\underline{x}' < \underline{x}' - \underline{x}|\hat{f}|\underline{x}' + \underline{x} > \exp\{2i\hbar^{-1}\underline{p}\cdot\underline{x}'\}.$$

As $\hat{f}$, in the Heisenberg picture, evolves into $\hat{f}(t)$, f_W evolves into $f_W(t)$, and it can be assumed that, as $\hbar \to 0$, $f_W(t)$ will tend in some suitably weak sense to the classical phase function corresponding to the given dynamical variable.

We consider the ACF

$$S(t) = \iint d\underline{x}d\underline{p}\bar{f}_W(\underline{x},\underline{p},o)f_W(\underline{x},\underline{p},t).$$

In view of the above, it can be assumed that, as $\hbar \to 0$, S(t) converges to the ACF of the given dynamical variable, evaluated through the classical equations of motion. For simple cases this can indeed be checked by simple computations.

Suppose now that the system has a pure point spectrum; let E_n be the eigenvalues and u_n the corresponding eigenfunctions. The time evolution of f_W is given by

$$f_W(\underline{x},\underline{p},t) = \sum_{n,m} c_{nm}\upsilon_{nm}(\underline{x},\underline{p})\exp\{-i\hbar^{-1}(E_n - E_m)t\}$$

where $\{\upsilon_{nm}\}$ is a complete orthonormal set of functions in $L^2(\mathbb{R}^4)$,

$$\upsilon_{nm}(\underline{x},\underline{p}) = 2\hbar^{-1}\sqrt{\pi}\int_{\mathbb{R}^2} d\underline{x}\ \exp\{2i\hbar^{-1}\underline{p}\cdot\underline{x}'\}\bar{u}_n(\underline{x}-\underline{x}')u_m(\underline{x}+\underline{x}')$$

and

$$c_{nm} = <u_n|\hat{f}|u_m> = \int d\underline{x}d\underline{p}\bar{\upsilon}_{nm}f_W. \qquad (1)$$

Therefore, one has

$$S(t) = \sum_{nm} |c_{nm}|^2 \exp\{-i\hbar^{-1}(E_m - E_n)t\} \tag{2}$$

so that S(t) is an almost periodic function, and <u>no</u> asymptotic decay is to be expected. However, for a classically chaotic system, S(t) will tend, as $\hbar \to 0$, to an exponentially decaying function. (More precisely, in order for this to be true, it is also necessary that f vanishes in a neighborhood of the origin.) The mechanism which makes this possible must be rooted in some peculiarities of the quantum Hamiltonian.

Suppose now that we are given an <u>ensemble</u> of Hamiltonians and let us average Equation (2) over this ensemble, with a fixed $\hbar \neq 0$,

$$\langle S(t)\rangle = \sum_{nm} \langle |c_{nm}|^2 \exp\{-i\hbar^{-1}(E_m - E_n)t\}. \tag{3}$$

It is not easy to work out Equation (3) without detailed information on the statistics. In an informal way, according to the program outlined at the beginning of this section, we look for statistical assumptions that can produce decay of ACF in the limit $\hbar \to 0$. Consider, for example, the following:

i) The statistics of the eigenvectors and that of the eigenvalues are independent.
ii) The distribution of the frequencies are absolutely continuous.

Assumption (i) is a very strong one. It holds in GOE (and GUE, also) as a consequence of rotational invariance. By using it, we get

$$\langle S(t)\rangle = \sum_{m,n} \langle |c_{nm}|^2\rangle \langle \exp\{-i\hbar^{-1}(E_m - E_n)t\}\rangle \tag{4}$$

and then, using (ii), we obtain

$$\lim_{t\to\infty} \langle S(t)\rangle = \sum_{m} \langle |c_{mm}|^2\rangle$$

that is, taking into account expression (1),

$$\langle S(\infty)\rangle = \sum_{E} |\langle u_E|\hat{f}|u_E\rangle|^2 = \sum_{E} |\int u_{E,\underline{w}} f_{\underline{w}} d\underline{x} d\underline{p}|^2. \tag{5}$$

It was plausibly conjectured by Berry that for ergodic systems $u_{E,w} \underset{\hbar\to 0}{\to} S(H - E)$ so that Equation (5) is a quantum version of the classical value

$$\langle S(\infty)\rangle_{c\ell} = \langle \int dE |f d\mu_E|^2\rangle$$

with μ_E the microcanonical measure on the energy surface $H = E$. $\langle S(\infty)\rangle_{c\ell}$ is thus the value entailed by classical mixing.

In the same informal way we may enquire about the influence of certain simple statistical parameters of the spectrum on $S(t)$, $t \to \infty$. In fact, the information at hand about the statistics of the spectrum is not usually so detailed as to include the joint distribution of all the eigenvalues: one usually has $\nu_n(E)$ with the probability that n eigenvalues will be found to be lesser than E; and the distribution $P(E,s)$, which has the probability that, given one eigenvalue at E, the next one will be found at a distance $\leqslant s$. From Equation (4) we get

$$\langle S(t)\rangle = 2 \sum_{\substack{m \\ p>0}} \langle |c_{m,m+p}|^2\rangle \langle \cos\{\hbar^{-1}(E_{m+p} - E_m)t\}\rangle .$$

It is now conceivable that, upon averaging the oscillating factors $\langle \cos\{\hbar^{-1}(E_{m+p} - E_m)t\}\rangle$ over different members of the ensemble, the terms with $p = 1$, being those with the longest periods, will dominate in the tail of $S(t)$. One has

$$M_m(t) = \langle \cos\{\hbar^{-1}(E_{m+1} - E_m)t\} = \int_0^\infty d\nu_m(E) \int_0^\infty dP(E,s)\cos(\hbar^{-1}st)$$

and, assuming for simplicity that $P(E,s) \equiv P(s)$ (as it is in the case, for example, for ensembles of 2-dimensional billiards),

$$M_m(t) \equiv M(t) = \int_0^\infty dP(s)\cos\{\hbar^{-1}st\}.$$

This formula shows the effect that different choices of $P(s)$ produce on the tail of $\langle S(t)\rangle$. Let for instance, $dP(s) \propto s^\nu e^{-\alpha s^2} ds$ (this formula makes it easy to investigate the effect of level repulsion, scaled by ν, on $S(t)$); then one has

$$M(t) \sim t^{-(\nu+1)} \sin \frac{\pi\nu}{2}\, \Gamma(\nu + 1);\quad t \to \infty.$$

Apart from even integral values of ν, this shows that an increasing level repulsion (i.e., an increasing value of ν) causes a faster decay of $\langle S(t)\rangle$ to its equilibrium value.

The physical relevance of this qualitative discussion is of course questionable, unless we can show that the ensemble average can be representative of individual members of the ensemble.

The property required is the above mentioned "self-averaging". By this we mean that, as $\hbar \to 0$, the ACF of individual systems of the ensemble should approach the averaged ACF, i.e., the statistical dispersion of the ACF should be vanishingly small in the limit $\hbar \to 0$.

Of course, this cannot be true uniformly in t, since, for any $\hbar \neq 0$, the averaged ACF is a decaying function of t, while individual ACFs are almost periodic. Therefore the behavior of the averaged ACF can be expected to be representative of the classical pattern only for a finite time interval, which, however, grows to infinity as $\hbar \to 0$.

For this reason, the above remarks on the algebraic decay of S(t) do not imply an algebraic classical decay of S(t). The classical decay as $t \to \infty$ is determined by behavior of the quantum <S(t)> over a different time scale. This behavior can be fully investigated when one assumes Gaussian ensembles as models.

4. GAUSSIAN ENSEMBLES AS MODELS OF QUANTUM CHAOS

To illustrate the reasons why we suggest that GOE, and related Gaussian ensembles, are in a sense "models" of quantum stochasticity, we now briefly describe the results that we have obtained [16] about ACFs in one such ensemble, namely the Gaussian Unitary Ensemble.

This ensemble is not fit to describe systems invariant under time reversal. The reason why we used it was that we found the analytical procedure to be simpler in this case; anyway, no qualitatively different results should be expected with GOE, so that the mentioned lack of invariance of GUE should not alter the heuristic value of our result.

Here, wave functions are N-dimensional complex vectors $\underline{c} \equiv \{c_i\}$, $i = 1,N$; in order to get a formal analog of the Wigner symbol, we used the matrix

$$\chi_W(k,\ell) = \frac{1}{\sqrt{N}} \sum_{n=1}^{N} \bar{c}_{k-n} c_{k+n} \exp\{4\pi i n\ell N^{-1}\}$$

where k − n, k + n are meant to be "modulo N" and χ_W is normalized according to:

$$\sum_{k,\ell=1}^{N} \chi_W(k,\ell) = 1.$$

Let the state vector evolve according to a Hamiltonian H, i.e., a random matrix in the N-dimensional GUE. Then, by considering the ACF of $\chi_W(k,\ell)$, between times 0 and t, and by averaging over H, one finds [16] that <S(t)> decays in time as $P(t)e^{-t^2}$, where P(t) is a polynomial. One useful fact here is that <S(t)> does not depend on the chosen initial state, because of rotationally invariant ensemble statistics.

If the limit $N \to \infty$ is taken, one finds that: (1) S(t) self-averages in this limit, and (2) $\langle S(t)\rangle \to 0$ for all $t \neq 0$. In other words, if the limit $N \to \infty$ is considered as a sort of "classical" limit, then GUE is a "good" ensemble!

Of course, this limit decay of $<S(t)>$ is too fast, no classical system can behave in this way. This is one more instance in which the over idealized character of Gaussian ensembles shows up. The problem then which so far remains open, is to find the way in which the strong assumptions which distinguish Gaussian ensembles can be relaxed, so that "realistic" ensembles can possess these weaker properties and still be good ensembles.

5. LEVELS REPULSION

In the previous section we have introduced the notion of level spacings distribution P(s). This distribution, being accessible to numerical and experimental computation, has played an important role in investigations about fluctuation properties of spectra of QCS and of nuclear spectra. For this reason we are now going to comment somewhat on it from the statistical ensemble viewpoint.

Consider an ensemble of (hermitean, real) matrices. Let P(s) be the statistical weight of those matrices that have two neighboring eigenvalues at a distance less than s. Suppose that P(s) has a density p(s).

We say that the ensemble has the level repulsion property, if $p(s) \to 0$ as $s \to 0$.

It is not clear whether this property, in itself, has any direct bearing on chaotic behavior. We know that GOE and GUE have this property; also, we know that the statistics of the spacings of levels of chaotic billiards like the stadium or Sinai's billiard, evaluated within the spectral sequence of <u>one</u> system, seem to possess the same property.

Here, we will give a simple argument to show that much more general ensembles than GOE and GUE exhibit level repulsion. This argument is basically a refinement of some ideas of Berry's [5].

Suppose that an ensemble of symmetric matrices is defined by a joint distribution function of matrix elements, which is absolutely continuous with a bounded density. We show that any such ensemble exhibits level repulsion.

First, we recall that in the space of symmetric real matrices, the set of matrices with degenerate eigenvalues is a manifold Σ of codimension two (e.g., for matrices of rank 2 it is a line). Now consider a matrix with two consecutive eigenvalues within ε of each other, with ε an infinitesimal. It is easily seen that this matrix, considered as a point in the space of matrices, is distant from Σ by one infinitesimal of order not greater than ε (consider, for example, a trace norm in the space of matrices).

So, this set of matrices has a volume which is infinitesimal, as $\varepsilon \to 0$, of an order not greater than that of a layer of thickness ε constructed around Σ: therefore, this volume is $O(\varepsilon^2)$ and, thanks to the assumptions on the distribution function, its statistical weight $P(\varepsilon)$ also will be $O(\varepsilon^2)$.

Therefore $p(\varepsilon)$, which is the derivative of $P(\varepsilon)$, will be $O(\varepsilon)$. However, to properly understand the practical importance of this result, the following remark is essential. If one performs a statistics of the spacing of neighboring levels in the spectrum of one matrix, chosen at random in the ensemble, one may expect that the resulting histogram will be an approximation of $p(s)$, and will therefore show level repulsion, only in the case that the ensemble has the self-averaging property for $p(s)$.

This is the case for GOE and GUE, but we do not know if this is true for general ensembles as we have considered here.

REFERENCES

1. T. A. Brody, J. Flores, J. B. French, P. A. Mello, A. Pandey and S. S. M. Wong, Rev. Mod. Phys., 53:385 (1981).
2. C. E. Porter, Statistical Theories of Spectra: Fluctuations, Acad., New York (1965).
3. S. S. M. Wong, in this volume.
4. O. Bohigas, M. J. Giannoni and C. Schmit, in this volume.
5. M. V. Berry, Lectures given at Les Houches Summer School 1981 (North Holland) and references therein.
6. G. Zaslavsky, Phys. Rep., 80:157 and references therein (1981).
7. G. Casati, F. Valz-Gris and I. Guarneri, Lett. Nuovo Cimento, 28:279 (1980).
8. J. Ford, Physics Today, April (1983).
9. P. W. Anderson, Phys. Rev., 109:1492 (1958); Rev. Mod. Phys., 50:191 (1978); see also R. E. Prange, D. R. Grempel and Shmuel Fishman, in this volume.
10. J. Bellisard and B. Souillard, in this volume.
11. G. Casati and I. Guarneri, "On the possibility of chaotic behavior in periodically perturbed Quantum Systems", Preprint.
12. G. Casati, Irreversibility and Chaos in Quantum System, Lect. Notes in Math., Springer Verlag (to appear).
13. A. I. Shnirelman, Ups. Mat. Nauk., 29:181 (1974).
14. G. Casati, I. Guarneri and F. Valz-Gris, Physica, 3D:664 (1981).
15. F. Vivaldi, G. Casati and I. Guarneri, Phys. Rev. Lett., 51: 727 (1983).
16. G. Casati and I. Guarneri, "Statistical properties of spectra and decay of correlations", Preprint.

ASPECTS OF DEGENERACY

M. V. Berry

HH Wills Physics Laboratory
Tyndall Avenue
Bristol BS8 1TL UK

1 INTRODUCTION

Without symmetry, degeneracies are considered to be 'accidential', reflecting the fact that for a typical Hamiltonian $\hat{H}$, representing a bound quantal system, no two of the energy levels E_n will coincide. But just as with road accidents the chance of a degeneracy can rise from negligible to inevitable if instead of considering individual Hamiltonians one embeds $\hat{H}$ in a population smoothly parameterised by variables $\underset{\sim}{R}=(X,Y,Z....)$. In section 2 I review an old argument of Von Neumann and Wigner (1929) indicating that for typical families $\hat{H}(\underset{\sim}{R})$ of real Hamiltonians, two parameters are necessary to produce a degeneracy, while if $\hat{H}(\underset{\sim}{R})$ is Hermitian (and not real) three parameters are necessary.

In a semiclassical context (Berry 1983a) where Planck's constant $\hbar$ is considered as a small parameter, quantal spectra exhibit a hierarchy of structures (Berry 1983b), characterised by energy ranges δE which are successively smaller in comparison with $\hbar$. In this hierarchy, degeneracies are the finest structures, with $\delta E=0$. Next come splittings due to barrier penetration, with $\delta E\sim O(\exp\{-\hbar^{-1}\})$. Then come spacings between adjacent and near-degenerate levels, which for systems with N freedoms correspond to $\delta E<O(\hbar^N)$. The mean spacings themselves have $\delta E\sim O(\hbar^N)$. Next come oscillatory clusterings involving many levels and associated with classical closed orbits, for which the 'energy wavelength' is $\delta E\sim O(\hbar)$. On the coarsest scale, $\delta E>O(\hbar)$ and only the average level density can be perceived (this depends only on the phase-space volume of the classical energy surfaces).

In section 3 I will discuss a different semiclassical aspect of degeneracies, and connect the avoided crossings of levels when a single parameter is varied with the emergence of chaotic behaviour when a classically integrable system is nonintegrably perturbed.

In section 4 I will give two versions of an argument connecting the number of parameters $\underset{\sim}{R}$ necessary to produce a degeneracy with the form of the distribution of spacings of neighbouring energy levels.

Finally, in section 5 I will summarise the curious phase behaviour of quantum states when continued in parameter space close to degeneracies.

2 HOW ACCIDENTAL IS A DEGENERACY?

Suppose that at some point $\underset{\sim}{R}^*$ in parameter space two states $|\psi_1^*\rangle$ and $|\psi_2^*\rangle$ are degenerate, with energy $E=E^*$, i.e.

$$\hat{H}(R^*)|\psi_1^*\rangle = E^*|\psi_1^*\rangle, \quad \hat{H}(R^*)|\psi_2^*\rangle = E^*|\psi_2^*\rangle. \tag{1}$$

If now the system is perturbed by varying $\underset{\sim}{R}^*$, does the degeneracy persist or do the levels split? According to degenerate perturba--tion theory (Von Neumann and Wigner 1929, translated in Knox and Gold 1964), the new states $|\psi_1(\underset{\sim}{R})\rangle$ and $|\psi_2(\underset{\sim}{R})\rangle$ are, to lowest order in $\underset{\sim}{R}-\underset{\sim}{R}^*$, linear combinations of the old ones, i.e.

$$|\psi_i(\underset{\sim}{R})\rangle \sum_{i=1}^{2} a_{ij}(\underset{\sim}{R})|\psi_j^*\rangle \quad (i=1,2) \tag{2}$$

To find the coefficients a_{ij} and the new energies $E_1(\underset{\sim}{R})$ and $E_2(\underset{\sim}{R})$, Schrodinger's equation is written in the $|\psi_j^*\rangle$ basis. This involves matrix elements

$$H'_{ij}(\underset{\sim}{R}) \equiv \langle\psi_i^*|\hat{H}(\underset{\sim}{R})-\hat{H}(\underset{\sim}{R}^*)|\psi_j\rangle \tag{3}$$

and gives, for the energy splitting,

$$\Delta E(\underset{\sim}{R}) \equiv E_2(\underset{\sim}{R})-E_1(\underset{\sim}{R}) = \{[H'_{11}(\underset{\sim}{R})-H'_{22}(\underset{\sim}{R})]^2 + 4|H'_{12}(R)|^2\}^{\frac{1}{2}} \tag{4}$$

At a degeneracy, ΔE must be zero, and this requires

$$H'_{11}(\underset{\sim}{R})=H'_{22}(\underset{\sim}{R}), \quad \mathrm{Re}H'_{12}(\underset{\sim}{R})=0, \mathrm{Im}\, H'_{12}(\underset{\sim}{R})=0. \tag{5}$$

Consider first the case where the family $\hat{H}(\underset{\sim}{R})$ consists of Hermitian Hamiltonians which are not restricted to be real. This represents quantal systems without time-reversal symmetry, such as a particle bound by a scalar potential and in an external magnetic field. Then the three equations in (5) give independent restrictions on $\hat{H}(\underset{\sim}{R})$ and so at least three parameters $\underset{\sim}{R}$ are required in order to satisfy them. If there are exactly three parameters, the degeneracy is isolated at the single point $\mathbf{R}^*$. If there are M > 3 parameters, then equations (5) are satisfied on an M-3 dimensional manifold including $\mathbf{R}^*$, so that the degeneracy is not isolated - for example in a four-parameter space a line of degenracy passes through $\mathbf{R}^*$. This conclusion can be expressed as follows: for Hermitian operators, degeneracies have codimension three.

Now consider the case where $\hat{H}(\mathbf{R})$ consists of <u>real</u> Hamiltonians (real symmetric matrices), representing quantal systems without time-reversal symmetry (no external magnetic fields). The last of equations (5) is now an identity, leaving just two independent conditions to be satisfied in order for a degeneracy to occur. If there are exactly two parameters $\mathbf{R}$, the degeneracy is isolated at the single point $\mathbf{R}^*$. For M > 2 parameters, the degeneracy is not isolated but lies on an M-2 dimensional manifold including $\mathbf{R}^*$ - for example in a three-parameter space a line of degeneracy passes through $\mathbf{R}^*$. This conclusion can be expressed as follows: for real symmetric operators, degeneracies have codimension two.

It is obvious that these arguments do not rely on there being any symmetry, and so really do establish conditions under which accidental degeneracies occur. If the family $\hat{H}(\mathbf{R})$ consists of Hamiltonians with geometric symmetry, then group theory enforces different conditions for degeneracy, as explained by Arnol'd (1972 and appendix 10 of 1978) for the case of real operators.

For $\mathbf{R}$ close to $\mathbf{R}^*$, the perturbation matrix elements H'_{ij} (equation 3) depend linearly on the components of $\mathbf{R}-\mathbf{R}^*$. If there are two parameters $\mathbf{R}=(X,Y)$ and the Hamiltonians $H(X,Y)$ are real, (4) implies that the level separation has the form

$$\Delta E(X,Y)=\{A(X-X^*)^2+ 2B(X-X^*)(Y-Y^*) + C(Y-Y^*)^2\}^{\frac{1}{2}} \tag{6}$$

where the quadratic form is positive-definite and A,B,C depend on components of $\nabla_R \hat{H}$ at $\mathbf{R}^*$. This is the equation of a double cone (Teller 1937) in the space ΔE,X,Y, and the levels $E_1(X,Y)$ and $E_2(X,Y)$ also intersect conically in E,X,Y space (fig.1). The diabolo geometry which organises degeneracies in these typical quantal systems with real Hamiltonians makes it natural to refer to the degeneracies themselves as <u>diabolical points.</u>

In molecular physics, the importance of diabolical points in electronic levels (where the parameters $\mathbf{R}$ are nuclear coordinates) has been emphasised by Vaz Pires et al (1978), Desouter-Lecomte et al (1979) and Dehareng et al (1983). In nuclear physics, the importance of diabolical points in nucleon spectra (where $\mathbf{R}$ para--meterises nuclear shape) was emphasized by Hill and Wheeler (1952).

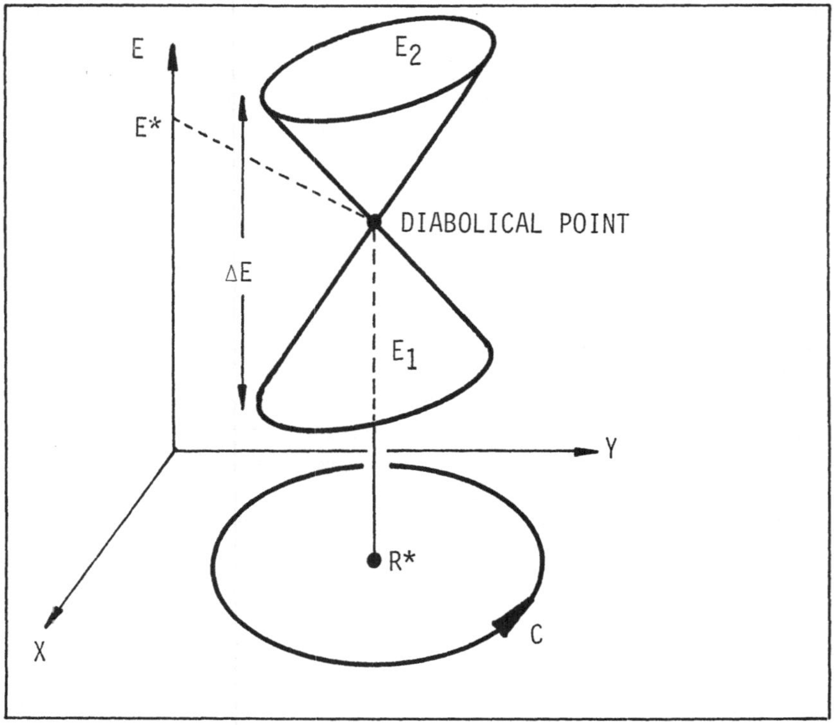

Fig.1. Geometry and notation near a diabolical point.

Berry and Wilkinson (1984) study diabolical points in the two-parameter family of'quantum triangles' (vibrations of triangular membranes) whose levels are determined by Schrödinger's equation for free motion within the triangles and vanishing wave function on the boundary. X and Y may correspond to any two angles (the third angle being determined and the size of the triangle causing trivial scaling of the levels). By exploring triangle space X,Y, a number of diabolical points were found numerically (their diabo--lical nature being established by a sign-change rule to be descri--bed in section 5). The lowest diabolical point involved levels 4 and 5 (the ground state being level 1) for the triangle 130.57°, 30.73°, 18.70°. This might be called the 'paradoxical triangle' on account of its being the non-special (generic) triangle with the special property that no other triangle has a lower pair of levels degenerating for reasons not connected with symmetry. (Of course there are lower degeneracies arising from symmetry such as levels 2, 3 for the equilaterial triangle, and levels 3,4 and 4,5 for certain isosceles triangles.)

For Hermitian Hamiltonians, the diabolical structure in E,X,Y space generalizes to a hyperconical intersection in the four-dimensional space E,X,Y,Z, where Z is the third parameter required to produce the degeneracy. For example, Z might be the strength of a magnetic field perpendicular to the plane of a quantum triangle; in this case I expect degeneracies for Z=0 to disappear when the field is switched on, but that new ones will appear for finite values of Z (for different triangles X,Y).

3 AVOIDED CROSSINGS AND CHAOS IN QUASI-INTEGRABLE SYSTEMS

It is clear that a family of generic Hamiltonians $\hat{H}(X)$ (Hermitian or real) depending on a single parameter X will not exhibit degeneracies: in the plane E,X, the eigenvalue curves $E_n(X)$ will not cross. Instead, the curves will display avoided crossings whose local form (which can be considered to arise from a close approach to a degeneracy in an augmented parameter space) is that of a pair of hyperbolas (slices of a cone or hypercone). Such avoided crossings are now familiar in calculations of the quantal spectra of families classically nonintegrable systems: in nonlinearly coupled oscillators (Noid et al 1980, Marcus 1980), in chaotic quantum billiards (Berry 1981), in (pseudointegrable) rational quantum billiards (Richens and Berry 1981) and in triangle quantum billiards (Berry and Wilkinson 1984).

If the family consists of Hamiltonians $\hat{H}(X)$ which are not generic but are special in some way, then degeneracies can be expected to occur. An important case is families of separable systems. For example in two dimensions a particle in a rectangular box with side ratio X has levels labelled by quantum numbers m,n with energies $E_{m,n} \sim m^2+X^2n^2$, which can degenerate when X^2 is rational; and a particle in a central potential V(r;X) may exhibit degeneracies between states of different angular momentum as X alters the strength or shape of V (these should not be confused with the two-fold degeneracy for nonzero angular-momentum states for all X, which is a direct consequence of the circular symmetry).

A slight generalization is to one-parameter families of classically integrable systems, whose phase-space trajectories are confined to tori (see for example the review by Berry 1978 or the book by Lichtenberg and Lieberman 1983). Then degeneracies will occur in levels approximated semiclassically by torus quantization (Maslov and Fedoriuk 1982, Berry 1983a) which for N freedoms can be formulated as follows: the state with quantum numbers $\mathbf{m}=(m_1 \ldots\ldots m_N)$ is associated with the torus whose actions $\mathbf{I}=(I_1 \ldots\ldots I_N)$ (corresponding to its N irreducible cycles) is

$$\mathbf{I}_m = (\mathbf{m} + \boldsymbol{\alpha}/4)\hbar \qquad (7)$$

where $\boldsymbol{\alpha}$ is a vector of integers (Maslov indices) depending on the embedding of the torus in phase space. The energy $E_m(X)$ of the state $\mathbf{m}$ is simply

$$E_{\mathbf{m}}(X)=H(\mathbf{I}_{\mathbf{m}};X), \tag{8}$$

where $H(\mathbf{I};X)$ is the Hamiltonian in action representation for the family of integrable systems.

Because torus quantization is only an approximation, the level crossings (degeneracies) predicted by (8) may in the exact quantal system be avoided crossings (near-degeneracies). A simple example occurs in one dimension for a particle in a double-well potential $V(x;X)$ whose form depends on X and is symmetric when X=0. In torus quantization (Bohr-Sommerfeld quantization in this case) the two wells are quantized separately, and there would be degeneracies when X=0; in reality, of course, these are split by barrier penetration, giving levels $E_n(X)$ with very narrow avoided crossings (splittings $O(\exp(-1/\hbar))$ as compared with spacings $O(\hbar)$).

A more interesting example is of families of multidimensional systems whose classical motion is quasi-integrable, i.e. perturba--tions of integrable systems. Then the torus quantization formula (8) is inexact in two senses: it is a semiclassical approximation to the quantum mechanics, and a torus approximation to the classical mechanics. For such cases, Marcus (1980), Noid et al (1980) and Ramaswamy and Marcus (1981) have associated multiple avoided cros--sings in the quantal spectrum $E_n(X)$ with chaos in the classical motion. This association is explained by the following simple argument, which was suggested to me by Ozorio de Almeida (private communication) and which does not appear explicitly in the papers cited although it is surely known to their authors.

On the basis of torus quantization (7 and 8), two states with quantum numbers $\mathbf{m}_1$ and $\mathbf{m}_2$ will degenerate for some parameter value X* if their tori $\mathbf{I}_{\mathbf{m}_1}$ and $\mathbf{I}_{\mathbf{m}_2}$ have the same energy, which if $\mathbf{m}_1$ and $\mathbf{m}_2$ are not too different implies

$$E_{\mathbf{m}_1}(X^*)-E_{\mathbf{m}_2}(X^*) \approx \hbar(\mathbf{m}_1 - \mathbf{m}_2)\cdot \nabla_I H(I_{\bar{\mathbf{m}}};X^*) = 0, \tag{9}$$

where $\bar{\mathbf{m}} \equiv (\mathbf{m}_1 + \mathbf{m}_2)/2$. Now

$$\nabla_I H(\mathbf{I}) = \boldsymbol{\omega}(\mathbf{I}) \tag{10}$$

which denotes the set of frequencies $\omega_1 \ldots \omega_N$ with which the trajectory winds round the torus. Because $\mathbf{m}_1 - \mathbf{m}_2$ is a vector whose components are integers, (9) implies that the degenerating tori have rationally dependent frequencies. For small $\hbar$, these degeneracies, which torus quantization predicts near those para--meters X* for which quantized tori are rational tori, will occur between many pairs of levels, i.e. those whose mean is $\bar{\underset{\sim}{m}}$ and whose difference $\underset{\sim}{m}_1 - \underset{\sim}{m}_2$ is small enough for the linear approximation (9) to hold.

But this prediction must fail in two ways. Firstly, torus quantization is not exact: because the family is generic (quasi-integrable), the exact quantal spectrum will exhibit not degeneracies but multiple avoided crossings. And secondly, the tori on which the prediction is based will not exist: because these tori have rationally dependent frequencies, they lie in narrow phase-space zones in which the nonintegrable perturbation destroys tori and replaces them (at least in part) by chaotic trajectories (see e.g. Berry 1978, Lichtenberg and Lieberman 1983). Thus nonexistent tori would predict nonexistent degeneracies, and actual chaos is associated with actual multiple avoided crossings.

The foregoing argument applies to quasi-integrable systems, and should not be misinterpreted to imply that multiple avoided crossings are always associated with classical chaos. In pseudo-integrable systems (rational billiards), for example, all trajectories are confined to N-dimensional invariant surfaces (multiple-handled spheres rather than tori), and there is no chaos in the sense of exponential separation; nevertheless, Richens and Berry (1981) found multiple avoided crossings in the spectrum of a family of such systems. (Isolated avoided crossings can occur even in integrable systems, as the earlier double-well example showed.)

4 DEGENERACIES AND LEVEL SPACINGS

In the semiclassical limit there is an intimate connection between the degeneracies of a family of systems and the distribution of level spacings for an individual system, which I will explain in two different ways. The spacings are defined so as to have the average value unity; this is achieved by magnifying the energy using the local smoothed level density $d\mathcal{N}(E)/dE$, where $\mathcal{N}(E)$ denotes the smoothed number of levels below E (Berry 1983a). For the system with parameters $\underset{\sim}{R}$, the n'th spacing is thus

$$S_n(\underset{\sim}{R}) \equiv (E_{n+1}(\underset{\sim}{R}) - E_n(\underset{\sim}{R}))\frac{d\mathcal{N}}{dE}(E_n;\underset{\sim}{R}) \qquad (11)$$

The level spacing distribution $P(S;E,\underset{\sim}{R})$ for the system $\underset{\sim}{R}$ at energy E is defined as

$$\left.\begin{array}{l} P(S;E,\underset{\sim}{R}) \equiv \left(\Delta E\dfrac{d\mathcal{N}}{dE}(E;\underset{\sim}{R})\right)^{-1} \displaystyle\sum_{n=\mathcal{N}(E-\Delta E/2;\underset{\sim}{R})}^{\mathcal{N}(E+\Delta E/2;\underset{\sim}{R})} \delta(S-S_n(\underset{\sim}{R})) \\ \left(\begin{array}{l}\Delta E\to 0 \\ \hbar^N/\Delta E\to 0\end{array}\right)\end{array}\right\}. \qquad (12)$$

This definition exploits the fact that semiclassically the level density is very large, so that in the limit it is possible to de- -fine a distribution over infinitely many levels (spacing $\sim \hbar^N$) in an infinitesimal range ΔE, for an individual system $\underset{\sim}{R}$.

Of particular interest is the form of P(S) for small spa- -cings. If P(S) vanishes as $S \to 0$, there is level repulsion; if P(S) tends to a constant as $S \to 0$, there is level clustering. Later I will show that these different types of behaviour occur for systems with different degeneracy structures.

In previous theories, developed to model the many-body physics of nuclei (Porter 1965), P(S) is defined not for an individual system but for an ensemble whose Hamiltonian matrix elements (real symmetric or Hermitian) are random (in fact with a Gaussian distribution). For semiclassical systems the connection between the ensemble and individual spacing distributions is made by realising that (by assumption) there is nothing particular about the system with parameters $\underset{\sim}{R}$, so that (12) would give the same P(S) for any other system close to $\underset{\sim}{R}$ in the family. Therefore (12) is unaffected by ensemble-averaging over a region of parameter space near $\underset{\sim}{R}$. Then, however, all terms in the summation over spa- -cings of levels n close to $\mathcal{N}(E)$ become the same, so that the prefactor cancels and

$$P(S;E) = \langle \delta(S - S_{\mathcal{N}(E)}(\underset{\sim}{R})\rangle_{\underset{\sim}{R}}, \qquad (13)$$

where $\langle\ \rangle_{\underset{\sim}{R}}$ denotes the ensemble average.

In the first way of understanding P(S), only small spacings are considered. The ensemble average (13) then contains contribu- -tions only from parameter-space regions where the surfaces $E_{\mathcal{N}}(\underset{\sim}{R})$ and $E_{\mathcal{N}+1}(\underset{\sim}{R})$ (fig.2), which correspond to the spacing $S_{\mathcal{N}}$, are close to diabolical points. In these regions, the energy surfaces are well approximated by cones (eq.6 or fig.1) for generic real Hamil- -tonians or hypercones for generic Hermitian Hamiltonians.

It is easiest to estimate not P(S) but its integral, namely

$$\left.\begin{aligned} \int_0^S dSP(S) &= \text{probability that level spacing} < S \\ &= \text{fraction of parameter space for which } S_{\mathcal{N}}(\underset{\sim}{R}) < S \\ &\propto \text{measure of parameter space for which sheets of cone have spacing} < S \\ &\propto S^2 \text{(real Hamiltonian, i.e. two parameters)} \\ &\propto S^3 \text{(Hermitian Hamiltonian, i.e. three parameters)}, \end{aligned}\right\} \qquad (14)$$

where the proportionalities S^2 and S^3 arise from the fact that when S is small the cones or hypercones expand linearly away from dege- -neracies.

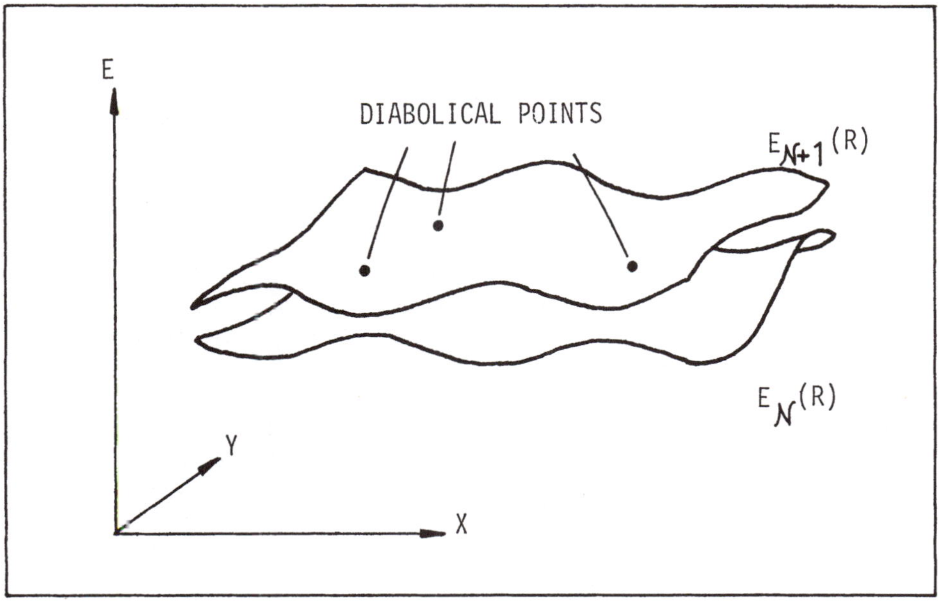

Fig.2. Neighbouring energy levels of generic real Hamiltonians, as a function of parameters $\underset{\sim}{R}=(X,Y)$. The surfaces are conically connected at diabolical points.

Thus

$$\left.\begin{array}{ll} P(S) \propto S & \text{(generic real Hamiltonian)} \\ \quad\;\; \propto S^2 & \text{(generic Hermitian Hamiltonian)} \end{array}\right\} (\text{as } S \to 0) \qquad (15)$$

These results indicate that generically there is level repul--sion, whose nature depends on whether the Hamiltonian is real or Hermitian. For real Hamiltonians, there is accumulating evidence to support the linear level repulsion predicted by (15), based on computations of spectra for several different nonintegrable systems. These include: the stadium billiard (McDonald and Kaufman 1979) and the Sinai billiard (Berry 1981), both of which are chaotic; the square torus billiard (Richens and Berry 1981), which is pseudo--integrable; and irrational triangle billiards (Berry and Wilkinson 1984), which are nonintegrable but neither pseudointegrable nor fully chaotic.

For Hermitian Hamiltonians, (15) predicts quadratic level repulsion. To my knowledge this prediction has not been tested by computation. One such test would be to calculate the levels for a classically nonintegrable planar billiard, with a uniform magnetic field perpendicular to its plane.

As I have already pointed out, separable systems, and

integrable systems with torus quantization, are not generic in that they exhibit degeneracies when a single parameter X is varied. This leads to a simple modification of the argument in equation (14): locally linear level crossings in the E,X plane imply $\int_0^S dSP(S) \propto S$. Thus P(S) tends to a constant as $S \to 0$ so that there is level cluste--ring - a result obtained in a different way, and illustrated by several computations, by Berry and Tabor (1977).

An extension of the scaling argument leading to (15) enabled Berry and Wilkinson (1984) to estimate that for generic planar billiards the semiclassical density of degeneracies in $\underset{\sim}{R}=(X,Y)$ space, between any pair of levels, increases as $\hbar^{-3}$.

Now we turn to the second way of understanding the form of P(S) for classically chaotic systems. This is due to Pechukas (1983), whose argument (which is based on an idea of Dyson 1962 - reprinted in the book by Porter 1965) I will now paraphrase. Let a single parameter vary; think of this as time and denote it accordingly by t rather than X. Regard the levels as a system of particles distributed along the energy axis, with coordinates $E_n(t)$. P(S) (and much more) will be determined by the dynamics of the levels under the time-dependent Hamiltonian $\hat{H}(t)$. Pechukas chooses a function of $\hbar$ for his parameter, which is unfortunate because it excludes quantum billiards (whose levels merely scale homogeneously with $\hbar$); I will work with an arbitrary parameter t.

The dynamics of the levels are determined by Schrodinger's equation, which for the state $|n(t)\rangle$ is

$$\hat{H}(t)|n(t)\rangle = E_n(t)|n(t)\rangle. \tag{16}$$

By elementary means, the velocity of the levels is obtained as

$$\dot{E}_n(t) = \dot{H}_{nn}(t) \tag{17}$$

and their accelerations as

$$\ddot{E}_n(t) = \ddot{H}_{nn}(t)+2\sum_{m\neq n}\frac{|\dot{H}_{mn}(t)|^2}{E_n(t)-E_m(t)}, \tag{18}$$

where dots denote d/dt and

$$\dot{H}_{mn}(t) \equiv \langle m(t)|\frac{d\hat{H}(t)}{dt}|n(t)\rangle . \tag{19}$$

The important equation is (18); its energy denominators show that the acceleration of the levels is dominated by strong interparticle repulsive forces which inhibit collisions (degeneracies).

The next step is to fix an energy E near which the spectrum

is to be studied, and introduce new coordinates $q_n(t)$ with E as origin and scaled so as to give a mean level spacing unity (cf 11 and its preceding paragraph); thus

$$E_n(t) \equiv E + \frac{q_n(t)}{d\mathcal{N}(E;t)/dE} \tag{20}$$

Now comes a crucial simplification. For classically chaotic systems, there is theoretical (Berry 1977, 1983a) and computational (McDonald and Kaufman 1979) evidence that quantum states resemble Gaussian random functions of their variables (e.g. position), which is a quantal reflection of classical exploration of the whole energy surface in phase space. This implies firstly that the matrix elements $\dot{H}_{mn}(t)$ in (18) couple the state n to many different states m with the same average strength, and secondly that $\dot{H}_{mn}(t)$ fluctuate rapidly with t. Dyson (1962), studying random matrices, treats these fluctuations as an underlying Brownian motion superposed on the interparticle repulsion. Pechukas (1983) makes the crucial simplification of simply replacing these fluctuating matrix elements by their t-averages. This amounts to the assumption that the time-scale of the fluctuations in $\dot{H}_{mn}$ is short in comparison with the time-scale of near-collisions resulting from interparticle bouncing; it is not clear how this assumption (which I expect to be valid in the semiclassical limit $\hbar\to 0$) might be justified.

With this simplification, the equations of motion become, in terms of the coordinates q_n (eq.20), and after subtracting any drift embodied in the average value of $\ddot{H}_{nn}$:

$$\ddot{q}_n(t) = \sum_{m\neq n} \frac{A_{mn}}{q_n(t)-q_m(t)} \tag{21}$$

where (with angle brackets now denoting an ensemble average over t)

$$A_{mn} \equiv 2\left(\frac{d\mathcal{N}(E)}{dE}\right)^2 \langle|\dot{H}_{mn}|^2\rangle_t . \tag{22}$$

This describes the classical mechanics of particles distributed (with unit mean density) along the q axis, whose 'Hamiltonian' is

$$\mathcal{H}(\{q_n\}\{p_n\} = \tfrac{1}{2}\sum_n p_n^2 + \sum\sum_{m<n} V_{mn}(q_m-q_n) \tag{23}$$

with interaction potential

$$V_{mn}(q_m-q_n) = - A_{mn}\ell n|q_m-q_n| . \tag{24}$$

The distribution of spacings $S_n (\equiv q_{n+1}-q_n)$, and other level statistics, is given by an average over t (cf 13).

Assuming that the interacting particles attain thermal equilibrium at some temperature $1/\beta$, this time average is also a thermal average, so that the joint probability distribution for all the levels is

$$P(\{q_n\} \propto e^{-\beta\sum_{m<n} V_{mn}(q_m-q_n)} = \prod_{m<n} |q_m-q_n|^{\gamma_{mn}} \qquad (25)$$

where

$$\gamma_{mn} \equiv A_{mn}\beta \qquad (26)$$

To find the temperature $1/\beta$ we realise from (23) that this is the mean value of the fluctuations in (momentum)^2, which from (17) and (20) is

$$\frac{1}{\beta} = \langle p_n^2\rangle_t = \langle \dot{q}_n^2\rangle_t = \left(\frac{d\mathcal{N}(E)}{dE}\right)^2 \left(\langle \dot{H}_{nn}\rangle_t - \langle \dot{H}_{nn}\rangle_t^2\right) \qquad (27)$$

Together with (26) and (22) this gives, for the exponent (25),

$$\gamma_{mn} = \frac{2\langle |\dot{H}_{mn}|^2\rangle_t}{\langle \dot{H}_{nn}^2\rangle_t - \langle \dot{H}_{nn}\rangle_t^2} \qquad (28)$$

For few-neighbour statistics such as P(S), only the <u>local</u> interparticle forces, as embodied in γ_{mn} for $|m-n|$ not large, are important. To find these γ_{mn}, use is again made of the semiclassi--cal chaos hypothesis, in the following explicit form: in the position representation the wavefunctions $\psi_m(\mathbf{x})$ and $\psi_n(\mathbf{x})$ of the states $|m\rangle$ and $|n\rangle$ are independent real or complex Gaussian random functions of $\mathbf{x}$ with the same statistics, the ensemble being parameterized by t. Under this hypothesis, the averages appearing in (28) of the matrix elements

$$\dot{H}_{mn} = \int\cdots\int d\mathbf{x}_1 d\mathbf{x}_2\ \psi_m^*(\mathbf{x}_1)\psi_n(\mathbf{x}_2)\langle \mathbf{x}_1|dH/dt|\mathbf{x}_2\rangle \qquad (29)$$

can be calculated (using elementary but lengthy algebra) and compared. The result is

$$\gamma_{mn} \begin{cases} = 1 \text{ (real } \hat{H}) \\ = 2 \text{ (Hermitian } \hat{H}) \end{cases}. \qquad (30)$$

The fact that γ_{mn} is independent of m and n is a time-averaged reflection of the fact that at any given t the operator $d\hat{H}/dt$ couples many semiclassically chaotic states with approximately equal strength. This behaviour contrasts with the coupling of semiclassical states in classically integrable systems (Pechukas 1983). The fact that the values of γ_{mn} in (30) are so simple reflects a curious feature of the statistical mechanics of energy levels: the strength of the potential is related to the temperature (in different ways for real and Hermitian Hamiltonians).

When combined with (25) and the constraint of unit density, these values of γ_{mn} give the joint distribution of the energy levels. It implies exactly the same few-neighbour statistics as the eigenvalues of random matrices (Porter 1965), in particular the level repulsion laws (15) for $P(S)$ as $S\to 0$ which we obtained by cone scaling. Bohigas and Giannoni (private communication) have found, for the levels of classically chaotic systems, strong computational evidence that few-neighbour statistics other than $P(S)$ are also in good agreement with the predictions of random-matrix theory.

The universality of few-neighbour statistics for generic semiclassical systems and random matrices cannot extend to the statistics of distant levels. Even in random matrix theory, the distant correlations are model-dependent, being different for the orthogonal matrices of the circular ensemble and the symmetric matrices of the Gaussian ensemble (Porter 1965). Semiclassically, the distant correlations between levels separated by $O(\hbar)$ (i.e. for groups of $O(\hbar^{-(N-1)})$ levels in an N-freedom system) must reflect the nature of the eventual decay of the interaction-strength γ_{mn} from the values (30); in any case it is known (Berry 1983ab, see also Hannay's paper in this volume) that such distant correlations take the form of oscillatory clustering associated with individual classical closed orbits.

5 PHASE CHANGES NEAR DEGENERACIES

Near a degeneracy in parameter space $\mathbf{R}$, energy levels $E_n(\mathbf{R})$ have diabolical connections as discussed in section 2 (see fig.1) which depend on whether the Hamiltonian $H(\mathbf{R})$ is real or Hermitian. Now I will discuss the nature of the stationary states $|n(\mathbf{R})\rangle$, by asking: how do the $|n(\mathbf{R})\rangle$ behave under continuation in $\mathbf{R}$ space? In particular, if $\mathbf{R}$ is varied round a circuit C, how are the states at the beginning and end of the loop related? Without further information the answer can only be that the states are related by an arbitrary phase factor because the only restriction on $|n(\mathbf{R})\rangle$ is that it is an eigenstate of $\hat{H}(\mathbf{R})$, and this leaves its phase undefined.

For real $\hat{H}(\mathbf{R})$, a natural restriction on $|n(\mathbf{R})\rangle$ is that in some basis (e.g. the position representation $\mathbf{r}$) wavefunctions $\psi_n(\mathbf{r};\mathbf{R})$ remain real. Only the sign of ψ_n remains undetermined, and this can be fixed by requiring $\psi_n(\mathbf{r};\mathbf{R})$ (which depends smoothly on

$\mathbf{x}$) to vary continuously with $\mathbf{R}$. With this (implicit) convention Herzberg and Longuet-Higgins (1963) (see also Longuet-Higgins 1975) found by considering the multipliers a_{ij} in (2) that the quantum state changes sign if and only if C encloses an odd number of diabolical points (as in fig.1). Such enclosure is possible because degeneracies have codimension two for real Hamiltonians.

The sign change is a test for diabolical points, and is employed by Berry and Wilkinson (1984) for quantum triangles whose angles were taken round a cycle C of changes so as to surround a triangle for which two states were degenerate: the normal derivative of the wavefunction of either state was seen to change sign round C. Globally, the sign change implies that the nodal lines (or more generally nodal hypersurfaces) in $\mathbf{x}$ space must move as C is traversed. Korsch (1983) shows a particularly simple example involving two degenerate states of a square quantum billiard; in this case the degeneracy occurs because of symmetry. Fig.3 shows the nodal patterns associated with a sign reversal for a circuit of a degeneracy (which is not symmetry-generated) in a quantum triangle; for details see Berry and Wilkinson (1984).

For Hermitian $\hat{H}(\mathbf{R})$, continuity as C is traversed is not sufficient to determine the phase change of a quantum state. A natural physical principle which does determine the phase change is to consider C being traversed very slowly in time and defined by $\mathbf{R}(t)$, so that $H(\mathbf{R}(t))$ is a time-dependent Hamiltonian governing the (adiabatic) evolution of quantum states $|\psi_n(t)\rangle$ according to the time-dependent Schrödinger equation. If the traversal takes a (long) time T (so that $\mathbf{R}(T)=\mathbf{R}(o)$), the states at the beginning and end of C are related by

$$\langle\Psi_n(o)|\Psi_n(T)\rangle = e^{i\gamma_n(C)}\, e^{-\frac{i}{\hbar}\int_o^T dt\, E_n(\mathbf{R}(t))} \tag{31}$$

The second factor contains the dynamical phase associated with the evolution of any quantum state; it can be regarded as the system's way of recording time. The first factor contains the Hermitian generalization of the sign reversal associated with degeneracies of real Hamiltonians - a generalization which must be subtle because Hermitian degeneracies have codimension three and so cannot be enclosed by circuits C. I call $\exp\{i\gamma_n(C)\}$ the geometrical phase factor; it can be regarded as the system's way of recording where it has been in parameter space - a sort of 'quantum learning'.

Exactly how $\gamma_n(C)$ is determined by the natural continuation defined by the time-dependent Schrödinger equation, is explained by Berry (1984); here I simply summarise the main result, for the case where the parameter space $\mathbf{R}=(X,Y,Z)$ is three-dimensional

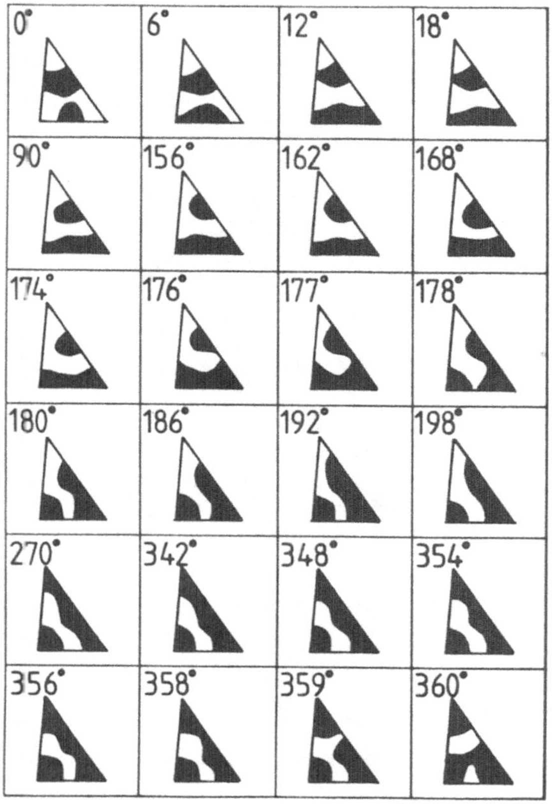

Fig.3. Sign reversal of a vibration mode for a triangle taken round a small circuit C, in angle space, surrounding a diabolical triangle; the indicated angles parameterise position on C.

(fig.4). It is that $\gamma_n(C)$ is determined by the flux through C of a vector field $\underset{\sim}{V}_n(\underset{\sim}{R})$ whose singularities are degeneracies involving the state $|n\rangle$.

Explicitly,

$$\gamma_n(C) = -\iint_S d\underset{\sim}{S}\cdot\underset{\sim}{V}_n(\underset{\sim}{R}) \tag{32}$$

where the integral is over any surface S spanning C and where

$$V_n(R) \equiv \mathrm{Im} \sum_{m\neq n} \frac{\langle n(\underset{\sim}{R})|\nabla_{\underset{\sim}{R}}\hat{H}(\underset{\sim}{R})|m(\underset{\sim}{R})\rangle \times \langle m(\underset{\sim}{R})|\nabla_{\underset{\sim}{R}}\hat{H}(\underset{\sim}{R})|n(\underset{\sim}{R})\rangle}{[E_m(\underset{\sim}{R}) - E_n(\underset{\sim}{R})]^2} \tag{33}$$

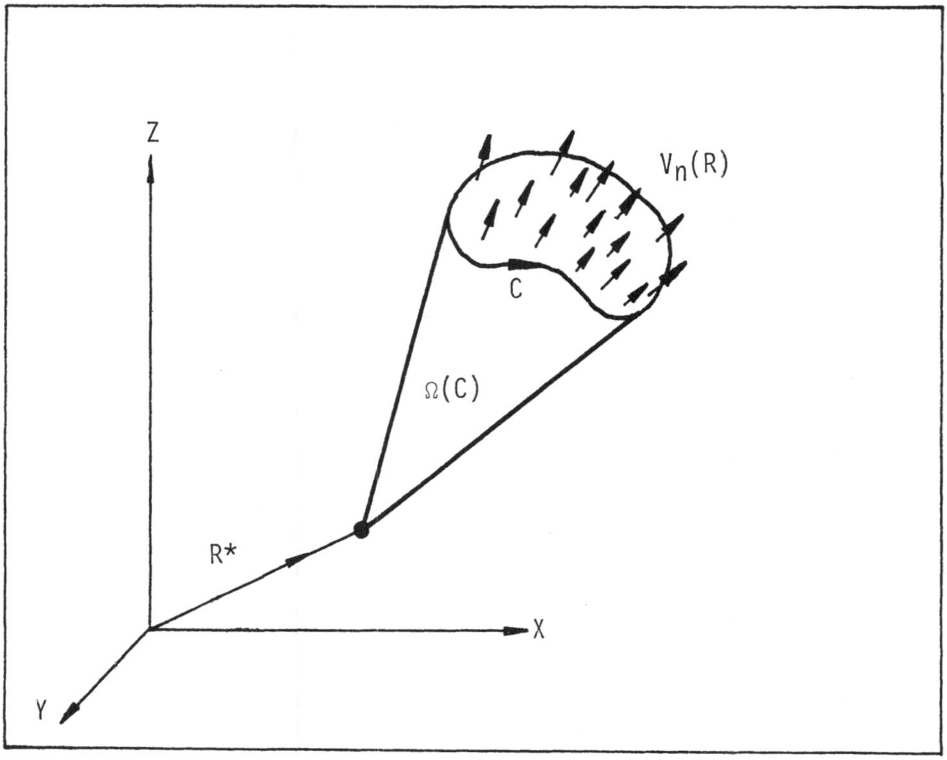

Fig.4 Circuit C near degeneracy $\mathbf{R}^*$ in $\mathbf{R}=(X,Y,Z)$ parameter space

In this formula, the arbitrary phases of the eigenstates $|n(\mathbf{R})\rangle$ cancel, so $\mathbf{V}_n(\mathbf{R})$ is manifestly independent of their choice.

The energy denominators make it obvious that $\mathbf{V}_n$ is singular at degeneracies $\mathbf{R}^*$. Close to $\mathbf{R}^*$, the sum involves only the state $|m\rangle=|n\pm 1\rangle$ with which $|n\rangle$ degenerates. In this case the form of the singularity is particularly simple in the parameterization for which the 2×2 Hamiltonian coupling the two degenerating states has the standard form

$$\hat{H}(\mathbf{R}) = \tfrac{1}{2}\begin{pmatrix} Z-Z^*, & X-X^* -i(Y-Y^*) \\ X-X^*+iY-Y^*, & Z^*-Z \end{pmatrix} \qquad (34)$$

Then $\mathbf{V}(\mathbf{R})$ is the magnetic field of a monopole located at $\mathbf{R}^*$, with strength $\pm$ 1/2 if $|n\rangle$ degenerates with $|n\mp 1\rangle$, and the phase factor associated with C is

$$e^{i\gamma_n(C)} = e^{\pm i\Omega(C)/2}, \qquad (35)$$

where $\Omega(C)$ is the solid angle (fig.4) subtended by C at the degeneracy.

For real Hamiltonians, $Y-Y^*=0$ in (34) and C lies in a plane and surrounds $\mathbf{R}^*$. Then S subtends $\Omega=\pm 2\pi$ and (35) reproduces the sign-reversal rule. By topological arguments not requiring explicit formulae for $\gamma_n(C)$, Stone (1976) proved that if C is expanded from one point $\mathbf{R}$ and contracted onto another so as to sweep out a closed surface enclosing $\mathbf{R}^*$, then the geometrical phase factor traverses a circle in its Argand plane; this property (which is easy to prove using (35)) is the Hermitian generalization of the sign-reversal test for degeneracy.

Apart from generalizing the sign-reversal rule near degene--racies, the geometrical phase factor defined by (31-33) has other interesting aspects. As described by Berry (1984), $\exp\{i\gamma_n(C)\}$ implies observable interference effects for spinning particles (bosons as well as fermions) in slowly-rotated magnetic fields, and contains the Aharonov-Bohm effect as a special case. Mead (1979, 1980,1983) and Mead and Truhlar (1979) describe how phase changes in molecular electronic states resulting from changes in nuclear configuration must affect the states and energies of nuclear motion.

REFERENCES

Arnol'd,V.I. 1972 Funct.Anal.Appl. **6** 94-101

Arnol'd V.I. 1978 Mathematical Methods of Classical Mechanics (Springer: New York)

Berry,M.V. 1977 J.Phys.A **10** 2083-2091

Berry,M.V. 1978 Regular and Irregular Motion in Topics in Nonlinear Mechanics (S.Jorna,ed) Am.Inst.Phys.Conf.Proc.**46** 16-120

Berry,M.V.1981 Ann.Phys.(N.Y) **131** 163-216

Berry,M.V. 1983a, Semiclassical Mechanics of Regular and Irregular Motion in Chaotic Behaviour of Deterministic Systems (Les Houches Lectures XXXVI,ed R.H.G.Helleman and G.Iooss) (North-Holland:Amsterdam) in press

Berry,M.V.1983b, Structures in Semiclassical Spectra: a Question of Scale in The Wave-Particle Dualism (ed.S.Diner,D.Fargue, G.Lochak and F.Selleri) (D.Reidel: Dordrecht) in press

Berry,M.V. 1984, Proc.Roy.Soc.Lond. A. submitted

Berry,M.V.and Tabor,M. 1977 Proc.Roy.Soc.Lond. **A356** 375-394

Berry,M.V. and Wilkinson,M 1984 Proc.Roy.Soc.Lond.A. submitted

Dehareng,D, Chapuisat ,X, Lorquet,J-C,Galloy,C and Raseev,G, 1983 J.Chem.Phys. **78** 1246-1264

Desouter-Lecomte,M, Galloy,C and Lorquet,J-C, 1979 J.Chem.Phys. **71** 3661-3672

Dyson,F.J, 1962 J.Math.Phys. **3** 1191-1198

Herzberg,G and Longuet-Higgins,H.C 1963 Disc.Far.Soc. **35** 77-82

Hill,D.L and Wheeler,J.A 1952 Phys.Rev. **89** 1102-1145

Knox,R.S and Gold,A 1964 Symmetry in the Solid State (Benjamin:N.Y)

Korsch,H.J 1983 Physics Letters A. In press

Lichtenberg,A.J and Lieberman,M.A 1983, Regular and Stochastic Motion (Springer:New York)
Longuet-Higgins,H.C 1975 Proc.Roy.Soc.Lond. A344 147-156
Marcus,R.A 1980 in Nonlinear Dynamics (ed.R.H.G.Helleman) Ann N.Y.Acad.Sci. 357 169-182
Maslov,V.P and Fedoriuk,M.V 1981 Semiclassical Approximation in Quantum Mechanics (D.Reidel:Dordrecht)
McDonald,S.W and Kaufman,A.N 1979 Phys.Rev.Lett.42 1189-1191
Mead,C.A 1979 J.Chem.Phys. 70 2276-2283
Mead,C.A 1980 Chem.Phys (Netherlands) 49 23-32, 33-38
Mead,C.A 1983 J.Chem.Phys. 78 807-814
Mead,C.A and Truhlar,D.G 1979 J.Chem.Phys. 70 2284-2296
Noid,D.W, Koszykowski,M.L, Tabor,M and Marcus,R.A 1980 J.Chem.Phys. 72 6167-6175
Pechukas,P 1983 Phys.Rev.Lett. In press
Porter,C.E 1965 Statistical Theories of Spectra:Fluctuations (Academic Press: New York)
Ramaswamy,R and Marcus,R.A 1981 J.Chem.Phys. 74 1379-1384,1385-1393
Richens,P.J and Berry,M.V 1981 Physica 1D 495-512
Stone,A.J 1976 Proc.Roy.Soc.Lond. A351 141-150
Teller,E 1937 J.Phys.Chem.41 109-116
Vaz Pires,M, Galloy,C and Lorquet,J-C 1978, J.Chem.Phys. 69 3242-3249
Von Neumann,J and Wigner,E 1929 Phys.Z. 30 467-470

PERIODIC ORBITS AND A CORRELATION FUNCTION FOR THE SEMICLASSICAL DENSITY OF STATES

J.H. Hannay

H.H.Wills Physics Laboratory
University of Bristol
Tyndall Avenue, Bristol BS8 1TL, U.K.

Through the rule that there is one quantum state per Planck cell in phase space, classical mechanics supplies an 'average' semiclassical density of states $\langle n(E)\rangle$ for a system with bound motion. Can it supply more refined 'averages' such as the correlation function for the density of states $\langle n(E)n(E+\Delta E)\rangle$? I shall argue that for the two extreme cases of integrable classical motion and ergodic classical motion information can indeed be obtained on not quite this, but a closely related correlation function. The results for the two types of motion are markedly different.

The starting point is the periodic orbit sum of Gutzwiller[1] (with subsequent important contributions by Balian and Bloch[2], and Berry and Tabor[3,4,5].) This is a semiclassical expression for the density of states

$$n(E) \equiv \sum_j \delta(E-E_j) = \langle n(E)\rangle + \tilde{n}(E) \qquad (1)$$

$$\tilde{n}(E) \underset{\hbar\to 0}{=} \sum_{\text{Periodic orbits}} \text{Oscillatory correction to } \langle n(E)\rangle \qquad (2)$$

where the idea is that if enough oscillatory corrections are included the expression will mimic the true spiky density of states with a δ function at each energy level (fig.1)

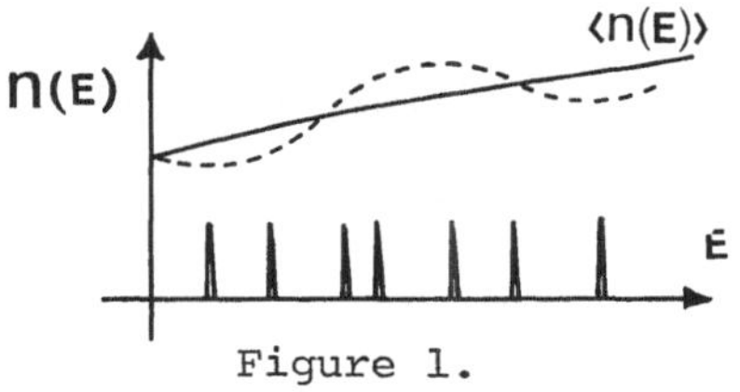

Figure 1.

The sum is over those special trajectories which close and retrace themselves - periodic orbits, with each contributing an amplitude times $\exp(iS(E)/\hbar)$ where S is the action $\oint \underset{\sim}{p}\cdot d\underset{\sim}{q}$ around the periodic orbit. For ergodic systems almost all periodic orbits are isolated paths but for integrable systems they come in continuous families - a stack of loops around a torus in phase space. In this case each torus of periodic orbits, rather than each individual periodic orbit gives an oscillatory correction.There are difficulties in applying the periodic orbit method to systems such as are described by KAM theory which are neither fully integrable nor fully ergodic. (It is not quite clear how to evaluate orbit amplitudes for the stable isolated periodic orbits which occur in such systems, in contrast to the unstable ones of ergodic systems). This is the reason for the present restriction to the two extremes.

It can seem strange that such detailed information on a system as its complete eigenvalue spectrum is available from such a special subset of orbits as periodic ones. It seems less so when it is recalled that (for non-special systems at least) periodic orbits are dense in phase space - the motion of any non-periodic phase trajectory can be matched with arbitrary accuracy for an arbitrarily long duration by motion on a nearby periodic orbit. Indeed this dense packing of periodic orbits is to be the basis of the present method.

To simplify this presentation I shall specialize to 'billiards' i.e. two dimensional enclosures with mirror reflection on the boundary (fig.2). Let the area of the enclosure be a and its perimeter l.

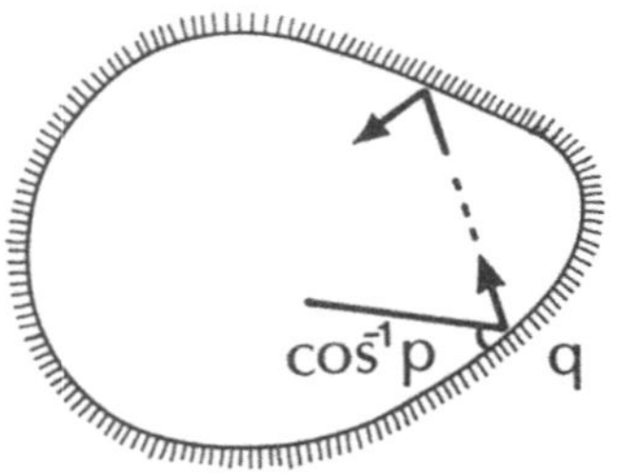

Figure 2.

The Schrödinger equation for such systems is $\nabla^2\psi + k^2\psi=0$ with $\psi=0$ on the boundary leading to eigenvalues k_j and a density of levels $n(k)\equiv\sum_j\delta(k-k_j)$. The average part is

$$\langle n(k)\rangle = ka/2\pi + \text{boundary terms (k independent)} \qquad (3)$$

The oscillatory part $\tilde{n}(k)$ is then given by the periodic orbit sum

$$\tilde{n}(k) = k^{\nu} \sum_{\text{periodic orbits } j} A_j \, e^{ikL_j} \, e^{i\alpha_j} \tag{4}$$

where $\nu=0$ for an ergodic billiard (with all periodic orbits isolated) and $\nu=1/2$ for an integrable billiard. The L_j are the lengths of the periodic orbits which will be assumed to be distinct (no symmetry) for simplicity. The α_j are phase factors independent of k which need not be specified as they will cancel out. The A_j are orbit amplitudes (also k independent) governed by the second variation of the length L_j with virtual disturbance (of the bounce points, say). They will be specified as required shortly. It should be noted that repetitions of periodic orbits are counted as seperate periodic orbits in (4). The repetition number runs through all positive and negative integers m except zero and L_j is m times the primitive orbit length.

All is set for introducing the correlation function, but what should be 'average' mean? To answer this one may ask the same question of $\langle n(k)\rangle$ (3). There one can write the following exact relation

$$\frac{a}{2\pi} = \left\langle \frac{n(k)}{k} \right\rangle \equiv \lim_{k\to\infty} \int_0^k \frac{n(k)}{k} \, dk \tag{5}$$

so that the average stands for an average over <u>all</u> k. The relation follows of course because of all k, almost all are infinite). Actually it will be convenient to extend the range of integration in (5) to -k to +k and divide by two. With the average so defined we can proceed analogously to obtain the correlation function. From (4)

$$\left\langle \frac{\tilde{n}(k)}{k^{\nu}} \; \frac{\tilde{n}(k+\Delta k)}{(k+\Delta k)^{\nu}} \right\rangle = \sum_{ij} A_i A_j \langle e^{ik(L_i - L_j)} \rangle \times e^{i\Delta k L_i} e^{i(\alpha_i - \alpha_j)} \tag{6}$$

$$= \sum_i A_i^2 \, e^{i\Delta k L_i}$$

since the average in(6) yields just the Kronecker delta. In words then we have the promising result that the correlation function is the Fourier transform of a weighted density of periodic orbit lengths $\sum_j A_j^2 \delta(L-L_j)$ (fig.3).

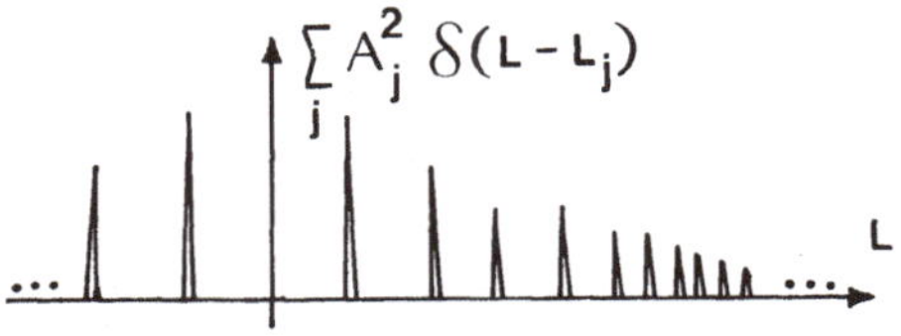

Figure 3.

Progress however depends on our extracting from classical mechanics some information on this weighted length density, for a priori one knows nothing of any generality about it. Happily a principle can be developed that extracts one particular feature at least. This 'uniformity' principle will yield the behaviour of the weighted length density function for large L rendering quantitative the qualitative observation that periodic orbits tend to become more numerous (increasing density) but weaker (decreasing amplitude) as L rises.

The basis of the principle is described below. Its consequence, and the result for the correlation function are these

$$\Sigma A_j^2 \delta(L-L_j) \underset{L\to\infty}{=} \begin{cases} |L| \times \text{Constant, Ergodic} \\ \text{constant, Integrable} \end{cases} \tag{7}$$

and by Fourier transformation the implication for the behaviour of the correlation function:

$$< \frac{\tilde{n}(k)\tilde{n}(k+\Delta k)}{k^{\nu}(k+\Delta k)^{\nu}} > \underset{\Delta k\to 0}{=} \begin{cases} -\frac{1}{\Delta k^2} \times \text{Constant, Ergodic} \\ 0 \text{ , Integrable} \end{cases} \tag{8}$$

In the integrable case this result of zero correlation is consistent with the Poisson distribution predicted by Berry and Tabor[3]. In the ergodic case the negative result is consistent with level repulsion (given a level at k it is less than averagely likely to find a level at $k+\Delta k$). It should be noted that although the results apply to small Δk they are not describing structure on the scale of the level spacing, about which a good deal is known[6]. This is because the correlation function averages over all k, and at high k the density of states $<n(k)>$ is high so that there are many levels in the range Δk. Indeed the ergodic result is reminiscent of that for the <u>tail</u> of the correlation function for the density of eigenvalues of a random matrix[7] which is used with increasing confidence to model that of ergodic Hamiltonian systems[8].

It remains to explain the uniformity principle and show how it justifies the relations (8). In general the uniformity principle says that, with appropriate (natural) weighting, the density of periodic orbits in phase space is uniform. For billiards the precise statement is considerably simplified by considering the 'bounce' map. In this the classical dynamics is fully specified as an area preserving map of a rectangle onto itself. The coordinates on the rectangle are (fig.2) (i) arc length q round the perimeter specifying the position of a bounce on the boundary and (ii) a measure of the angle of incidence with the normal on bouncing - its sine; p, (for area preservation). Knowing these two quantities one obviously knows, from the shape of the enclosure, the q and p of the next bounce - the shape determines the mapping of the rectangle. For an integrable billiard successive iterates of a typical initial point jump along a curve eventually filling it out, while for an ergodic map the point jumps about everywhere within the rectangle filling it evenly.

I shall treat only this latter, ergodic, case where the explanation of the uniformity principle is maximally simple. One way to express the ergodic jumping is the following, with $\underline{r} \equiv (q,p)$:

$$\frac{1}{2N}\sum_{|n|=1}^{N} \delta(\underline{r}-\underline{r}_n) \underset{N\to\infty}{=} \text{constant } c \text{ (independent of } \underline{r}) \tag{9}$$

where $\underline{r}_n$ is the nth iterate of an initial point $\underline{r}_o$ (with n taking negative as well as positive values for subsequent convenience). The value of the constant c is easily found by integrating both sides of (9),to be the reciprocal of the rectangle area (that is, 2l since the range of p is two).

The result's independence of $\underline{r}$, and indeed $\underline{r}_o$, holds rigorously if the δ function is smoothed by an arbitrarily small amount and the implicit assumption is that the limit of zero smoothing can be taken.

The key step is now easy - we merely evaluate the function of $\underline{r}$ (i.e. the constant c) at $\underline{r}_o$ to get

$$\frac{1}{2N}\sum_{|n|=1}^{N} \delta(\underline{r}_o-\underline{r}_n) \underset{N\to\infty}{=} c \tag{10}$$

But as a function of $\underline{r}_o$ for a particular n, $\delta(\underline{r}_o-\underline{r}_n)$ is itself a set of δ functions - isolated δ spikes in the rectangle sitting on those points $\underline{r}_o$ which return to themselves after n iterations - the 'periodic points of period n' (fig.4).

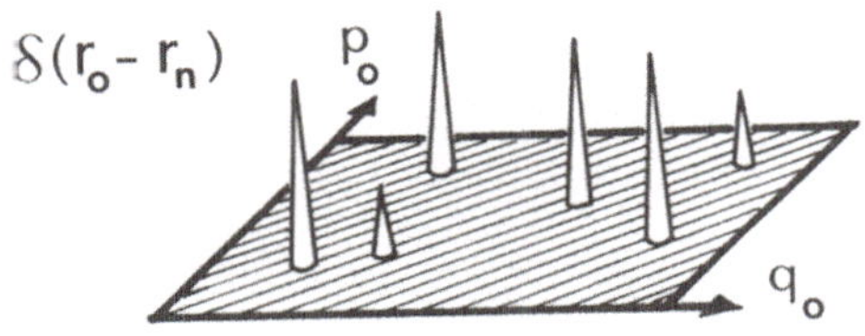

Figure 4.

These correspond to the <u>periodic orbits</u> of the billiard which close after n bounces on the boundary. Of course each iterate of a periodic point is also a periodic point so that for one periodic orbit there is a cycle of periodic points.

Each δ spike in the set $\delta(\underline{r}_o - \underline{r}_n))$ evidently has its own weight - the derivative Jacobian $|\det^n(\partial\underline{r}_n/\partial\underline{r}_o - 1)|^{-1}$. For every periodic point within the same cycle the weight is the same but different cycles generally have different weights. Now it is not difficult to show that the amplitude A of a periodic orbit appearing in the path sum (4) is related to the corresponding periodic point weight by

$$A^2 = \frac{L^2}{(2\pi m)^2} \left|\det(\partial\underline{r}_n/\partial\underline{r}_o - \underline{\underline{1}})\right|^{-1} \tag{11}$$

where m is the repetition number of the orbit (so that L/m is the primitive length). By integrating (10) over the rectangle we obtain a sum rule for the amplitudes of periodic orbits

$$\frac{1}{2N} \sum_{\substack{\text{Periodic orbits } j \\ \text{with } |n_j| \leqslant N \text{ bounces}}} \frac{|n_j| A_j^2 m_j^2}{|m_j| L_j^2} \underset{N\to\infty}{=} C/(2\pi)^2 \tag{12}$$

where the $|n_j/m_j|$ factor is required because a single periodic orbit corresponds to a cycle of $|n_j/m_j|$ periodic points.

Finally the sum (12) can be simplified by recognizing that it is dominated as $N\to\infty$ by long orbits (indeed any finite subset of orbits is irrelevant). Of infinitely long orbits in an ergodic system almost all are primitive ($m=\pm 1$); this must be so if periodic orbits proliferate rapidly (and that they do so is indeed implied by the result below). So the first simplification is to set $|m_j|=1$. The other is that the average length of a long orbit is proportional to its number of bounces. (The constant of proportionality is $\pi a/l$, the mean length of a chord in the enclosure, though this is not actually needed.) So the $|n_j|$ inside the sum and the N outside it can be converted respectively to $|L_j|$ and L. Thus (12) becomes

$$\frac{1}{2L} \sum_{\substack{\text{Periodic orbits } j \\ \text{with } |L_j| \leqslant L}} \frac{A_j^2}{|L_j|} \underset{L\to\infty}{=} C/(2\pi)^2 \tag{13}$$

This implies the stated $|L|$ dependence of the weighted length density of periodic orbits (7) for an ergodic system and by

Fourier transformation the - $1/\Delta k^2$ dependence of the correlation function.

The argument appropriate to integrable motion and indeed the derivation of the correlation function for general Hamiltonian systems rather than just billiards is given in the work on which the present contribution is based:

J.H.Hannay & A.M.Ozorio de Almeida
'Periodic orbits and a correlation function for the semiclassical density of states' (to be submitted to J.Phys.A).

In this general case the role of wavenumber k is played not by the energy E but by the parameter $\hbar^{-1}$. With the density of states considered as a function $n(E,\hbar^{-1})$ of $\hbar^{-1}$ for fixed E, the results are analogous. The correlation function, with its average over all $\hbar^{-1}$, goes like $-(\Delta\hbar^{-1})^{-2}$ for the ergodic case and zero for the integrable one.

REFERENCES

1 M.C.Gutzwiller, J.Math.Phys.8 (1967) 1979-2000
J.Math.Phys.10 (1969) 1004-1020
J.Math.Phys.11 (1970) 1791-1806
J.Math.Phys.12 (1971) 343-358

2 R.Balian and C.Bloch, Ann.Phys.N.Y.60 (1970) 401-447
Ann.Phys.N.Y.64 (1971) 271-307
Ann.Phys.N.Y.69 (1972) 76-160
Ann.Phys.N.Y.85 (1974) 514-545

3 M.V.Berry and M.Tabor, Proc.Roy.Soc.A349 (1976) 101-123

4 M.V.Berry and M.Tabor, Proc.Roy.Soc. A356 (1977) 375-394

5 M.V.Berry and M.Tabor, J.Phys.A10 (1977) 371-379

6 M.V.Berry, 'Semiclassical Mechanics of Regular and Irregular Motion'. Lectures given at Les Houches 1981, (North-Holland, R.H.G.Helleman, ed.).

7 C.F.Porter (ed.), 'Statistical Theory of Spectra: Fluctuations' (Academic Press, N.Y.1965).

8 P.Pechukas, Phys.Rev.Lett. (to appear)

MILD CHAOS

Martin Gutzwiller

IBM T. J. Watson Research Center
Yorktown Heights, New York 10598

Abstract: The different roles of chaos in physics are discussed, and 2 examples of a smooth kind of chaos are given. First, the proliferation of small terms and the occurrence of small denominators is shown to cause serious difficulties in the practical calculations of the moon's motion through the sky. Second, the scattering of an electron on a rigid molecule is reduced to the motion of a particle on a surface of negative curvature. The resulting phase shift is an analytic, and yet chaotic function of the wave vector.

The word "chaos" is not a technical term for the time being, in contrast to words such as ergodicity, mixing and entropy, which can be defined even if there are different kinds of each. Chaos, at this point, describes rather a vague feeling of malaise among scientists, an embarrassing state of affairs whose precise symptoms are not clear. This applies in particular to chaos in quantum mechanics; but in any situation, our recourse to chaos is almost an admission of failure about our efforts to find order and beauty in the description of nature.

One is forced to ask (paraphrasing the words of I. I. Rabi): Who ordered chaos? More specifically, what is it good for? Where is it needed? Can it be avoided? The answer is almost contradictory in terms: Chaos is necessary to unify the following four basic areas of physics,

1) Newtonian Mechanics,
2) Thermodynamics and Equilibrium Statistical Mechanics,
3) Atomic and Molecular Physics,
4) Electromagnetism including Optics.

Each of these separate fields was discovered and developed independently of the others. They have been found in agreement with every theoretical prediction imaginable, to almost arbitrary precision. Nobody has any serious doubts about our understanding of these four (by now) classical areas on the basis of the underlying mathematical principles, even if not all problems have been solved as yet. This situation is quite different from what prevails, e.g., in hydrodynamics, physics of condensed matter, nuclear and elementary particle physics.

What happens when the 4 classical areas come in contact with one another? Evidently, different concepts, pictures, mathematical structures have to be tied together, such as ordinary non-linear differential equations and parabolic linear partial differential equations, in the case of classical and quantal mechanics. This process of establishing the connection works in some very special circumstances, but in general, it runs into great difficulties. With only slight exaggeration, one always finds some kind of chaotic behavior at the root of our troubles.

Best known in this regard is the ergodic hypothesis of Boltzmann which is supposed to reduce thermodynamics to Newtonian mechanics. Although the importance of this idea has been recognized for roughly 100 years, it was tested for the first time (and found to fail) only 30 years ago by Fermi, Pasta and Ulam.[1] A lot of progress has been made since then, but it is not the topic of this conference.

A much less clear picture emerges from all the efforts to see classical mechanics as some limiting case of quantum mechanics. The difficulty here was recognized by Einstein in 1917, but his paper[2] was totally ignored for 40 years. The fundamental problem he posed about the transition from classical to quantal mechanics has yet to find a general treatment. The main obstacle is our inability to provide an effective description for the chaotic trajectories in classical Hamiltonian systems.

The main topic of this conference, however, is the bridge from quantum mechanics to statistical mechanics, the third, and possibly most difficult step in the unification of the first three among the four basic areas. More precisely, we would like to know how chaotic behavior manifests itself in a quantum mechanical system.

This is an entirely new field of inquiry. Before we get our feet wet, and our hands dirty, we should realize that we are dealing with something completely new. Classical chaos always seems to produce fractal sets,[3] and its lesson is by now fairly clear. Every time a phenomenon is examined on a finer scale, it reveals the same structure as on the rougher scale. An infinite descent is possible whose essence is self-similarity.

Quantum mechanics is specifically designed to describe the atom as the smallest constituent in nature. Its purpose is to stop the infinite descent to an ever finer scale, and this goal is achieved at a considerable price in terms of mathematical complexity as well as philosophical complications. Thus, not only do we face more difficult technical problems, but we have to ask ourselves to what extent our calculations can be experimentally tested in principle, let alone in practice.

Looking at nodal patterns in wave functions may be informative as far as the mathematics goes, but it is doubtful whether any experiment can be designed to investigate this matter. On the other hand, subjecting a simple quantum pendulum to periodic kicks and measuring its increase in kinetic energy appears quite feasible. When Casati, Chirikov, Ford and Israelev[4] found a qualitative difference between classical and quantal behavior in this system, a true characteristic of quantum chaos had been established. It may take a while, however, before its mathematical explanation does more than relate this mystery to the enigma of Anderson localization in a crystal with random potential[5] (although to have such connections is very valuable).

The virtual identity of quantum mechanics with atomic physics should give us some clues. The extreme chaos of classical mechanics gets smoothed over on the atomic scale. But then, to what extent can a smooth object ever be called chaotic? The classical measures such as entropy seem much too rough to give any useful criteria. It is imperative to study particular examples and to become familiar with their quantal behavior, before constructing sophisticated mathematical theories and deriving general theorems.

Some mental pictures are necessary to provide guidance. One would like to find appropriate qualifiers to describe the different kinds of chaos, something like severe, sharp or harsh for classical motions, and perhaps gentle, benign, or mild in quantum mechanics. At the same time, even some presumably well-behaved systems in classical mechanics should be recognized for what they are in practice, that is mildly chaotic, and therefore perhaps representative of typical quantal behavior.

An example of this last kind from the author's own experience will be discussed first, namely, the motion of the moon around the earth under the disturbing influence of the sun. This is the first problem in non-linear mechanics which mankind has tried to describe in a scientific manner, and with significant success as far back as 3000 years ago. It has also turned out to require the most advanced numerical tools for a satisfactory practical solution. But a complete mathematical treatment is not in sight, possibly because there is a subtle form of chaos, as yet not understood.

The other example in this paper takes off from some of the author's recent work[6]. The quantum scattering of an electron on a molecule will be shown to resemble the scattering on a surface of constant negative curvature. The latter gave rather surprising results. Its phase shift although analytic demonstrates a peculiar kind of chaos which may be typical of quantum systems in general.

Gentle Chaos in the motion of the moon

The satellites of planets are supposed to move around their parent exactly as the planets move around the sun. Our own moon is a glaring exception to this rule. Her motion is strongly perturbed by the sun; so much so that all other disturbances (non-spherical shape of the earth, the neighboring planets, etc.) are smaller by a factor 10^{-4}. The following simplified three-body problem gives therefore the essence of the lunar motions.

The center of mass Γ of the earth and the moon moves around the sun in a fixed plane, the ecliptic, along a fixed Kepler ellipse of semi-major axis a' and eccentricity e'. The mean angular speed n' of Γ around the sun is connected with the masses (S for sun, E for earth, M for moon) by Kepler's third law, $n'^2 a'^3 = G(S + E + M)$, with the gravitational constant G.

The main events in the lunar motion have to do with her position relative to the sun as seen from the earth. Therefore, the Cartesian vector (x,y,z) which points from the center of the earth to the center of the moon is referred to a special moving frame. The x and y directions lie in the ecliptic in such a manner that the mean position of the sun is always on the negative x-axis at a distance a'. Thus, the moon's frame of reference is geocentric and turning at the uniform rate n'.

The Hamiltonian H, in terms of the momenta (u,v,w) conjugate to (x,y,z), is given by

$$H = 1/2(u^2 + v^2 + w^2) + n'(uy - vx) + n'^2(x^2 + y^2)$$
$$- \frac{G(E+M)}{r} - \frac{GS}{r'} \sum_{j=2}^{\infty} \frac{E^{j-1} - (-M)^{j-1}}{(E+M)^{j-1}} \left(\frac{r}{r'}\right)^j P_j(\cos\gamma),$$

where the momenta have already been divided by the reduced mass $EM/(E + M)$. The distance between the moon and the earth is $r = (x^2 + y^2 + z^2)^{1/2}$, and the distance of the sun from Γ is r'. The angle between the moon and the sun as seen from the earth is γ which occurs in the Legendre polynomials $P_j(\cos\gamma)$. The mean position of the sun in the rotating framework is $x = -a'$, and the true position is known explicitly as a function of time with the polar coordinates (r',ϕ') in the (x,y) plane

$$r' = a'[1 + \frac{1}{2}e'^2 - e' \cos \ell' - \frac{1}{2}e'^2 \cos 2\ell' + ...],$$

$$\phi' = \pi + 2e' \sin \ell' + \frac{5}{4}e'^2 \sin 2\ell' + ...,$$

where $\ell' = n't + \ell'_0$. Therefore, one has $r \cos \gamma = x \cos \phi' + y \sin \phi'$.

The individual terms in this Hamiltonian can be compared with the various energies of an electron in an atom under the combined influence of magnetic and electric fields. The second term is the interaction with a constant magnetic field of moderate size, since the Larmor frequency $n' = 2\pi/\text{year}$ is about 1/13 of the atomic frequency $n = 2\pi/$ sidereal month. The third term contains both diamagnetism and the centrifugal potential. Both of these combine with the lowest term $j = 2$ in the multipole expansion which is the quadrupole field of the sun in the neighborhood of the earth, and has the opposite sign $-n'^2(3x^2-r^2)/2$. The sun destabilizes the moon, but only along the x-direction in lowest order.

The Hamiltonian has now been completely defined, and the results from its trajectories can be discussed. As long as the sun's eccentricity e' is neglected, there is no explicit time dependence in H. With 3 degrees of freedom one expects 3 frequencies corresponding to a motion on a torus in phase space. The 3 independent periodic motions were well known to the Babylonians as far as 3000 years ago. They ought to be given full credit for discovering the existence of tori in phase space.

The 3 freedoms take the moon around the earth in a sidereal month with the frequency $\lambda \sim 2\pi/27.32158$ days, toward and away from the earth in an anomalistic month with $\mu \sim 2\pi/27.55455$ days, above and below the ecliptic in a draconitic month with $\nu \sim 2\pi/27.21222$ days. There is a slow perturbation from the sun as the year passes with $\kappa \sim 2\pi/356.256$ days, the sidereal frequency. It makes the torus expand and contract adiabatically, the first instance of this phenomenon in the history of the sciences. The new and full moons recur in a synodic month with $\sigma = \lambda - \kappa \sim 2\pi/29.53059$ days. The above values for the various periods are approximate; their values were known to Hipparchus correctly to the 5 significant figures given above (~1 second). They constitute the first high precision data in the sciences. There is no reason to believe that they are rationally related to one another.

One can now write the general expression for the coordinates of the moon relative to the earth

$$x = a_0 \cos \alpha_0 + a_1 \cos \alpha_1 + a_2 \cos \alpha_2 + \ldots,$$

$$y = a_0 \sin \alpha_0 + a_1 \sin \alpha_1 + a_2 \sin \alpha_2 + \ldots,$$

$$z = c_0 \sin \gamma_0 + c_2 \sin \gamma_1 + c_2 \sin \alpha_2 + \ldots.$$

The coefficients a_i and c_i are real numbers which depend on the 6 initial conditions, or other 6 convenient parameters. The angular variables α_i and γ_i are all of the form $\alpha = \omega t + \phi$ where both the frequency ω and the phase ϕ are linear combination with integer coefficients of the four basic frequencies and their associated phases. The set of α's and the set of γ's differ in that α contains only even multiples of ν while γ has only odd multiples of ν. The above expression for x,y,z describes the Greek idea of epicycles, although it was given in this form only by Copernicus. It is still the most effective formula for tori in phase space, and is used almost exclusively by astronomers and physicists.

The Babylonians knew the first two terms in x and y with frequencies σ and μ, as well as the first term in z with frequency ν. That is all they needed as astrologers in order to predict new moons, full moons, and, in particular, eclipses. The Greeks, being scientists, also measured the positions of the half moons in the sky, and made a capital discovery which is the essence of non-linear mechanics. Ptolemy established the existence of a third term in x and y whose frequency is the linear combination $2\sigma - \mu$. He obtained the value of its coefficient correctly from the observations, but that value can also be obtained from celestial mechanics once the "trivial" Babylonian terms are known. Tycho Brahe added two more non-trivial terms in x and y, and one in z.

By the year 1600, well before Galileo, Kepler, and Newton the description of the lunar trajectory had been given its modern form. What has been done since? The expansion for x,y and z has been carried to many thousands of terms with the help of Newtonian mechanics. The most elaborate results are given by Schmidt, Gutzwiller, and Eckert[7] with the help of Bellesheim and Levitan, using the above Hamiltonian.

The requirements of modern observations demand a precision of 10^{-10} in the coordinates of the moon. Occultations of fixed stars by the moon can be timed with the precision of the apparent star diameter, $\sim 10^{-5}$ arc seconds. Echoes of light pulses from the reflectors on the surface of the moon give distances correct to several centimeters. But at this point, one enters the realm of slight chaos.

First, there is the problem of proliferating small terms, as discussed by Gutzwiller.[8] The root of the sum of squares for all coefficients smaller than 10^{-10} turns out to be larger than 10^{-9}. It becomes therefore necessary to compute all terms down to a precision of 10^{-12}. Since even such small coefficients should be known with 2 decimals accuracy, one has to go all the way to 10^{-14} just to manage the noise from the small terms in the expansion.

Second, even in the rather benign case of the moon, resonances are bound to show up at this extreme level of precision. In particular, the combination $\mu - 3\lambda + 2\nu$ yields a period of 180 years or a denominator of 1/2000. This special term can only be controlled if all calculations are done to an accuracy of 10^{-17}, including, of course, collecting all the coefficients of this size in the expansion of x,y and z. Even with this effort, the resonant term can only be secured to 10^{-12}, i.e., it can be shown to be insignificant at the price of much sweat and toil.

There is no reason to believe that the moon's motion is not given by the epicycle expansion of Copernicus. On the other hand, it is very unlikely that a mathematical convergence proof can be put together in the interesting range of parameters. In that sense, we don't know whether the lunar trajectory is chaotic, although we can be confident of the moon's company for many more millions if not billions of years. In that time span, a lot of other effects come into play, most significantly the tides and the secular acceleration.

What can we learn from our lunar experience? Even a simple system with 3 degrees of freedom and 1 adiabatically slow perturbation may be difficult to categorize mathematically as either integrable or chaotic. The practical description becomes unwieldy and the original physical simplicity gets lost. The suppression of noise becomes a problem even in the calculations, and the presence of resonances may ultimately defeat our computing skill.

The benign chaos of quantum scattering

Geometrical constructions are helpful when complicated mathematical arguments are used to explain a problem in physics. Shapes are as much part of our experience as are mechanical contraptions. Understanding means that different instances in our life are brought together.

The free motion of a particle on a curved surface in our 3-dimensional Euclidean space can be visualized particularly well. It follows a line on that surface which is locally the shortest, i.e., compared to other lines which deviate by small amounts and over only a limited section of the whole line. Such shortest lines are called geodesics. Their equations are of second order in each of the two parameters which describe the surface, and can be derived from an appropriate Hamiltonian, if necessary.

In order to study chaotic motion on a 2-dimensional surface, we need an ingredient which guarantees a certain amount of instability and disorder in the behavior of the geodesics. That ingredient is undoubtedly negative curvature for the underlying surface. The neighborhood of any point then looks like a saddle point, i.e., the tangent plane to the surface intersects the surface, quite in contrast to the tangent plane for any point on the sphere. Again in contrast to any sphere-like surface, a surface with largely negative curvature cannot easily be closed.

One practical realization is an ordinary (x,y) plane which is distorted in certain isolated regions by pulling out some sort of a chimney. The latter may reach all the way to infinity either in the positive or in the negative z-direction, and may or may not have a non-vanishing diameter in the limit of large $|z|$. But such a chimney could also have a finite height, provided there is a sharp conical tip at the end. This idea of a Euclidean plane which is deformed into a surface of negative curvature was first discussed by Hadamard.[9] The very same picture arises when one investigates the motion of an electron through a molecule. The latter will at first be treated as a 2-dimensional object which consists of a finite number of atoms, called A, B, C, ... etc. Each one has its nucleus rigidly attached to a fixed position in the (x,y) plane which will be given the same name as the atom.

The incoming electron feels in the neighborhood of each atom the attraction of a screened Coulomb potential, which is assumed circularly symmetric around the position of the nucleus. Thus, the atom A creates around the location A the potential $V_A(r) = -(e^2/r).\ \xi_A(r)$ where r is the distance from A, and the function ξ_A decreases monotonically from some fixed positive integer Z_A, the nuclear charge of A at r=0, to 0 at some distance R_A, the radius of the atom A. The net electric charge inside a circle of radius r around A is given by $[\xi_A(r) - r\, d\xi_A/dr]$.

The model of a 2-dimensional molecule can be further simplified without any loss of significant features by assuming that the various circular regions of radius R around each nucleus don't overlap. The overall potential V(x,y) which acts on the electron is made up of separate pieces, $V_A(r_A)$, $V_B(r_B)$, etc. Now, any particular trajectory for the electron follows from the variational principle of Euler and Maupertuis: The first variation of the integral

$$\int_1^2 \sqrt{2m(E-V(x,y))}\ \sqrt{dx^2+dy^2}$$

between fixed endpoints (x_1,y_1) and (x_2,y_2) vanishes provided the electron has the energy E and mass m. The stationary value of the integral is called $S(x_2,y_2;x_1,y_1)$, the action integral for the particular trajectory from (x_1,y_1) to (x_2,y_2).

The trajectory of the electron is identical with the geodesic on a curved space which is described by two parameters x and y, and where the length of any curve is measured by the integral in the variational principle of Euler and Maupertuis. A pictorial representation of this space can be found for the potential energy V(x, y) of our 2-dimensional molecule.

Since the electron comes in from infinitely far with a given kinetic energy E, the length measurements outside the molecule agree with the Euclidean definition up to a factor $\sqrt{2mE}$. As the electron enters the circle of radius R around a particular atom, the Euclidean measurement is stretched by a factor $[2m(E - V)]^{1/2}$ where V is negative. This stretching of the flat space is equivalent to creating a hill inside the atom such that the length of its slope is in the ratio $[2m(E - V)]^{1/2}$ to $(2mE)^{1/2}$ compared to its projection onto the (x,y) plane. The curvature of this hill is negative for the screened Coulomb potential, and it has a conical tip of angle 60°.

The last detail is characteristic of the pure Coulomb nature of the potential in the neighborhood of the nucleus. A cone of this type can be cut open along one of its generators, and its surface can then be spread out in a plane. The two sides of the cut now form exactly the two opposing rays of one straight line. Thus, when the electron runs straight into the nucleus, or equivalently, straight up the corresponding hill, it comes out (or down) precisely on the same path on which it came in (or up). Physically speaking, its angular momentum with respect to the particular nucleus vanishes in this situation.

When the angular momentum is not zero, the electron goes around the hill without reaching the top. Its trajectory can be obtained by putting down a string on the surface, loose at first, and then pulling the two ends in the ingoing and in the outgoing directions of the electron. The negative curvature of the hill guarantees the existence of exactly one shortest path which, of course, always stays on the surface. There is a one-to-one relation between the angle of deflection (going from 0 to π) and the angular momentum (going from $R(2mE)^{1/2}$ to 0).

The classical trajectory of the electron through the molecule now becomes a sequence of shortest paths, around the various atoms. Such a sequence can be described by a sequence of letters, a word, where each letter describes the electron going around a particular atom. A counter-clockwise path around A is named a, and a clockwise path is named $\bar{a}$; similarly, B gives rise to b and $\bar{b}$; etc.. The alphabet has as many letters as there are atoms in the molecule.

Any word in this alphabet describes a particular trajectory provided two rules are obeyed. 1.) Any letter must always be preceded and succeeded by a different letter, and it cannot be preceded nor succeeded by its inverse. 2) The preceding and the succeeding letters must be different, and cannot be inverses of each other. The first rule precludes the trajectory from winding around the tip of any one hill, because the conical nature of the tip would make any such winding contract to zero. The second rule prevents the trajectory from entering an atom with vanishing angular momentum, making a U-turn. This last restriction is peculiar to the screened Coulomb potential.

The negative curvature of the surface again guarantees the existence of exactly one trajectory corresponding to each admissible word. There is a great proliferation of trajectories if the molecule is large enough. In a rough manner of speaking, their number increases exponentially with their length, in striking contrast to the situation in an integrable system where the number of trajectories increases with a power of their length, the exponent equal to the number of freedoms. The scattering of electrons from a molecule is chaotic!

The transition to quantum mechanics is relatively easy. It suffices to make a summation over all classical paths for the phase function $\exp(iS(x_2y_2;x_1y_1)/\hbar)$. The initial and the final points are chosen very far from the molecule, and can move out to infinity in a carefully controlled manner. The phase function gets multiplied with an amplitude which decreases rapidly as the trajectory gets more complicated. The results is a scattering crossection in function of E and the asymptotic directions of motion.

What will be the likely result of such a calculation? There are two things to keep in mind. As described above, quantum mechanics was only included to the extent which classical mechanics allows it. The presence of chaos can only be inferred from the classical behavior, and the latter makes any computation exceedingly difficult. Therefore, a more amenable, but mathematically more abstract, example of quantum scattering seems desirable. With some luck, the semi-classical result coincides with the fully quantum mechanical.

Such an example was worked out by the author.[6] The similarities with the scattering of an electron on a molecule are quite obvious. The underlying space has negative curvature which has, moreover, everywhere the same value, and is normalized to -1. The classical paths are described by all the words in four letters, a, $\bar{a}$, b, and $\bar{b}$. The only restriction is that a letter and its inverse cannot be next to each other, but repetitions of the same letter are allowed. The motion at large distance from the scatterer is essentially 1-dimensional, so that one can use the momentum $(2mE)^{1/2}$ as the only parameter, or even more appropriately, the wave vector $k = (2mE)^{1/2}/\hbar$.

The high symmetry of the underlying space makes the semiclassical result coincide with the quantal, i.e., with the complete solution of Schrödinger's equation. Since the incoming and outgoing particles use the same channel, the only quantity of interest is the phase shift $\beta(k)$. It consists of two parts, a monotonically increasing part which is easily calculated and of no interest, and a non-monotonic function of k with average value 0 whose excursions from zero increase very slowly with increasing k while becoming more jagged. The latter function is given by the logarithm of the ratio $\zeta(1 + 2ik)/\zeta(1-2ik)$ in terms of Riemann's famous zeta-function $\zeta(s)$. It is plotted in figure 1 with k varying in the arbitrary range $10000<k<10020$.

The "size" of the scatterer is given by the radius of curvature which was normalized to 1. Therefore, the parameter k is already given in natural units. A value such as $k\sim 10000$ should be well into the classical region by usual standards.

The fascinating element in figure 1 is the incredible amount of apparent structure. Since this particular plot is preceded by 500 similar ones, and succeeded by an as yet unknown number of others of the same character, the unsuspecting viewer guesses that the scatterer must be exceedingly complicated. But on the contrary, the underlying space is totally featureless, exactly as a sphere, i.e., a surface of constant positive curvature, which has no characteristic details whatever. Figure 1 represents pure quantum chaos!

There are good reasons to believe that this plot does not go on indefinitely. Since one looks at an analytic function at a distance of 1/2 from its singularities (on the critical line $Re(s)=1/2$ if Riemann's conjecture is correct), one expects the structure to get wiped out when the singularities move much closer to one another than 1/2. This starts happening for $k\sim 10^7$, another factor 1000 from figure 1. From the numerical evidence in favor of Riemann's Conjecture, one believes the phase shift to remain analytic out to $k\to\infty$ with a strip of analyticity of width 1/2, a truly smooth function all the way!

The last statement can be sharpened. The phase shift $\beta(k)$ has a discrete Fourier spectrum whose frequencies are multiples (but not linear combinations) of the logarithms of the rational prime numbers. Thus, there is a

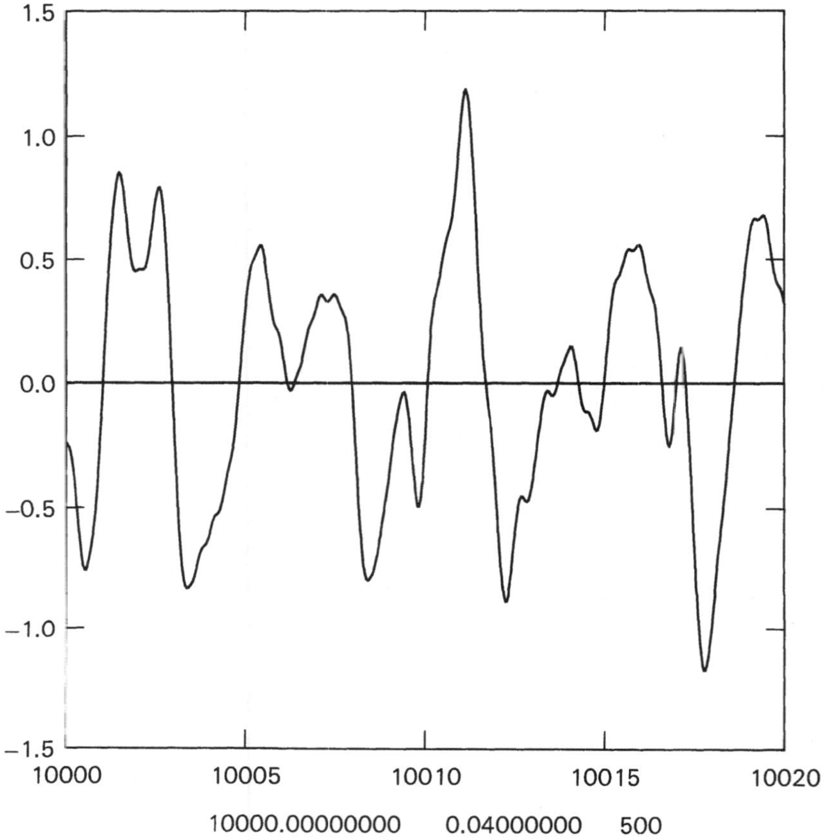

Figure 1.Phase shift B (radians) as function of wave vector k in the interval from k=10000 to 10020.

tremendous storehouse of independent frequencies. Enough to permit $\beta(k)$ to mimic any other analytic function over a finite interval, as was shown by Reich.[10] This kind of chaos does not involve discontinuous objects such as fractals. Yet, its benign nature manages to choose indifferently from among all smooth objects, like a good actor who can play any role in the repertory without even letting you know which one he is doing at any one time.

In conclusion, there seem to exist manifestations of chaos which are more elusive than the somewhat unanticipated, but then rather brazen appearance of fractals. Quantum mechanics, if not certain familiar problems in classical mechanics, seems to present examples of this new type which one might call mild chaos for the lack of a better word.

References

1. I. E. Fermi, J. Pasta and S. Ulam, Studies in Nonlinear Problems, in Enrico Fermi's Collected Papers, The University of Chicago Press, 1965 Vol. II, p. 977-988.

2. A. Einstein, Verhandlungen der Deutschen Physikalischen Gesellschaft 19 (1917) 82-92.

3. It is hard to find an explicit reference to the fractal nature of classical Hamiltonian systems, although that result is implicit in some recent work such as C. P. Kadanoff, Phys. Rev. Letters 47 (1981) 1641 and J. Stat. Phys. 27 (1982) 631. Cf. G. Schmidt and J. Bialek, Physica 5D (1982) 397.

4. G. Casati, B. V. Chirikov, F. M. Izraelev, and J. Ford, Stochastic Behavior of a Quantum Pendulum Under a Periodic Perturbation, in Lecture Notes in Physics 93, Springer Verlag 1979, p. 334-352.

5. S. Fishman, D. R. Grempel and R. E. Prange, Chaos, Quantum Recurrences and Anderson Localization, Phys. Rev. Letters 49 (1982) 509.

6. M. C. Gutzwiller, Stochastic Behavior in Quantum Scattering, Physics 7D (1983).

7. D. Schmidt, M. C. Gutzwiller, W. J. Eckert, The Main Problem of Lunar Theory according to Hill and Brown, to be published.

8. M. C. Gutzwiller, The Numerical Evaluation of Eckert's Lunar Ephemeris, Astronomical Journal 84 (1979) 889-899.

9. J. Hadamard, J. Math Pure Appl. 4 (1898) 27.

10. A. Reich, Arch. Math. 34 (1980) 440-451.

REGULAR AND IRREGULAR ENERGY SPECTRA

Asher Peres

Department of Physics
Technion - Israel Institute of Technology
32 000 Haifa, Israel

Einstein[1] was the first to show how the Bohr-Sommerfeld quantization rules could be extended to systems which are not separable but are nevertheless integrable. Brillouin[2] and Keller[3] later refined his method (torus quantization). Gutzwiller[4] and Percival[5] suggested that the "regular" spectrum of an integrable system ought to be qualitatively different from the "irregular" spectrum of a nonintegrable one. In particular, there should be no natural labelling of the irregular energy levels by a set of quantum numbers. Pomphrey[6] found that the energy levels of an irregular spectrum are more sensitive to small perturbations than those of a regular spectrum; his results, however, are not conclusive, as they might be due to the presence of avoided crossings.[7] Several authors investigated the statistical distribution of energy differences of *consecutive* levels and found, in various numerical experiments, that regular spectra had a tendency to clustering[8] while irregular ones showed "level repulsion."[9-12] Similar results will indeed be reported below. However, there seems to be nothing fundamental in the consecutiveness of energy levels.

In this paper, I show that there is a much simpler and fundamental distinction between regular and irregular spectra. If a system is classically integrable, its quantum energy spectrum consists of sets of nearly equidistant levels (in the semi-classical regime of small $\hbar$, i.e., when quantum numbers are large and in particular when there are many levels in a narrow energy domain). There is no such regularity for a classically chaotic system.

This property, which was empirically found by several authors[7,13,14] follows at once from the fact that classical orbits of integrable systems are multiply periodic in time.[15] Indeed, the only way a small wave packet (in the limit $\hbar\to 0$) can follow a classical trajectory[14] is if the energy levels are arranged in sets with

$$\Delta E = \hbar \sum n_k \omega_k \quad , \tag{1}$$

where the n_k are integers and the ω_k are the classical frequencies.

The same result follows from the EBK quantization scheme[1-3] which gives very reliable results in the limit of large quantum numbers. We write the Hamiltonian as $H(I_1 \ldots I_N)$, and take the action variables I_k as $I_k=(m_k+\alpha_k)\hbar$, where the m_k are large integers and the α_k are constants. The energy levels are given by[8,14]

$$E_{m_1 \ldots m_N} = H(\alpha_1\hbar \ldots \alpha_N\hbar)+\hbar\sum m_k\omega_k+\tfrac{1}{2}\hbar^2\sum m_k m_\ell \,\partial\omega_k/\partial I_\ell + \ldots \tag{2}$$

where $\omega_k=\partial H/\partial I_k$ are the classical frequencies. If we neglect the terms of order $\hbar^2$ and higher terms, the energy levels are equidistant, with spacing given by Eq. (1). Any corrections to the equal spacing law are of order $\hbar^2$ (or higher) and due to anharmonicity (dependence of ω_k on I_ℓ). In other words, the equal spacing law is a local one (in energy space) valid if there are many energy levels in a small energy range.

This can be tested by a very simple algorithm. It follows at once from (1) that if

$$\Delta E = E_1 - E_0 \tag{3}$$

is the spacing between *any* two levels, arbitrarily called E_1 and E_0, there should be additional levels located at $E_n=E_0+n\Delta E$, where n is (very close to) any (not too large) positive or negative integer. This is easily seen by considering the integers n_k in Eq. (1) as a unit lattice in N-dimensional space (N is the number of degrees of freedom). If we draw a vector between any two points of the lattice, that vector, suitably displaced, will connect further pairs of points in the lattice. (There is however a difficulty: Very close levels *may* have widely different quantum numbers and therefore not belong to an equidistant set, because of anharmonic effects.)

Let D be the deviation of $(E_n-E_0)/(E_1-E_0)$ from an integer, due to anharmonicity. An elementary computer program can tabulate D for all energy levels E_n (for given E_1 and E_0). This is considerably simpler than using fast Fourier transforms. All the levels for which D is less than some arbitrarily chosen "width" W are considered to belong to the same equidistant set. The result, plotted as a function of W, gives a kind of "line shape" for the energy difference ΔE.

As an example, this test can be applied to the Hénon-Heiles system[7,13,14] and one easily sees a clustering of ΔE around the classical frequency $\omega=1$. The second classical frequency is more difficult to detect (unless one goes to *extremely* low $\hbar$) because low lying energy levels of the Hénon-Heiles system are almost degenerate. Surprisingly, the regular sequences of eigenvalues are found to continue smoothly up to energies which are classically chaotic. Quantum chaos thus appears more remote than classical chaos. This can be explained by the existence of tori remnants.[16,17] As long as the missing parts of these "vague tori" are small compared to $(2\pi\hbar)^N$, the EBK quantization remains roughly valid. Here too, one would need an extremely small $\hbar$ to distinguish between the onset of stochasticity around E=0.125 and the dissociation limit E=0.166.

Another difficulty with the Hénon-Heiles system is the necessity to truncate the Hamiltonian matrix to a finite number of energy levels. There is no obvious semi-classical limit for such a *truncated* Hamiltonian. There exists however another very simple model[18] which overcomes this difficulty. It consists of two coupled rotators with Hamiltonian

$$H = L_z + M_z + L_x M_x \ . \tag{4}$$

Here, L_m and M_n have the same commutation relations as the components of two independent angular momenta. This system has two constants of motion besides the Hamiltonian, namely $L^2=\hbar^2\ell(\ell+1)$ and $M^2=\hbar^2 m(m+1)$. For given ℓ and m, we thus have to deal with *finite* matrices only. The energy levels were computed by Mario Feingold for several values of ℓ and m (details will appear in his thesis). When ℓ=m, there are four symmetry classes, depending on whether the wavefunction $\psi(\ell_z, m_z)$ is even or odd under the exchange of ℓ_z and m_z, and whether ℓ_z+m_z is even or odd.

Fig. 1 shows the 121 energy levels of the even-even class for ℓ=m=10, as functions of L=10.488$\hbar$, on a scale such that E_{max}=1, where

$$\begin{aligned} E_{max} &= \sqrt{(\ell+1)/\ell}\ 2L \qquad \text{if} \qquad L<1 \ , \\ &= \sqrt{(\ell+1)/\ell}\ (L^2+1) \qquad \text{if} \qquad L>1 \ . \end{aligned} \tag{5}$$

(Apart from the square root which is a quantum effect, E_{max} is the maximum value of the energy for the *classical* system described by (4), with Poisson brackets in lieu of commutators.)[18]

As shown in ref. 18, most classical orbits are regular for L<0.8, and most are chaotic for L>1.3 (they again become regular for L>40000, but this does not concern us here). Likewise, the quantum spectrum shows a clear regularity for L<0.8 and no apparent regularity for L>1.3. It should be noted that there are no level crossings in the regular region. There are only "avoided crossings" but the gaps between the lines are extremely narrow, much smaller than the line thickness of the graph. It follows from Eq. (1) that these

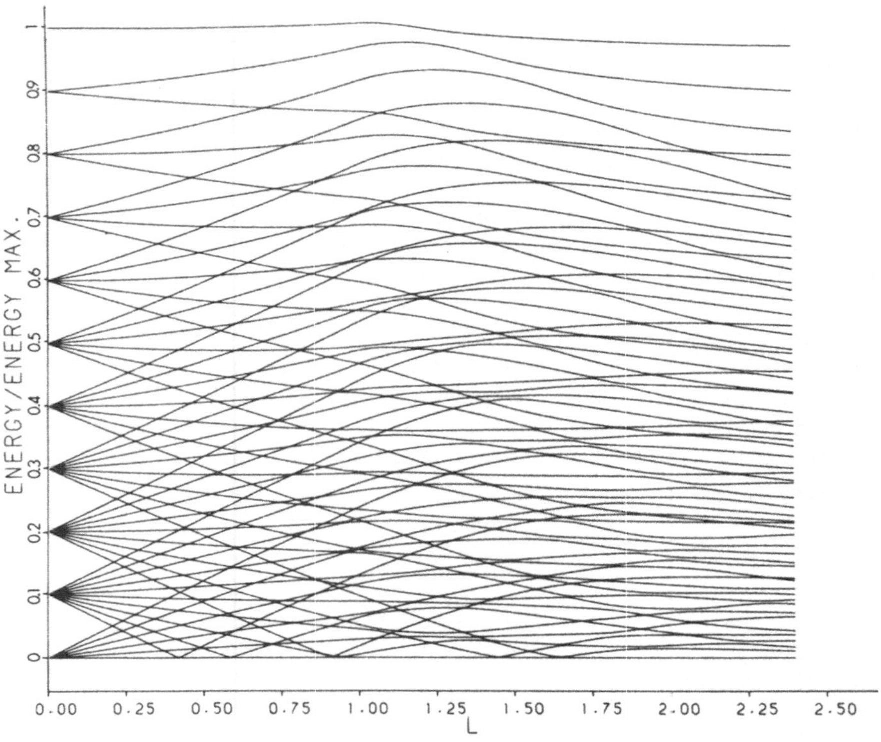

FIG.1. Energy levels of coupled rotators, with the Hamiltonian of Eq. (4), as functions of L=M. The spectrum is symmetric around E=0 and only the positive levels have been drawn.

narrowly avoided crossings are related to classical resonances: $\sum n_k \omega_k = 0$. These classical resonances are in turn related to tiny chaotic domains which must exist in the regular region of phase space.

When we move on to the chaotic region, the gaps of the avoided crossings are much broader. This level repulsion agrees with the empirical findings reported in refs. 9-12. As these broadly avoided crossings tend to overlap, it is impossible to assign unambiguous quantum numbers to a given energy level.[5] The situation is similar to the quadratic Zeeman effect.[19]

Fig. 2 shows the result of the equidistance test in the regular region L=0.5, for ℓ=m=20 (there are 441 even-even levels) and 0.57< E<1. A similar test, performed in the classically chaotic region L=2, showed only a quasi-uniform background.

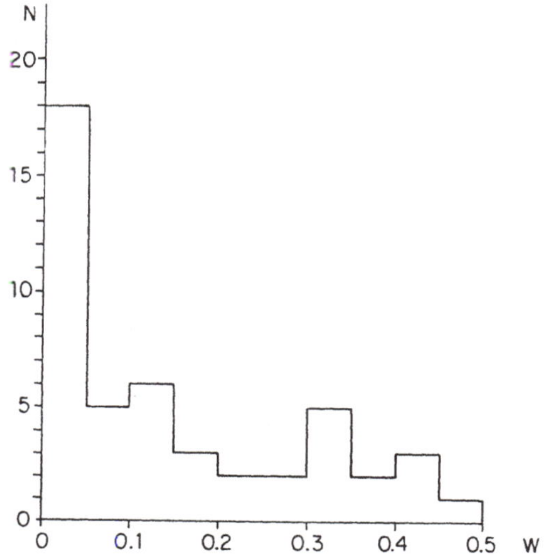

FIG. 2. Arrangement of energy levels of the double rotator in equidistant sets, for L=M=0.5. There are 47 even-even levels in the range 0.57<E<1. Taking E_0=0.710562 and E_1= 0.730530, we find that 18 levels deviate from the equidistant law by less than W=5%. The 29 other levels, which belong to different sets, are a kind of uniform background.

In summary, the quantum energy spectrum of a system which is classically integrable consists of sets of nearly equidistant levels. This rule is valid in the semi-classical regime, namely when $\hbar$ is small and quantum numbers are large. The rule still applies to quasi-integrable systems having small chaotic regions in phase space, namely regions smaller than $(2\pi\hbar)^N$. However, the equidistance rule breaks down for a classically chaotic system. A very simple numerical algorithm can therefore distinguish regular from irregular spectra. This algorithm, however, becomes inefficient in the *extreme* semi-classical limit $\hbar\to0$, because the density of energy levels increases as $\hbar^{-N}$ and it is then difficult to distinguish the levels belonging to an equidistant set from the neighboring ones. (This is *not* a practical difficulty, because nobody would seriously try to test the regularity of a classical system by examining the $\hbar \to 0$ limit of the corresponding quantum system!)

ACKNOWLEDGMENTS

I am grateful to M. V. Berry for several clarifying comments and to G. Casati for the hospitality of the University of Milan, where the final version of this paper was written. I am indebted to M. Feingold for supplying Fig. 1 and the energy spectrum for Fig. 2.

This work was supported by the Israel Academy of Sciences and Humanities, the Lawrence Deutsch Research Fund, the Gerard Swope Fund and the Fund for Encouragement of Research at Technion.

REFERENCES

1. A. Einstein, Verhandl. Deut. Phys. Ges. 19:82 (1917).
2. L. Brillouin, J. Phys. Rad. 7:353 (1926).
3. J. B. Keller, Ann. Phys. (N. Y.) 4:180 (1958).
4. M. C. Gutzwiller, J. Math. Phys. 12:347 (1971).
5. I. C. Percival, J. Phys. B6:L229 (1973).
6. N. Pomphrey, J. Phys. B7:1909 (1974).
7. D. W. Noid, M. L. Koszykowski, M. Tabor and R. A. Marcus, J. Chem. Phys. 72:6169 (1980).
8. M. V. Berry and M. Tabor, Proc. Roy. Soc. A349:101 (1976); A356: 375 (1977).
9. S. W. McDonald and A. N. Kaufman, Phys. Rev. Letters 42:1189 (1979).
10. G. Casati, F. Valz-Gris and I. Guarneri, Lett. Nuovo Cim. 28: 279 (1980).
11. M. V. Berry, Ann. Phys. (N. Y.) 131:163 (1981).
12. H. J. Korsch and M. V. Berry, Physica 3D:627 (1981).
13. R. E. Wyatt, G. Hose, H. S. Taylor and M. J. Davis, Phys. Rev. A (1983, in press).
14. N. Moiseyev and A. Peres, J. Chem. Phys. (1983, in press).
15. H. Goldstein, Classical Mechanics (2nd ed., Addison-Wesley, Reading, Mass. 1980) p. 466.
16. C. Jaffé and W. P. Reinhardt, J. Chem. Phys. 77:5191 (1982).
17. R. B. Shirts and W. P. Reinhardt, J. Chem. Phys. 77:5204 (1982).
18. M. Feingold and A. Peres, Physica D (1983, in press).
19. J. B. Delos, S. K. Knudson and D. W. Noid, Phys. Rev. Letters 50:579 (1983); Phys. Rev. A (1983, in press).

STOCHASTIC IONIZATION OF BOUND ELECTRONS

Roderick V. Jensen

Mason Laboratory, Yale University
New Haven, CT 06520 U.S.A.

Abstract

Recent experimental measurments of the ionization of Rydberg atoms in low frequency electromagnetic fields are discused. These observations of ionization rates which depend strongly on the intensity of the oscillating fields and only weakly on the frequency provide clear evidence for chaotic behavior in a quantum system. In an attempt to understand this ionization mechanism an analagous, one dimensional system is considered consisting of a surface state electron bound to the surface of liquid helium by its image charge. A classical analysis of the behavior of this nonlinear oscillator in a microwave field shows that, above a critical field, the electron diffuses in energy until it ionizes. Since the microwave frequencies and field strengths required for stochastic ionization are readily available, this system provides an opportunity to thoroughly explore the manifestations of classical chaos in a quantum system.

I. Introduction

In the last ten years deterministic classical systems with chaotic dynamics have been the subject of extensive research; however, little progress has been made on the question of whether the chaotic dynamics persist in a quantum mechanical description of these systems. This question is important in a variety of problems, such as the calculation of the vibrational and rotational spectra of polyatomic molecules and the determination of the response of atoms and molecules to time-dependent electro-magnetic fields[1], which have significant applications in chemistry, laser isotope separation, and the development of short wavelength lasers. Both the n-body problems and the driven oscillators correspond to non-integrable, classical systems which exhibit chaotic behavior.

To date, the attempts to develop a quantum description of these systems remain incomplete and controversial. These problems first arose in the early development of quantum mechanics. In 1917 Einstein[2] remarked that the semi-classical quantization procedure fails for non-integrable classical systems such as the three-body problem. More recently, numerical investigations[3] of non-integrable systems which exhibit chaotic behavior in the classical limit suggest that the quantum dynamics may also be stochastic. However, theoretical studies of some time-dependent Hamiltonians with discrete quasi-energy spectra[4] indicate that the quantum dynamics of such systems are always quasi-periodic and never chaotic.

The question of whether a quantum system can be chaotic must, ultimately, be resolved by experimental investigations of real physical systems which are amenable to theoretical analysis. Although the majority of papers at this conference have dealt with the quantum description of time-independent, n-body systems like polyatomic molecules, I will consider the equally important and closely related problems of time-dependent quantum systems which correspond to driven, nonlinear oscillators in the classical limit. Specifically, I will examine the behavior of bound electrons in the presence of intense electromagnetic fields. The nonlinear behavior of these systems is easier to study, experimentally, since the degree of nonlinearity is determined by the amplitudes of the externally applied fields. Experimental and theoretical studies of these systems are essential for an understanding of the variety of novel features exhibited by atomic physics in large amplitude laser and microwave fields.

This paper is divided into two parts:

First, I will discuss some very suggestive experiments on the ionization of highly excited hydrogen and helium atoms in microwave fields performed by Bayfield, Koch and Mariani[5-7] at Yale University over the last ten years. These observations of ionization rates which depend strongly on the intensity of the microwave fields and only weakly on the frequency provide strong evidence for stochastic behavior in a quantum system.

Second, in an attempt to understand this ionization process I will consider a simpler quantum system for theoretical and experimental study consisting of surface state electrons (SSE) which are bound to the surface of liquid helium by their image charge.[8] A classical analysis of the behavior of these one-dimensional "hydrogen" atoms in a microwave field predicts a microwave power threshold for the onset of stochastic ionization as well as an ionization rate.[9] Because this system is one dimensional, the semi-classical and quantum analyses are greatly simplified as well. Consequently, this non-integrable dynamical system provides a unique opportunity to test the predictions of the classical, semi-classical, and quantum theories with a real experiment which should resolve much of the controversy surrounding the problem of quantum chaos.

II. Microwave Ionization of Rydberg Atoms

In their pioneering work Bayfield and Koch[5] measured the ionization rate of a beam of neutral hydrogen atoms, carefully prepared in highly excited Rydberg states (n ~ 66), as they passed through a microwave cavity. Although the microwave freqency (9.9 GHz) is only ~ 40 % of the resonant frequency for single photon excitation to the n = 67 level and ~ 1 % of the photon frequency for excitation to the continuum, they observed significant ionization above a critical microwave field of ~ 20 V/cm. Their early data for the ionization probability as a function of microwave field, F, normalized to the binding Coulomb field ($1/n^4$ in atomic units) is shown in Fig. 1 by the solid dots fitted by the solid curve. For low intensity microwave fields no ionization is observed; however, as the fields are increased the entire beam is ionized.

These remarkable observations require an explanation. Since the microwave frequency is less than the classical orbital frequency of the electron, Stark ionization at the peak electric field is a prime canidate for the ionization mechanism. (The interaction with the magnetic field is much weaker and can be neglected.) However, if the highly excited electron is treated as a classical particle orbiting the nucleus (a good approximation in this case because of the large n values), then theoretical calculations by Banks and Leopold[10] of the static electric field strengths required to dissociate the electron from the atom give an ionization threshold which is too high. This classical ionization threshold is shown in Fig. 1

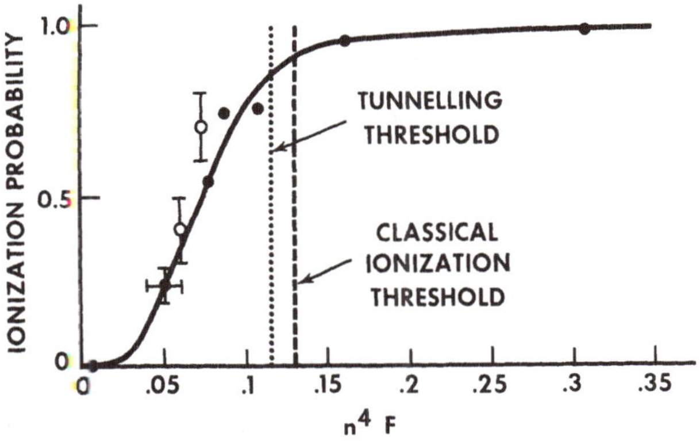

Fig. 1. Ionization probability as a function of peak microwave field, F, for highly excited hydrogen atoms, n ~ 66. The solid circles are experimental data points[5] which are fitted by the solid curve. The open circles are the results of a classical, Monte Carlo simulation of the experiment.[14]

by the dashed line. Although the classical threshold coincides with the saturation of the ionization signal, significant ionization occurs well below this level.

Estimates of the of ionization due to tunnelling in a static field based on the quantum mechanical calculations of Damburg and Kolosov[11] and the experimental measurments of Koch and Mariani[12] also give rates which are to slow to explain the experimental results. In Fig. 1 the dotted line shows the peak microwave field required for significant tunnelling during the beam transit time through the cavity ($\tau_{tr} \sim 10^{-7}$ sec).

Clearly, the observed ionization must be a dynamic effect. Another likely explanation of the ionization at low field strengths is multi-photon absorption. However, since nearly 100 photons must be absorbed to reach the continuum, the theoretical investigation of this mechanism has been intractable, requiring one hundred orders of perturbation theory and complicated by the effects of intermediate resonances. Consequently, quantum mechanics has thus far been unable to account for these experimental observations.

The only successful theoretical analysis involved a classical treatment of the orbiting electron in a plane polarized, oscillating electric field. The electron dynamics are described in atomic units[13] by the Hamiltonian

$$H(\vec{r}, \vec{p}, t) = p^2/2 - Z/r + xF \cos\Omega t \qquad (1)$$

where $Z = 1$ is the charge of the hydrogen nucleus, F is the peak electric field strength, and Ω is the angular frequency of the oscillating field. Leopold and Percival[14] integrated the classical equations of motion for this nonlinear oscillator in a time-dependent field and demonstrated the possibility of classical excitation and ionization of the electron. Using a Monte Carlo model of Bayfield and Koch's expriment, they were able to calculate ionization rates which agreed well with the experimental measurments. Their results are indicated by the open circles in Fig. 1. The accompaning error bars were due to uncertainty in the effective interaction time in the microwave cavity.

The physical ionization mechanism evoked by this classical calculation for a driven nonlinear oscillator is stochastic diffusion of the electrons in phase space. Ionization results when the chaotic electron trajectories wander into the continuum. Since the Kolmogorov-Arnold-Moser theorem[15] guarantees that for small microwave fields most of the trajectories will remain regular, the microwave fields must exceed a critical level before the orbits become sufficiently chaotic to cause significant ionization. Unfortunately, because the electrons move around in a six-dimensional phase space, detailed numerical and analytical investigations of the transition to global stochasticity has proven difficult. In fact the numerical calculations[14] were so time consuming that only two values of the microwave field were considered. Moreover, analytical treatments of the transition to global stochasticity[16-18] due to resonance overlap[19] in the classical phase space of the electron remain incomplete.

Because of the large quantum numbers in these original experiments it is not suprising that the classical treatment provides an adequate description. More recently, Koch and Mariani[20] have studied the ionization of lower Rydberg levels ($n \sim 29$) in an attempt to observe the modifications of the classically chaotic electron dynamics due to quantum effects. In addition they have been able to make accurate measurments of the beam ionization as a continuous function of the microwave field. Some typical data for a hydrogen beam initially prepared in the $n = 29$ state is shown in Fig. 2. The ionization probability is a much steeper function of the microwave intensity than for the non-adiabatic, $n \sim 66$, experiments.

In Fig. 2 I have also indicated estimates for the static field thresholds for ionization both by quantum mechanical tunnelling[11,12] and classical dissociation.[10] The appearance of the ionization signal roughly coincides with the peak field level for significant tunnelling and the signal saturates at the field level for classical dissociation. Since the frequency of the microwave source is fixed at 9.9 GHz, the ratio of the perturbation frequency to to the electron orbital frequency is now $\sim$ .04. Consequently, it would appear in this case that the observed ionization is simply field ionization in the quasi-static microwave field.

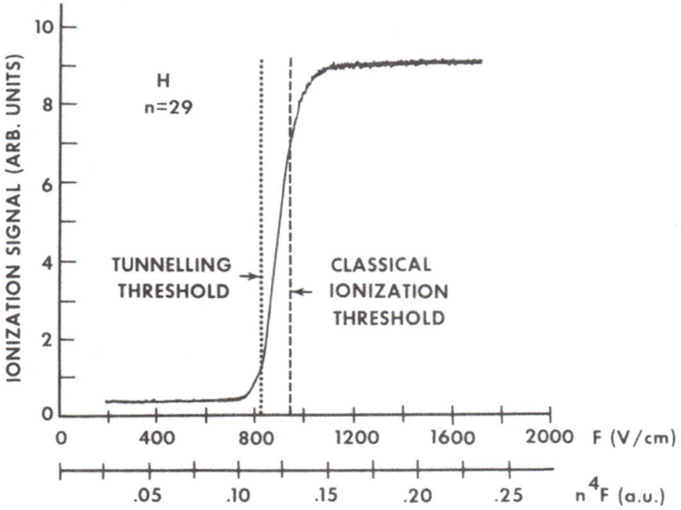

Fig. 2. Ionization signal as a continuous function of the peak microwave field, F, for a beam of hydrogen atoms in the $n = 29$ state.[7,20]

However, the static ionization thresholds shown in Fig. 2 only apply to the easiest to ionize Stark sublevel in the principal level n = 29. Both theoretical and experimental studies of hydrogen atoms in a static electric field conclude that the classical thresholds for dissociation and the quantum thresholds for significant tunnelling for the n^2 different, Stark shifted sublevels range from $n^4F \sim .12$ to .3 .[10-12] Although the hydrogen beam was prepared with a broad distribution of sublevels[21], the experimental data indicates that all of the atoms field ionize at the electric field levels for the easiest to ionize sublevels.

Although these results for lower quantum levels and quasi-static fields fail to exhibit the ionization due to global stochasticity seen in the earlier non-adiabatic experiments for $n \sim 66$, the data does suggest that the perturbing microwave field may cause diffusion in the action space corresponding to the parabolic quantum numbers characterizing the Stark sublevels. Then all of the sublevels will ionize through the escape channel provided by the easiest to ionize sublevel. Both theoretical and experimental investigations are in progress that should further elucidate the behavior of the lower principal quantum levels in the presence of an oscillating electric field.

Mariani and Koch[7,20] have also examined the microwave ionization of beams of highly excited helium atoms. Their experimental data for the ionization probability for helium atoms with one electron excited in the n = 29 state is shown in Fig. 3 along with the corresponding data for hydrogen atoms in the n = 29 state. Because the helium electron is so highly excited one might expect that the electron dynamics would be similar to that of hydrogen. However, the experimental measurments demonstrate that the field threshold for ionization is significantly lowered due to the interaction with the inner electron. Above threshold the ionization probability exhibits a remarkable plateau corresponding to an ionization rate which is independent of the amplitude of the perturbing field. Finally, when the microwave fields reach the hydrogen ionization threshold the helium beam is fully ionized in accordance with our expectations. This lowering of the ionization threshold has also been observed for Rydberg states of sodium by Pillet et al[22].

Mariani and Koch and their co-workers van de Water and Bergeman have developed a quasi-static quantum analysis of their experiment which shows that the observed ionization threshold in helium corresponds to the appearance of the first avoided crossings in the static Stark map of the perturbed energy levels.[7] However, the subsequent behavior of the ionization signal remains a mystery. As in the hydrogen case, these fascinating experimental results provide evidence for a variety of nonlinear phenomena in quantum systems which require a theoretical explanation.

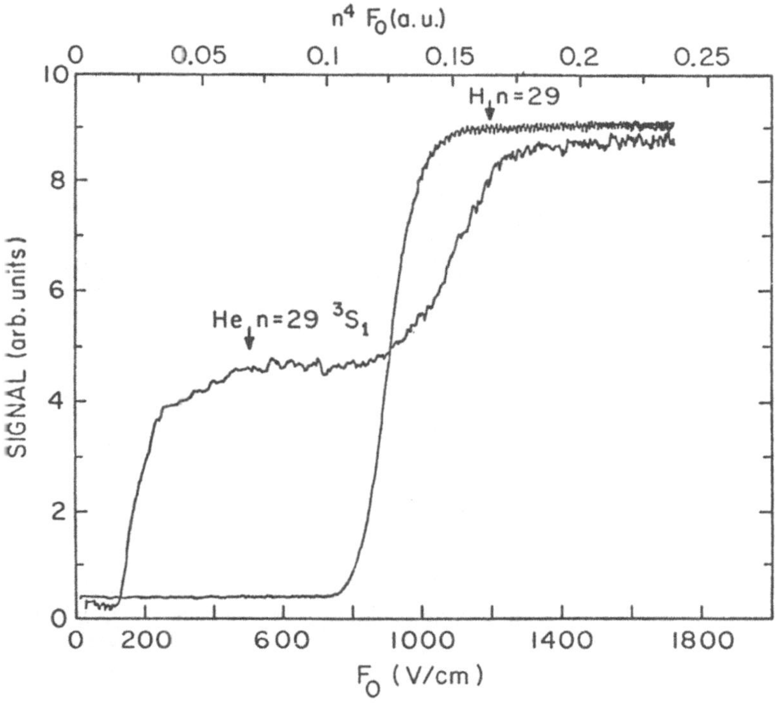

Fig. 3. Ionization signal as a function of peak microwave field, F_0, for both hydrogen and helium Rydberg atoms with one electron excited in the n = 29 state.[7]

III. Stochastic Ionization of Surface State Electrons

In an attempt to understand the physical mechanism responsible for the ionization of highly excited hydrogen atoms in a oscillating electric field I proposed[9] a different quantum system for theoretical and experimental study consisting of surface state electrons (SSE) which are weakly bound to the surface of liquid helium by their image charge. Spectroscopic studies by Grimes et al[8] and by Lambert and Richards[23] found that the energy levels of the SSE are given by the hydrogenic formula

$$E_n = - Z^2R/n^2 \quad ,n = 1,2,3,... \tag{2}$$

where R = 13.6 ev. These energy levels result from a one-dimensional quantum mechanical treatment of the SSE which assumes an attracting Ze^2/x potential due to the image charge and a repulsive barrier at the surface due to Pauli exclusion. In the classical limit the electron bounces back and forth between the helium surface and the one-dimensional Coulomb potential.

The dynamical behavior of this one-dimensional "hydrogen" atom in a spatially uniform, oscillating electric field normal to the liquid helium surface is described by the Hamiltonian

$$H(x,p,t) = p^2/2 + \begin{cases} -Z/x & , x > 0 \\ \infty & , x \leq 0 \end{cases} + xF \cos\Omega t \qquad (3)$$

where F and Ω are the amplitude and oscillation frequency of the perturbing field in atomic units[13]. The theoretical analysis of the behavior of this perturbed, nonlinear oscillator is much more tractable than that for the full three dimensional hydrogen atom. Furthermore, since liquid helium is a poor dielectric, the image charge is very small, $Z \sim 7.1\text{x}10^{-3}$, and the binding energies and characteristic frequencies of this one-dimensional "hydrogen" atom are four orders of magnitude smaller than those for real hydrogen atoms. Consequently, available microwave sources can be used to investigate the dynamics of the perturbed SSE in both the classical ($n \gg 1$) and quantum ($n \sim 1$) limits. Therefore, the SSE in an oscillating field provides an ideal system to test the predictions of the classical, semi-classical, and quantum theories with a real experiment.

The remainder of this paper presents a classical treatment of the dynamics of the SSE in a microwave field which shows that ionization results from stochastic diffusion of chaotic electron orbits and provides analytic estimates for the field threshold for stochastic ionization as well as an ionization rate.[9] These results, which are based on the resonance overlap criterion for the onset of global stochasticity[19], should be valid for large quantum numbers, $n \gg 1$, corresponding to the classical limit. Morever, this calculation predicts the microwave powers and frequencies required to explore the manifestations of classical chaos in the quantum regime (low n). More detailed semi-classical and quantum treatments of this system will be considered in future work for comparison with experimental results which will, hopefully, be forthcoming.

First, I consider the integrable dynamics generated by the unperturbed Hamiltonian. A bound electron with energy -E bounces back and forth in the potential well between $x = 0$ and $x = a = Z/E$ (a.u.) with frequency $\Omega_0 = (8Z/a^3)^{1/2}$ (a.u.) .

A canonical transformation to action-angle variables,

$$I = \sqrt{Za/2} \qquad (4)$$

$$\theta = \begin{cases} 2[\sin^{-1}(\sqrt{x/a}\,) - \sqrt{x/a(1-x/a)}\,] & , p \geq 0 \\ 2\pi - 2[\sin^{-1}(\sqrt{x/a}\,) - \sqrt{x/a(1-x/a)}\,] & , p < 0 , \end{cases} \qquad (5)$$

reduces the unperturbed dynamics to straight line trajectories in action-angle space. The new, unperturbed Hamiltonian is

$$H_0(I) = -Z^2/2I^2 \qquad (6)$$

which gives a constant angular velocity

$$\Omega_0(I) = dH_0/dI = Z^2/I^3 \quad . \tag{7}$$

Since the standing wavelengths of the microwave radiation are long compared with the maximum excursion, a, of the SSE from the liquid helium surface, the spatial variation of the perturbing electric fields can be neglected. In addition, for nonrelativistic electron velocities the interaction with the oscillating magnetic field is also negligable. We therefore consider a perturbation of the form

$$V(x,t) = xF \cos\Omega t \tag{8}$$

due to externally applied microwave fields.

For sufficiently small electric fields the Kolmogorov-Arnold-Moser (KAM) theorem[15] guarantees that most of the straight line trajectories in action-angle space will be only slightly distorted by the perturbation. If I expand the perturbation in a Fourier series[19] in Θ, the perturbed Hamiltonian can be written as

$$H(\Theta,I,t) = H_0(I) + F \sum_{m=-\infty}^{\infty} V_m(I)\cos(m\Theta + \Omega t) \quad . \tag{9}$$

The maximum distortion of the orbits in action-angle space will occur at resonances where the phase, $m\Theta + \Omega t$, is stationary[19]. The resonant frequencies and actions are therefore determined by the relation[19]

$$m\Omega_0(I) + \Omega = 0 \quad . \tag{10}$$

Then using Eq.'s (7) and (10), the action resonant with the mth subharmonic of the perturbation is given by

$$I_m = (mZ^2/\Omega)^{1/3} \quad . \tag{11}$$

For small perturbations the Hamiltonian can be approximated in the vicinity of each resonance by the Hamiltonian of a pendulum. The electron trajectories near the resonances are confined in narrow island chains in action-angle space. The electrons gain and lose energy as they ride the perturbation but no net change in the energy occurs. The island chains corresponding to the three lowest resonances are shown in Fig. 4. (Note that Z = 1/8 in the dimensionless units which were used in making this plot.)

As the perturbation increases the islands grow wider in action. When the islands are sufficiently large the electron can diffuse in action (or energy) by wandering from one island chain to another. These transitions occur when the orbit of the electron is so distorted by one resonance that its oscillation frequency becomes resonant with another resonance corresponding to a subharmonic of the oscillating microwave field. This qualitative picture provides a means of estimating the size of the perturbation required to make the transition from

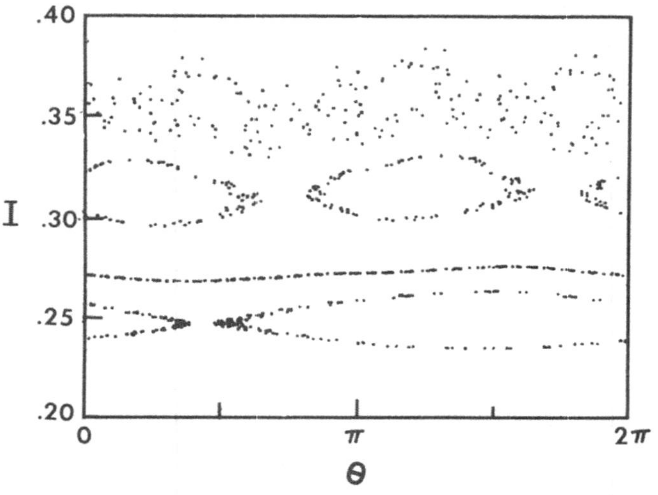

Fig. 4. Island chains for the m=1,2,3 resonances for a perturbation with Ω = 320 GHz and F = 14 V/cm. The m=3 islands already exhibit large stochastic regions. Also shown is a confining KAM surface between the m=1 and m=2 resonances.

regular to stochastic behavior. Roughly speaking, this occurs when the islands generated at the resonances overlap. The "resonance overlap criterion" has been the subject of extensive numerical investigations [24-25] which indicate that it provides a good estimate for the onset of stochasticity.

The separation of the resonances is easily computed using Eq. (11)

$$\delta_m = I_{m+1} - I_m = (Z^2/\Omega)^{1/3} \left[(m+1)^{1/3} - m^{1/3}\right] \tag{12}$$

$$\sim I_m/3m \quad \text{for large m.}$$

The widths of the islands are determined by the corresponding Fourier amplitudes, V_m, of the perturbation. Following Ref. 19, I estimate the island widths by approximating the Hamiltonian in the vicinity of the resonance by the Hamiltonian for a pendulum. The island width corresponds to the width of the trapping (libration) region[26]

$$W_m = 4\,(FV_m/\Omega_0')^{1/2}\,|_{I=I_m} \tag{13}$$

where

$$\Omega_0'\,|_{I=I_m} = d\Omega_0/dI\,|_{I=I_m} = 3Z^2/I_m^{\,4}. \tag{14}$$

The Fourier components, V_m, of the oscillating microwave potential, Eq. (8), are determined by evaluating the integrals

$$V_m(I) = (1/2\pi)\int_0^{2\pi} d\Theta\, e^{im\Theta}\, x(\Theta,I) \equiv aJ_m'(m)/2m \tag{15}$$

$$\sim\ .325\ I^2/(Zm^{3/2})\quad \text{for large m}$$

using a trick suggested by Landau and Lifschitz[27], where J_m' is the derivative of the ordinary Bessel function of order m. Then combining Eq.'s (12)-(15), the width of the mth resonance is

$$W_m = 4\,I_m^{\,3}(FJ_m'(m)/3mZ^3)^{1/2}\ . \tag{16}$$

The 0th order islands overlap when the ratio of the island width to the separation is greater than one,

$$1 < .5(W_{m+1} + W_m)/\delta_m \sim\ 4\,F^{1/2}m^{11/12}/(Z^{1/6}\Omega^{2/3})\quad \text{for large m.} \tag{17}$$

This criterion determines the critical microwave field required for excitation from one classical resonance to another

$$F >\ .06\ Z^{1/3}\Omega^{4/3}/m^{22/12}\quad \text{(a.u.)}\ . \tag{18}$$

If the SSE has an initial energy corresponding to an action which lies between I_m and I_{m+1}, then the application of microwave fields in excess of the threshold defined by Eq. (18) will cause the electron to diffuse to higher values of the action. Moreover, since the island overlap, Eq. (17) , increases with m, once the microwave field exceeds the threshold for stochastic diffusion for electrons with action I_m then these electrons can diffuse to still larger actions (or energies) until they ionize.

Since this calculation assumes that the electron has an initial action, I(0), which is greater than $I_1 = (Z^2/\Omega)^{1/3}$, this prediction of the critical field for the onset of global stochasticity only applies for perturbing microwave fields with frequencies, Ω, larger than the initial, electron bounce frequency, $\Omega_0(I(0))$ given by Eq. (7) in atomic units.[13] If $I(0) < I_1$ then the primary island in action-angle space, centered at I_1, must expand until it reaches I(0) in addition to overlapping the m = 2 island chain before ionization can occur. An explicit calculation of the critical field in this case is presented in Ref. 28.

When the islands overlap, the classical excitation rate can be estimated using random walk arguments. A quasi-linear[29] treatment of the distribution of trajectories in action-angle space leads a Fokker-Planck type diffusion equation with

$$D_{ql}(I) = \Delta I^2/(2\Delta t) \sim 1.65\mathrm{x}10^3\ I^4F^2/\Omega \quad \text{(a.u.)}. \tag{19}$$

Then the characteristic diffusion rate for an electron to random walk from the m to the m+1 resonance can be estimated by

$$\nu_e \sim 2D_{ql}/\delta_m{}^2 \sim 40\ m^{8/3}F^2/\Omega^{5/3} \quad \text{(a.u.)}. \tag{20}$$

Since Eq. (20) is a rapidly increasing function of m, this provides a convenient order of magnitude estimate of the stochastic ionization rate.

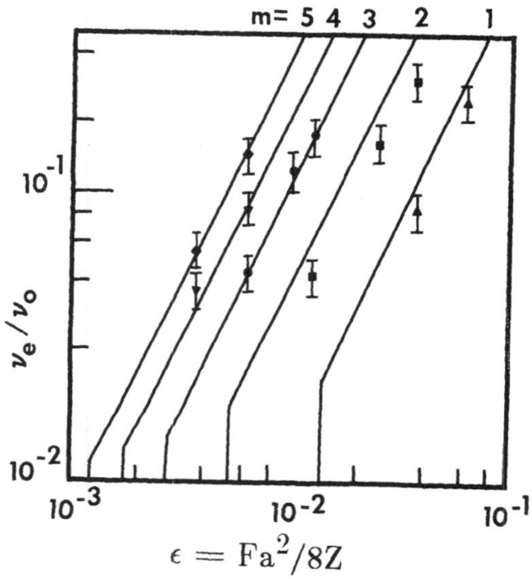

Fig. 5. Theoretical, Eq. (13), and numerical excitation rates from the m to m+1 resonances are plotted as a function of the dimensionless electric field strength, ϵ, for the first 5 resonances with $\Omega = 320$ GHz. The five different symbols for the numerical data points correspond to the different values of m. The error bars represent an estimate of statistical errors in the Monte Carlo calculation. No excitations were observed, numerically, for $\epsilon < \epsilon_c$ as predicted by Eq. (18).

Our analytic estimates for the stochasticity threshold, Eq. (18), and the excitation rate, Eq. (20), have been verified by numerical integrations of the perturbed equations of motion. For small electric fields the electrons remained confined near their initial action (See Fig. 4); however, as the field was increased above the threshold the trajectories spanned several resonances indicating the breakup of confining KAM[15] surfaces. The numerical results for both the stochastic threshold and excitation rate are compared with the analytic predictions as functions of microwave field in Fig. 5. Although the resonance overlap criterion is only expected to be accurate to within a factor of 2,[26] the quasi-linear diffusion estimate is very reliable once the dynamics have become fully chaotic. These remarks are illustrated by the excellent agreement between the theoretical and numerical excitation rates.

If I apply this theory to a SSE initially in the ground state, $E_0 = Z^2R = 6.8x10^{-4}$ ev, the feasibility of the proposed experiment can be assessed. The maximum excursion and oscillation frequency of the corresponding classical electron are

$$a \sim 1.51x10^{-6} \text{ cm} , \tag{21}$$

$$\nu_0 = \Omega_0/2\pi \sim 320 \text{ GHz} , \tag{22}$$

and the binding electric field at maximum excursion is

$$E_0 = Z/a^2 \sim 450 \text{ V/cm} \quad . \tag{23}$$

If the microwave frequency is $\sim$ 320 GHz, then $I(0) \sim I_1$ and stochastic diffusion occurs when the $m = 1$ and $m = 2$ island chains overlap. In this case an oscillating electric field of $F \sim 115$ V/cm is sufficient to classically ionize the ground state SSE at a rate $\nu_e \sim 14$ GHz. Moreover, since the nth quantum level has action $I \equiv n$, it corresponds to the mth classical resonance with $m = n^3\Omega/Z^2$. Therefore, the critical field decreases significantly as $(n^{66/12}\Omega^{1/2})^{-1}$ for SSE's in excited states. The electric field threshold also decreases for higher frequency perturbations since the initial action of the SSE then coreresponds to classical resonances with $m > 1$.

These estimates show that the effects of time-dependent perturbations on the quantum mechanical bound states of surface state electrons can be studied using available microwave sources. The classical analysis predicts that modest electric field strengths, $\sim$ 25% of the binding field, lead to stochastic electron trajectories which ionize rapidly within $\sim$ 25 bounce periods. If the classical stochasticity of this driven oscillator persists in the quantum regime, experimentalists should observe both enhanced line widths for the quantum levels and measurable ionization rates which increase as functions of microwave power.

IV. Discussion

The classical treatment of the surface state electron in a microwave field gives ionization thresholds and rates which are in qualitative agreement with the experimental measurements by Bayfield and Koch[5] performed with real hydrogen atoms. This suggests that the physical ionization mechanism for large quantum numbers in the experiments[5] and the numerical simulations[14] is stochastic diffusion due to overlap of classical resonances in action-angle space.

The same procedure used to analyze the perturbed SSE can also be extended to the Hamiltonian, Eq. (1), which describes the dynamics of a classical electron orbiting the nucleus in the presence of an oscillating electric field. Although the classical, six-dimensional, action-angle phase space can be reduced to four dimensions by exploiting the cylindrical symmetry about the direction of the electric field, the analysis is considerably more complicated than that for the perturbed SSE because of the interplay of the additional degrees of freedom. For example, in higher dimensions the KAM surfaces no longer confine the phase space trajectories. So the electron can always wander to higher actions or energies by "going around" these two dimensional surfaces in the four-dimensional space. Consequently, there is no well defined threshold for the onset of stochastic diffusion. However, this "Arnold diffusion",[26] is usually extremely slow and rapid ionization due to stochastic diffusion will not occur until most of the KAM surfaces between the classical resonances are destroyed.

Meerson et al.[16-17] and Zaslavskii[18] have attempted to calculate the electric field threshold for the destruction of KAM surfaces due to island overlap by transforming the classical equations of motion to appropriate action-angle variables. Although a number of details remain to be worked out, rough estimates of the stochasticity threshold when the microwave frequency is close to the oscillator frequency are in good agreement with the experimental results for hydrogen atoms in the $n \sim 66$ level. However, these preliminary calculations are not adequate to explain the experimental results for the quasi-static experiments ($n \sim 29$) or the low ionization thresholds in helium. Extensive theoretical and experimental studies of bound electrons in oscillating electric fields are currently being pursued in an effort to explain these remarkable phenomena and to further explore the nonlinear quantum dynamics of these systems.

The perturbed SSE, in particular, provides an ideal system for the investigation of quantum chaos. The behavior of this quantum oscillator is properly described by a one-dimensional Schroedinger equation with a time-dependent potential. Although this partial differential equation with variable coefficients does not appear to be amenable to analytic solution, it can be studied, numerically, using efficient algorithms for the solution of partial differential equations. Moreover, since the low lying quantum levels of the SSE are readily accesible to experiment, the predictions of both the classical and quantum treatments can be tested. Some preliminary numerical calculations for bound electrons in one and two dimensions in an oscillating electric field were presented at this conference by Lamb.[30] These results indicate that the quantum mechanical description of these systems also exhibits ionizing solutions. More extensive numerical calculations to examine the connection with the classical predictions of stochastic ionization remain to be completed.

In conclusion, experimental and numerical investigations of non-integrable quantum systems consisting of bound electrons in oscillating fields show a variety of novel features. For large quantum numbers these phenomena can be attributed to the onset of chaotic dynamics in the classical description. These observations therefore provide clear evidence for chaos in a quantum system. Moreover, further investigations for low quantum numbers of both Rydberg atoms and surface state electrons in large amplitude electromagnetic fields promise to provide a detailed picture of the manifestations of chaos in quantum systems.

Acknowledgments

The author thanks D. Mariani and P. Koch for extensive discussions of their experimental measurements and for providing the data used in Fig.'s 1-3, and I.B. Bernstein for useful theoretical discussions. This work was supported by the National Science Foundation.

References

1. Special Issue on Laser Chemistry, Physics Today 33, No.11, 27-59 (1980).
2. A. Einstein, Verh. Dtsch. Phys. Ges. 19, 82 (1917).
3. G. Casati, B.V. Chirikov, F.M. Israelev, and J. Ford, in Stochastic Behavior in Classical and Quantum Systems, ed. G. Casati and J. Ford, Lect. Notes in Physics 93, (Springer, New York, 1979).
4. T. Hogg and B.A. Huberman, Phys. Rev. Lett. 48, 711 (1982).
5. J E. Bayfield and P.M. Koch, Phys. Rev. Lett. 33, 258 (1974).
6. J E. Bayfield, L.D. Gardner and P.M. Koch, Phys. Rev. Lett. 39, 76 (1977).
7. D.R. Mariani, W. Van de Water, P.M. Koch, and T. Bergeman, Phys. Rev. Lett. 50, 1261 (1983).
8. C.C. Grimes, T.R. Brown, M.L. Burns and C.L. Zipfel, Phys. Rev. B13, 140 (1976).
9. R.V. Jensen, Phys. Rev. Lett. 49, 1365 (1982).
10. D. Banks and J.G. Leopold, J. Phys. B11, 37 (1978).
11. R.J. Damburg and V.V. Kolosov, J. Phys. B12, 2637 (1979).
12. P.M. Koch and D.R. Mariani, Phys. Rev. Lett. 46, 1275 (1981).
13. H.A. Bethe and E.E. Salpeter, Quantum Mechanics of One- and Two-Electron Atoms, (Plenum, New York, 1977) p. 3.
14. J.G. Leopold and I.C. Percival, J. Phys. B 12, 709 (1979).
15. V.I. Arnold, Mathematical Methods of Classical Mechanics, (Springer, New York,1978).
16. B.I. Meerson, E.A. Oks, and P.V. Sasorov, JETP Lett. 29, 72 (1979).
17. B.I. Meerson, E.A. Oks, and P.V. Sasorov, J. Phys. B15, 3599 (1982).
18. G.M. Zaslavskii, Phys. Rep. 80,157 (1981).

19. G.M. Zaslavskii and B.V. Chirikov, Sov. Phys. Usp. 14, 549 (1971).
20. P.M. Koch, J. Phys. (Paris), Colloq. 43, C2-187 (1982).
21. Private communication, D. R. Mariani.
22. P. Pillet, W.W. Smith, R. Kachru, N.H. Tran, and T. F. Gallagher, Phys. Rev. Lett. 50, 1042 (1983).
23. D.K. Lambert and P.L. Richards, Phys. Rev. B23, 3282 (1981).
24. B.V. Chirikov, Phys. Rep. 52, 263 (1979).
25. J.M. Greene, J. Math. Phys. 20, 1183 (1979).
26. A.J. Lichtenberg and M.A. Lieberman, Regular and Stochastic Motion, (Springer, New York, 1983).
27. L.D. Landau and E.M. Lifschitz, Classical Field Theory, (Pergamon, New York 1975) pp. 181-182.
28. R.V. Jensen (submitted to Phys. Rev. A).
29. G.M. Zaslavskii and N.N. Filonenko, Sov. Phys. JETP 25, 851 (1968).
30. W. Lamb, see paper in this volume.

QUANTUM DIFFUSION LIMITATION AT EXCITATION OF RYDBERG ATOM IN VARIABLE FIELD

D. L. Shepelyansky

Institute of Nuclear Physics
630090, Novosibirsk 90, USSR

Abstract

A computer simulation method is used to investigate excitation of a hydrogen-like atom in the field of an electromagnetic wave from the states with $n \sim 50$ and parabolic quantum numbers $n_1 \gg n_2$ (or $n_1 \ll n_2$). It is shown that diffusion over levels is much slower than in the classical case. A significant part in atom excitation is played by multiphoton resonances with the number of photons $k \gtrsim \gtrsim 10$.

1. INTRODUCTION

For the last few years a number of interesting experiments was carried out on ionization and excitation of hydrogen atoms by the microwave field from the states with the main quantum number $n \sim 50$ [1-3]. The first of these experiments showed at the frequency of $\omega/2\pi = 9.9$ GHz, a strong ionization of atoms with $n \approx 65$ at the electric field $\varepsilon \gtrsim 0.06\ n^{-4}$ and adiabatic tunneling parameter $\gamma = \omega/\varepsilon\ n \approx 7$ [4] ($\omega \approx 0.43\ n^{-3}$)*. To explain the results of the experiment a diffusion mechanism of atom ionization was offered in [5]. The cause of electron diffusion arising in an atom due to the purely monochromatic field effect can be understood in the following way. It is quite possible to assume [6] that at $n \approx 50 \gg 1$ the quasiclassical approximation is good enough and to pass on to investigation of the classical system. The latter is substantially nonlinear and at field $\varepsilon > \varepsilon_{cr}$ there arises a stochasticity leading to

* Here and below we use atomic units; n, ℓ, m are principal, orbital and magnetic quantum numbers.

a diffusion excitation of an electron [7]. In case of linear and circular polarizations and $\omega \approx \Omega = n^{-3}$ the estimation for ε_{cr} was obtained in [7,8] on the grounds of the overlap criterium [9]. Dependences of critical field and diffusion rate on frequency are given in [10,11]. Despite all attractiveness of this classical approach, its only justification is the coincidence of the relative portion of ionized atoms at $\varepsilon\, n \approx 0.06$ [1] with the probability of ionization obtained through computer simulation of classical system's dynamics [6]. The accuracy of coincidence is about 30%.

At the same time a number of numerical experiments with simple quantum systems [12,13] have shown that statistical properties of quantum dynamics are much weaker than those of classical ones. Moreover, in the course of time the effect of quantum corrections increases [14] and leads to diffusion's slowing down and its final almost complete stopping. First this effect of quantum diffusion limitation was observed in the model of quantum rotator [12]. The same phenomenon can take place at diffusion ionization of atoms with $n \gg 1$. Preliminary estimates [11] show that relative magnitude of the quantum corrections can be compared to unity and so there may be a significant distinction between quantum dynamics of atom excitation and classical ones.

2. THE MODEL OF SURFACE-STATE ELECTRONS

When investigating quantum and classical dynamics we employed the computer simulation method. We chose an external field linear polarized along the axis z. The atom was initially excited into states with quantum numbers which satisfied the following conditions $n \sim 50$, $m = 0$, $n_1 \gg n_2$ (or $1 \lesssim \ell \ll n^{2/3}$). For these states the ratio of cross-size to longitudinal one is small and that is why it is possible to describe the dynamics of excitation in this case by the one-dimensional Hamiltonian

$$H = \frac{p^2}{2} - \frac{1}{|z|} + \varepsilon\, z \cos \omega t \qquad (1)$$

with the boundary condition $\Psi(0) = 0$, which corresponds to an infinite potential wall at $z = 0$. This Hamiltonian also describes the exitation of states of atom with $\ell = 0$ in the external field with the potential $V = \varepsilon |r| \cos\omega t$. Consequently for Equation (1) matrix elements $z_{nn'} = r_{nn'}$. To prove the fact that such a modification of the initial Hamiltonian will not lead to any significant change of the dynamics we can present the following arguments:

1. When $1 \leqslant \ell \ll n^{2/3}$ the matrix elements $z^{2\ell\pm 1}_{nn'}$ do not depend on ℓ [15], as a result of which their sum z_ε that determines the probability of transition from one shell into another in dipole approximation to within several per cent is equal to $r_{nn'}$ ($n, n' \sim \sim 50$). So, for $n = 40$, $n' = 41$, $\ell = 2$, the ratio $z_\varepsilon / r_{nn'} = -0.996$ (here we use for z_ε its quasiclassical value).

2. Due to Coulomb degeneration the transition frequency between adjoining sublevels of one and the same shell prove much smaller than Kepler frequency: $3\varepsilon n/2\Omega \lesssim 0.1 \ll 1$ for $\varepsilon_0 = \varepsilon n^4 \lesssim 0.06$. So, despite the fact that the width of Stark multiplet exceeds the distance between the shells $\Delta\Omega \gtrsim \Omega$, the motion of sublevels proves synchronized and it is possible to substitute the whole shell for one level. A proof of such approximation is given by the experiments [2,3], where the observed resonant picture of excitation in the field with $\varepsilon_0 \approx 0.06$ corresponded to the unperturbed spectrum of atom levels.

3. In the classical system when $\ell \ll n$ the dependence of ε_{cr} and the rate of diffusion on ℓ proves insignificant [11]. The Hamiltonian (1) describes precisely the motion for the orbits with eccentricity $e = 1$ stretched along the field. In the quantum system the states with parabolic quantum numbers $n_1 = n - 1$, $n_2 = 0$ (or vice versa) and close to them ($n_1 \lessgtr n_2$) correspond to such classical orbits. Let us see if this condition will be violated in the course of time. To do this we shall consider unperturbed ($\varepsilon = 0$) phase-action variables corresponding to parabolic quantum numbers. For these variables

$$Z = \frac{3}{2} n(n_1 - n_2) + \sum_{\substack{m_1, m_2 \\ m_1 + m_2 \neq 0}} B_{m_1 m_2} \exp[i(m_1\lambda_1 + m_2\lambda_2)]$$

$$B_{m_1 m_2} = \frac{1}{m_1 + m_2} [\mu_2 \mathcal{J}_{m_1}(\mu_1(m_1 + m_2)) \mathcal{J}'_{m_2}(\mu_2(m_1 + m_2)) - \mu_1 \mathcal{J}'_{m_1}(\mu_1(m_1 + m_2)) \mathcal{J}_{m_2}(\mu_2(m_1 + m_2))], \tag{2}$$

$\mu_{1,2} = [n_{1,2}(n - n_{2,1})/n^2]^{1/2}$; λ_1, λ_2 - phases conjugated to n_1, n_2; where the derivative is taken over the argument of Bessel function. According to the correspondence principle, the values $B_{m_1 m_2}$ determine matrix elements $z_{n_1 n_2}^{n_1' n_2'}$ of the transition between the unperturbed levels with $m_{1,2} = n'_{1,2} - n_{1,2}$. When $n_1 \gg n_2$ ($m = 0$) matrix elements for the transitions with the change of n_2 contain a supplementary small parameter n_2/n. The change magnitude of n_2 can be estimated in the course of time in the following way. For $\omega \approx \Omega$ and $\varepsilon > \varepsilon_{cr}$ there starts a diffusion electron excitation. According to the theoretical estimates [11] the rate of diffusion over n and the value of ε_{cr} are:

$$D_n = \frac{d(\Delta n)^2}{dt} \approx \frac{1}{3} \varepsilon^2 n^7; \quad \varepsilon_{cr} \approx \frac{1}{84 n^4}. \tag{3}$$

When $\varepsilon > \varepsilon_{cr}$ the phase λ_1 changes randomly in the course of time, which makes it possible to estimate the rate of diffusion over n_2 : : $D_{n_2} \sim (n_2/n)^2 D_n$. From the above we see that in the course of a numerical experiment $\tau = \omega t/2\pi \lesssim 50$ at $\varepsilon n^4 \approx 0.04$, $n_2 \sim 1$ the change

$\Delta n_2 \sim 1$ and so the one-dimensional approximation is not violated. It is interesting to note that when $\varepsilon n^4 n_2 \ll 1$ the dynamics over the second degree of freedom (n_2, λ_2) prove substantially quantum, since in this case the perturbation theory can be used [16]: $B_{m_1 m_2}/\Omega \ll 1$ $(m_2 \neq 0)$. Let us point out the fact that when $m_1 + m_2 = \text{const} \sim 1$ the matrix elements $B_{m_1 m_2}$ with $m_1 - m_2 \gg 1$ are exponentially small. This proves still more rigorously the effect observed in the experiments where the shell shows itself as one level [2,3].

To determine in quantative terms how good is the selected one-dimension approximation, we made a comparison with the results of numerical experiments for a real atom in a linear polarized field carried out by F. M. Izrailev. The initial condition was given by parabolic quantum numbers $n_1 = n_0 - 1$, $n_2 = 0$. The comparison was made for $n_0 = 10, 9$ and showed a good agreement with the results obtained in a one-dimensional approximaton. So, for instance, the probabilities of excitation $W_{n>n_0}$ into states with $n > n_0$ coincide with the accuracy of 10 - 20%.

Hence, proceeding from the above arguments we can conclude that the system (1) will describe in qualitative and quantitative terms the dynamics of Rydberg atom excitation from states $n_1 \gg n_2$ or $n_1 \ll n_2$ (and qualitatively from $1 \lesssim \ell \ll n^{2/3}$). Besides, this model describes exactly the dynamics of surface-state electrons excitation [17] in a variable field, directed perpendicular to the surface of liquid helium and that is why its investigation is of interest in itself. It was offered to carry out an experiment with such a system in reference [18]*.

3. NUMERICAL EXPERIMENTS

Simulation of quantum dynamics of system (1) was carried out in the following way. At the initial moment of time one level of the unperturbed system with $n_0 \sim 50$ was excited. The field frequency was close to the Kepler one: $\omega n_0^2 = \omega_0 \sim 1$ and the field $\varepsilon n_0^4 = \varepsilon_0 \sim 0.06$. Further on we carried out a numerical integration of equations for amplitudes C_n of the unperturbed Hamiltonian's states,

$$i\dot{C}_n = -\frac{1}{2n^2} C_n + \sum_{n'=n_{min}}^{n_{max}} V_{nn'} C_{n'}. \tag{4}$$

Thus, levels with $n_{min} \leqslant n \leqslant n_{max}$ were only involved in the dynamics. Since at the cited values of parameters ε and ω diffusion in the classical system moves up along n [11], levels with $n < n_0$ are weakly

* We note that the functional dependence $\varepsilon_{cr}(\omega)$ in [18] was not found correctly (there is a mistake in f.(8)). See for a correct answer [11].

excited and that is why the accepted $n_0 - n_{min} < 10$ proved sufficient. During all the time the probability $W_{n_{min}} = |C_{n_{min}}|^2$ was at the level 10^{-4}, and a further decrease of n_{min} did not influence the dynamics of excitation. The working value of n_{max} equal to $\simeq 2n_0$ was selected in most cases. At the same time a number of control experiments with $n_{max} \simeq 2.7n_0$ was carried out. A good agreement of working and control points (see below) shows that in order to investigate electron excitation into states of the discrete spectrum within the time $\tau \simeq 40$ periods of an external field, we can restrict ourselves to a finite number of levels with $n_{max} \simeq 2n_0$. At values $n_{max} \simeq 2n_0$, $n_0 \simeq 50$ and $\omega \simeq \Omega$ to come out into a continuous spectrum it is necessary to absorb five photons more as compared to states $n \simeq n_{max}$. That is why it is quite natural to expect that the probability of multiphoton ionization will be relatively small. Besides, the main aim of the experiments was to investigate the diffusion mechanism of excitation, occurring for discrete states only.

In the process of numerical integration probabilities $W_n = |C_n|^2$ were found, according to which the first $M_1 = \langle n - n_o\rangle/n_o$ and the second $M_2 = \langle (n-n_0)^2\rangle/n_0^2$ moments of distribution were determined as well as $W_{n>n_0+4}$ - the probability of excitation in states with $n > n_0 + 4$ and $W_{n \geqslant [1.5n_0]}$ - in states with $n \geqslant [1.5n_0]$ where brackets indicate an integer part. The step of integration Δt was determined from the condition $\varepsilon z_{n_{max}n_{max}}\Delta t \lesssim 0.3$ and it was usually $\simeq 200$ times less than an external field period. When it was reduced two times the relative change of cited values and probabilities $W_n > 10^{-2}$ during the time $\tau = 40$ was less than 1% (for $10^{-5} < W_n < 10^{-2}$ the change reached 10%). The accuracy of conservation of the complete probability $W = \Sigma_n W_n = 1$ was not worse than 0.1%. The calculation of matrix elements $z_{nn'} = r_{nn'}$ was done numerically according to the formules given in [19]. At $n \simeq 40$ their agreement with classical values was 2 - 3%. The time of integration of one period of external field at $n_0 = 30$, $n_{max} = 63$, $\varepsilon_0 = 0.04$, $\omega_0 = 1$ was $\simeq 15$ seconds of computer ES-1060.

Together with the quantum dynamics simulation a numerical investigation of the classical system was carried out. To exclude the peculiarity at the point $z = 0$, it turned out to be suitable to pass on from unperturbed phase-action variables (n,λ) to new variables (n,ξ) and to a new "time" η, where the equations of motion are:

$$\frac{dn}{d\eta} = -\varepsilon n^2 \cos\omega t \sin\xi, \quad \frac{dt}{d\eta} = 1 - \cos\xi$$
$$\frac{d\xi}{d\eta} = -n^{-3} + 2\varepsilon n \cos\omega t(1-\cos\xi), \quad \lambda = \xi - \sin\xi. \qquad (5)$$

The initial distribution of classical trajectories in the phase space was a line: $n = n_0$ and equipartition distribution along the phase variable λ, which corresponded exactly to the initial conditions in the quantum case (one level with $n = n$). The complete number of trajectories was $N = 1000$. In view of the fact that in the classical system there was a finite number of levels a reflection boundary condition at the corresponding n_{max} was introduced in the classical model. However, this had a weak effect on the values $W_{n>n_0+4}$ and $W_{n\geqslant[1.5n_0]}$. So, for instance for $\varepsilon_0 = 0.04$, $\omega_0 = 1$, $\tau = 40$ when reflection was replaced by absorption, their change proved less than one standard deviation. The value of the latter was found by four groups, consisting of 250 classical trajectories equally distributed along λ. At the selected integration step the accuracy of conservation of energy in a constant field, i.e., at $\omega = 0$, $t \leqslant 80\cdot\pi n_0^3$, $\varepsilon_0 = 0.06$ was better than 10^{-5}. It should be mentioned that the classical dynamics depend on scale variables only $\varepsilon_0 = \varepsilon n_0^4$, $\omega_0 = \omega n_0^3$ and at fixed values of the latter ones, they do not depend on n_0 [6,20].

4. THE RESULTS OF THE EXPERIMENTS

a) One-Frequency Excitation

The main group of experiments was conducted for values $n_0 = 30$, 45, 66 in the range of parameters ε_0, ω_0 where a strong stochasticity occurs, which leads to a diffusion excitation of a classical electron. However, despite the fact $n_0 >> 1$ the obtained results show that the exitation of an atom is of a fundamentally quantum nature. In fact, as one can see in Figure 1, quantum averages are close to classical ones only for a very short time $\tau < \tau^* \approx 5$ periods of an external field. When $\tau > \tau^*$ a quantum diffusion limitation is observed, because of which the excitation of a quantum atom proves much weaker than that of a classical one. We note that the given effect is not related to the finite number of selected levels. Indeed, for $n_0 = 45$ when $n_{max} - n_0$ is reduced one time and a half (to $n_{max} = 80$), the relative change M_2 increases from 4% when $\tau =$ $= 10$ to 24% when $\tau = 40$ and the ratio of the classical value M_2 to the quantum one changes from 2 to 6. Besides, the effective width of distribution $\Delta n \approx 5 << n_{max} - n_0 = 52$ and the probability on levels with $n \approx n_{max} = 97$ is $\sim 10^{-6}$. The classical value $(\Delta n)^2$ agrees well with the theoretical estimate (3) $(\Delta n)^2 \approx 2\varepsilon_0^2 n_0^2 \tau$. But to check it in a more detailed way, it is necessary to calculate the local coefficient D_n which goes beyond our paper.

It should be noted, however, that the proximity of the first and the second moments of quantum distribution to their classical values does not at all mean that the excitation occurs due to diffusion. An example of this is the excitation from $n_0 = 30$ at

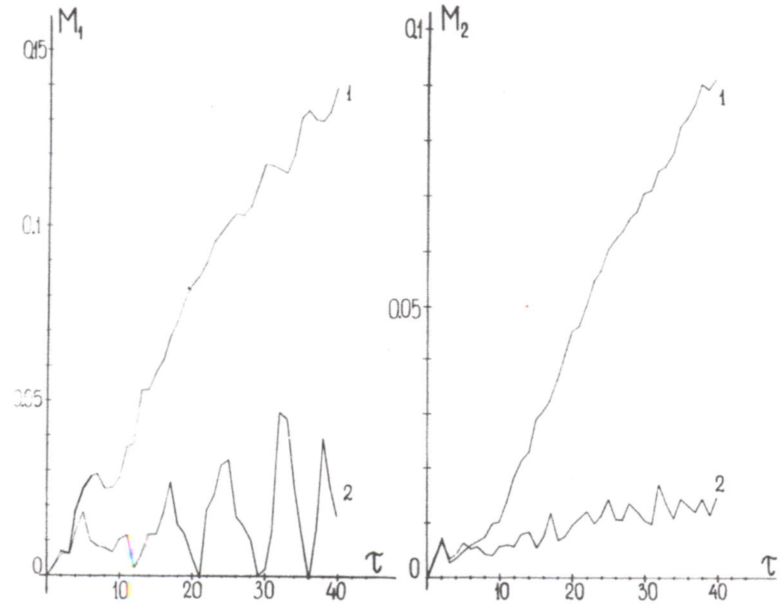

Fig. 1. Dependence of the moments of distribution on time for classical (1) and quantum (2) atoms with $\varepsilon_0 = \varepsilon n_0^4 = 0.04$, $\omega_0 = \omega n_0^3 = 1$, $n_0 = 45$, $n_{max} = 97$.

$\varepsilon_0 = 0.04$, $\omega_0 = 1$. In this case for $\tau \leqslant 30$ the difference between quantum moments and classical ones is smaller than 20% (except the values $\tau = 4,5,6,7$ where it comes up to 100% for M_1). Nevertheless, in the quantum case the dependence of excitation probabilities $W_{n>n_o+4}$, $W_{n\geqslant[1.5n_0]}$ on frequency ω_0 is of a pronounced resonance character and differs fundamentally from the classical one (Figure 2a,c). For $\omega_0 = 0.9975$, $\tau = 30$, $n > 34$ maximus of probability distribution fall on levels with $n = 36,37,39,41,44,47,49,52,58$ and 66. These values ($n \neq 36$) approximately correspond to unperturbed resonant transitions from the initial level $n_0 = 30$. For the last three values of n the necessary number of photons k is equal correspondingly to 10,11, and 12 (see Figure 2). For $\omega_0 = 1.002$, $\tau = 30$ the main contribution (63%) into the probability $W_{n>n_0+4}$ comes from levels with $n = 35$ (33%), 36 (18%) and 37 (12%). Despite the fact that close to the resonance $W_{n\geqslant[1.5n_0]}$ makes up about 20% of the complete probability, control experiments have shown that when $n_{max} - n_c$ increased one time and a half, the change of probabilities of excitation proves insignificant (see Figure 2a,c).

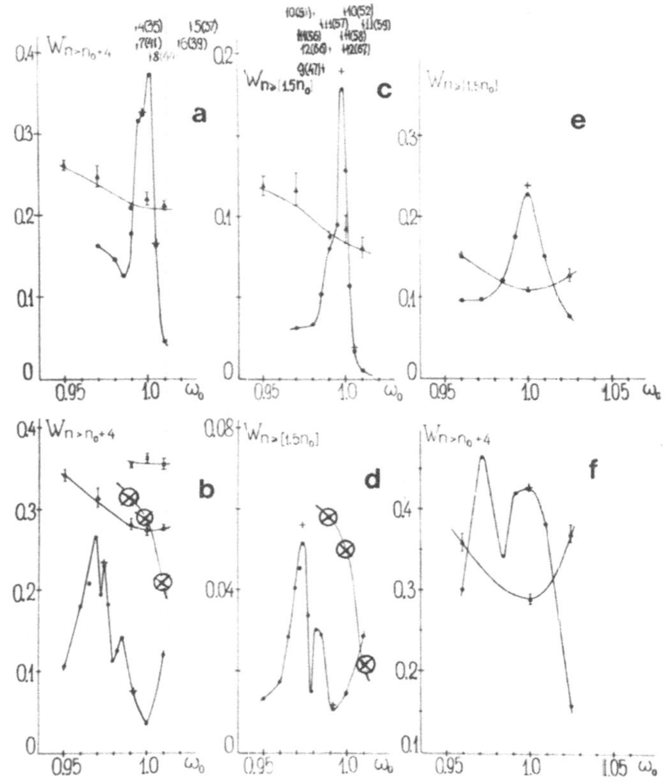

Fig. 2. Dependence of probabilities of atom excitation $W_{n>n_0+4}$ and $W_{n\geqslant[1.5n_0]}$ on frequency ω_0 in the quantum (●) and classical (▲) case. For (a),(c): $\varepsilon_0 = 0.04$, $n_0 = 30$, $n_{max} = 63$ ((+)-control points for the quantum system with $n_{max} = 80$); (b),(d): $\varepsilon_0 = 0.04$, $n_0 = 45$, $n_{max} = 97$ ((+) - $n_{max} = 122$); (⊕) - quantum and (x) - classical values for $\varepsilon_0 = 0.04$, $n_0 = 66$, $n_{max} = 143$ (for $W_{n\geqslant[1.5n_0]}$ classical values (▲) are given in Figure 2c); the values of probabilities are taken at the moment of time $\tau = \omega t/2\pi = 30$. For (e),(f): $\varepsilon_0 = 0.06$, $n_0 = 45$, $\tau = 12$, $n_{max} = 97$ ((+) - the control point with $n_{max} = 122$). Arrows point to the position of resonances and figures alongside to the number of photons and the number of a resonant level. The curves are given to complete a graphic picture.

Thus, in the investigated case the mechanism of excitation proves not a diffusion but a multiphoton one. At that, up to $k = 10$ photons can be effectively absorbed. For $\omega_0 = 0.9975$, $n_0 = 30$,

$\tau = 30$, $0.03 \leqslant \varepsilon_o \leqslant 0.037$ the experimental value $k = 9.973 \pm 0.006$.

For $n_o = 45$, $\varepsilon_o = 0.04$ excitation also is of a pronounced resonant character (Figure 2b,d). However, we cannot connect the observed resonances with multiphoton transitions between unperturbed levels. For $\omega_o = 0.974$, $n > 49$, $\tau = 30$ levels with n = 50-55,57,59, 63,67,72,86 and 93 are the most excited. Even in resonance, absolute values of probabilities of excitation happen to be much smaller than their classical values (Figure 2). However, so far as the field ε_o grows, this difference decreases and at $\varepsilon_o \approx 0.06$ the probabilities become comparable (Figure 3). Nevertheless, as is seen in Figure 2e,f, even if $\varepsilon_o = 0.06$, the mechanism of excitation is of a resonant nature and results in a more effective excitation than classical diffusion does.

The dependence of the probability of excitation on the frequency of an external field proves rather strong. So, for $n_o = 45$, $\varepsilon_o = 0.03$, $\omega_o = 1$, 1.3 for $\tau = 40$, the value $W_{n \geqslant [1.5n_o]} < 2 \cdot 10^{-3}$ and the probability of excitation is two orders of magnitude higher for

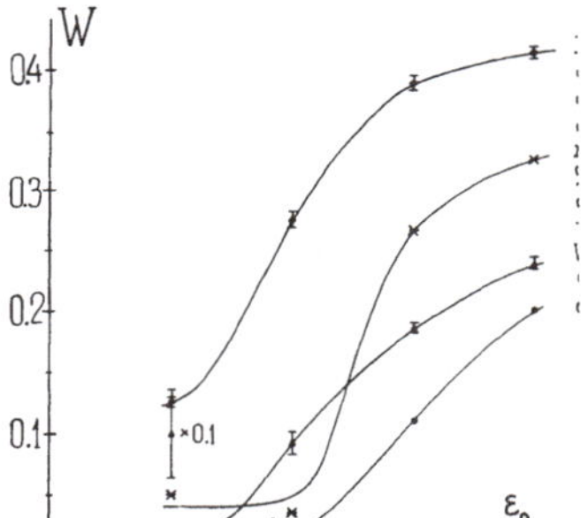

Fig. 3. Dependence of the probability of atom excitation on field ε_o for $\omega_o = 1$, $n_o = 45$, $\tau = 30$, $n_{max} = 97$. Symbols (•), (*) correspond to quantum probabilities $W_{n \geqslant [1.5n_o]}$, $W_{n > n_o+4}$ and (▲), (♦) to classical ones.

$\omega_0 = 1\sqrt{2} \approx 0.707$, $\tau = 30$. In the classical system $W_{n\geqslant[1.5n_0]}$ changes in order of magnitude from 0.016 at $\omega_0 = 1$, $\tau = 40$ to 0.12 at $\omega_0 \approx 0.707$, $\tau = 30$. However, despite the fact that at $\omega_0 \approx 0.707$ the probabilities of excitation and the moments of distribution M_1 and M_2 are close to their classical values (for M_1, M_2 at $\tau \leqslant 30$ the difference $\leqslant 30\%$), the dependence $W_{n>n_0+4}$, $W_{n\geqslant[1.5n_0]}$ on the frequency ω_0 differs greatly from the classical one. As it is evident in Figure 4 the excitation of levels with $n \geqslant 67$ occurs in a resonant way, levels with n = 69,76,78,79,80,81 and 82 for $\omega_0 \approx 0.707$ being the most excited. All of them except n = 76 ($W_{76} \approx 0.03$) have about the same probability $W_n \approx 0.014$. For the frequency $\omega_0 = 0.697$ the probabilities of excitation of these levels are one order less and do not correspond to the positions of maximums W_n. An emergence of such a group of strongly excited close states can be qualitatively explained in the following way. After a small number of periods ($\tau \approx 5$) the initial distribution spreads quickly. The width of spreading is equal $\Delta n \approx 7$, and $\langle n\rangle \approx 49$. As when neighboring levels are nearly equidistant for all of them a simultaneous

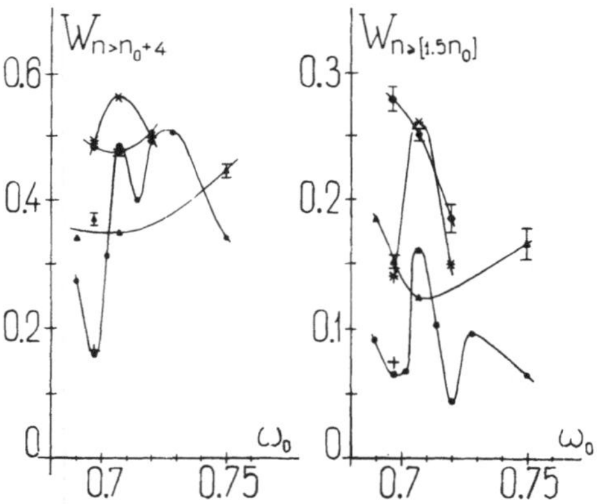

Fig. 4. Dependence of probabilities of atom excitation on the frequency ω_0 when $n_0 = 45$, $n_{max} = 97$. Points (●) are quantum values for $\varepsilon_0 = 0.03$, $\tau = 30$, (▲) - classical values. The control point (+) corresponds to $n_{max} = 122$. Quantum (*) and classical (♦) values correspond to $\varepsilon_0 = 0.04$, $\tau = 15$.

multiphoton excitation can take place. However, it should be noted that the number of photons necessary for that is rather great: $k = (E_{80} - E_{45})/\omega \approx 22$.

As the field increases up to $\varepsilon_0 = 0.04$ the probability of excitation $W_{n \geqslant [1.5n_0]}$ grows sharply from 0.062 to 0.26 ($\omega_0 = 0.707$, $\tau = 15$). When the frequency changes the values $W_{n>n_0+4}$ are close to classical ones, but for $W_{n \geqslant [1.5n_0]}$ a resonance dependence on ω_0 is still observed, and the difference from the classical value of probability is 100% (Figure 4).

As the frequency grows, the probability of quantum system excitation becomes smaller than in the classical case. So, when $n_0 = 45$, $\varepsilon_0 = 0.04$, $\omega_0 = 1.3$, $\tau = 50$ in the classical system $W_{n>n_0+4} = 0.359 \pm \pm 0.013$, $W_{n \geqslant [1.5n_0]} = 0.136 \pm 0.003$ and in the quantum one ($n_{max} = 97$) $- 0.181$ and 0.051 correspondingly. Quantum moments of distribution M_1 and M_2 are 2.5 times smaller than classical ones. The same is valid for $\omega_0 = 1.25$.

The growth of the number of initially excited levels up to $n_0 = 66$ does not lead to a qualitative change of dynamics. So, for $\varepsilon_0 = 0.04$, $\omega_0 = 1.01$, $\tau = 35$ the moments of quantum distribution are rather (≈ 4 times) smaller than the classical ones. Besides, the quantum distribution breaks sharply at $n > 76$. This is its fundamental distinction from the classical one, which diminishes very slowly at great n. As a result the classical value of $W_{n \geqslant [1.5n_0]}$ is about 5 times greater than the quantum one (Figure 2d) and $W_{n>76} = 0.23$ is 5.4 times greater. The dependence of the probability of excitation on frequency also differs from the classical one (Figure 2b,d). A significant role belongs here to the multiphoton mechanism of excitation. So, for $\omega_0 = 1$, $\tau = 20$ maximums of distribution W_n at $n = 60,76$ and 89 approximately correspond to $k = 7,8$ and 15-photon transitions from initial level (Figure 5).

b) Two-Frequency Excitation

A number of experiments was carried out for the case when an external field has two harmonics: $\varepsilon(t) = \varepsilon(\cos\omega t + \cos\nu t)$. One frequency has been fixed $\omega n_0^3 = \omega_0 = 1$ and the second was being changed in a small range near the values $\omega_1 = \nu n_0^3 = 0.7, 1.3$. For the former one a visible excitation of the classical system occurs at $\varepsilon_0 = \varepsilon n_0^4 \geqslant \geqslant 0.015$. For smaller values spreading of the classical packet during the time $\tau = \omega t/2\pi \leqslant 40$ proves insignificant (Figure 6a). But at $\varepsilon_0 = 0.02$, $\tau = 20$ there appears already a developed stochastisity, leading to a substantial excitation of the system (Figure 6b,c). In the quantum case when ε grows the width of distribution grows as

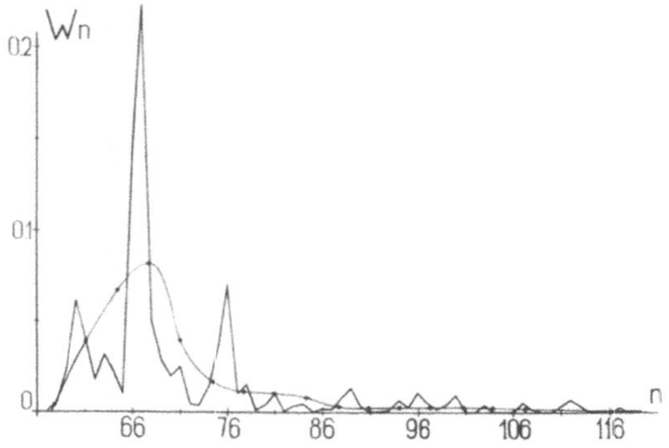

Fig. 5. Distribution over levels in the classical and quantum cases when $\varepsilon_0 = 0.04$, $\omega_0 = 1$, $n_0 = 66$, $\tau = 20$, $n_{max} = 143$. The broken line is the quantum distribution; points correspond to cells of histogram of classical distribution with $\Delta n/n_0 = 0.05$, the smooth curve is given to complete a graphic picture.

well (Figure 6) which agrees qualitatively with the idea of the diffusion mechanism of excitation. The probability of excitation of levels close to n_0 ($W_{n>n_0+4}$) is close to the classical one (see Figure 7a), and the probability of excitation of high levels of $W_{n\geqslant[1.5n_0]}$ is much greater than that and is of a marked resonance nature (Figure 7c). The latter points to the fact that multiphoton excitation may also be rather substantial in the case of a wide distribution over levels (Figure 6b). In all the investigated cases the quantum distribution turned out to be more cutting than the classical one, peaks W_n in a weak field correspond to resonance transitions (e.g., in Figure 6a the peak at n = 51 corresponds to 5-photon resonance with $n_0 = 45$).

For $\omega_1 \approx 1.3$, $n_0 = 45$, $\varepsilon_0 = 0.02$ the quantum system is less excited than the classical one (Figure 7b,d). So, for instance, for $\omega_1 = 1.27$, $n_0 = 45$, $\varepsilon_0 = 0.02$ the classical value M_2 is 3 times less than the classical one, and M_1 is 300 times less. When n_0 comes up to 66 probabilities of excitation are close to classical ones (see Figure 7b,d for $\omega_1 = 1.31$, $\tau = 40$), but the first moment of quantum

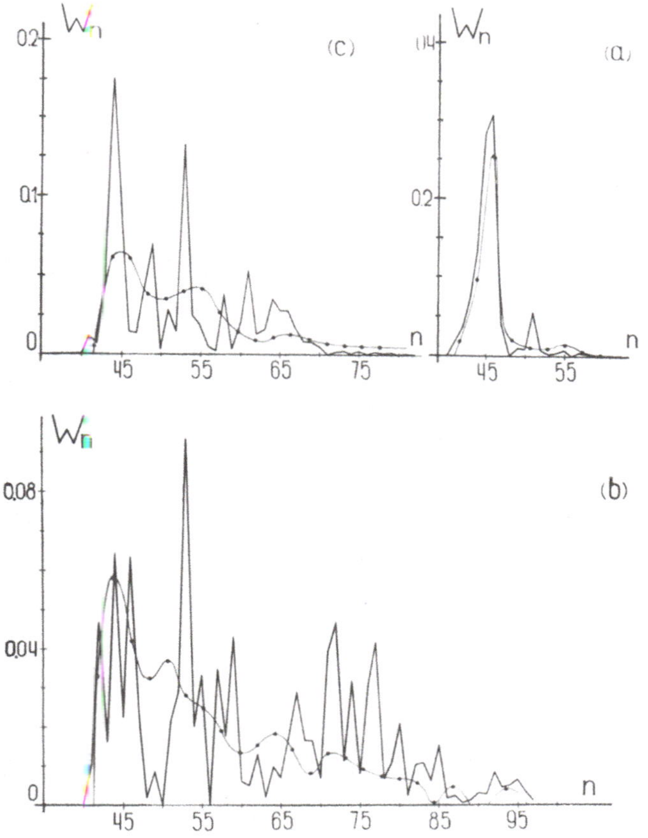

Fig. 6. The same as In Figure 5 but in case of the two-frequency excitation for $n_0 = 45$, $\omega_0 = 1$, $\tau = \omega t/2\pi = 20$, $n_{max} = 97$. (a) $\varepsilon_0 = 0.01$, $\omega_1 = 0.70711$; (b) $\varepsilon_0 = 0.02$, $\omega_1 = 0.735$; (c) $\varepsilon_0 = 0.02$, $\omega_1 = 0.70711$.

distribution at $\tau > 10$ differs 2 or 4 times from the classical ones and at $10 < \tau < 15$ has another sign. Sharp peaks ($0.08 < W_n < 0.12$ for $n = 61,67,70$ and 71) about 1.5 - 3 times exceeding the classical value W_n are observed in distribution for $\tau = 40$. So, despite the fact that in the quantum case a substantial spreading of the packet occurs (the excited number of levels $\Delta n \approx 20$, $M_2^2/M_2^{c\ell} \approx 0.8$) quantative characteristics of distribution somehow differ from classical ones.

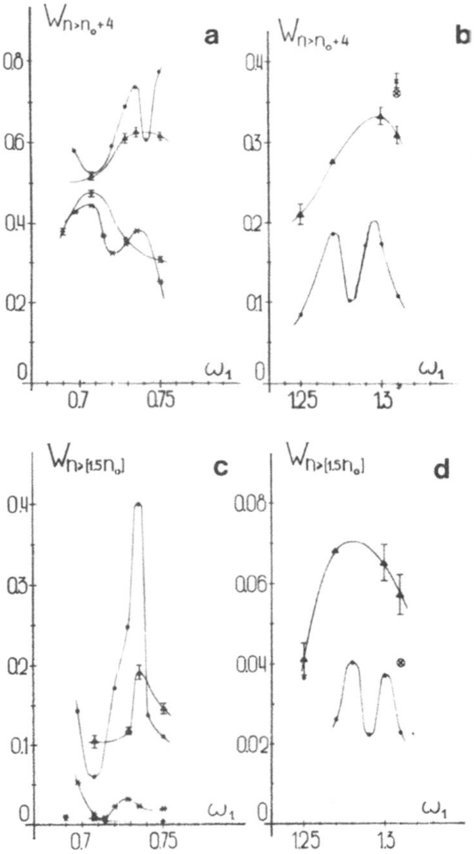

Fig. 7. Dependence of the probability of atom excitation on frequency in case of two-frequency excitation when $n_0 = 45$, $\omega_0 = 1$, $n_{max} = 97$. (a),(c): for $\varepsilon_0 = 0.015$ with quantum (*) and classical (♦) values; for $\varepsilon_0 = 0.02$ correspondingly (●) and (▲), $\tau = 20$; (b),(d): the same as (a),(c) but for $\tau = 40$. Symbols (⊕), (x) correspond to the quantum and classical systems when $n_0 = 66$, $\varepsilon_0 = 0.02$, $\tau = 40$, $n_{max} = 143$. For $W_{n\geqslant[1.5n_0]}$ the classical value is (▲).

5. DISCUSSION

Thus the carried out experiments have shown that the diffusion excitation of atoms in a field of the monochromatic wave from states with $n \sim 50$ is rather suppressed. A simple estimate of values n, starting with that which on the diffusion excitation will be close to the classical one, can be obtained in the following way [13]. According to Equation (3) the number of diffusively excited levels

Δn grows with the number of periods τ ($\omega \approx \Omega$) according to the law $(\Delta n)^2 \approx D_o\tau \approx 2(\varepsilon_o n_o)^2\tau$. At the same time Δn determines an effective number of lines in the spectrum of the wave function $N_L \sim \Delta n$. The spectrum can be considered continuous as far as $N_L \geqslant \tau$. As diffusion needs a continuous spectrum the time of quantum diffusion limitation is determined by the condition $N_L(\tau^*) \sim \tau^*$ which gives

$$\tau^* \sim 2(\varepsilon_o n_o)^2 \sim D_o. \tag{6}$$

At $\varepsilon_o \sim \varepsilon_{cr}$ this time $\tau^* \propto n_o^2$, which agrees with the estimate given in [11]. Since $\tau \gtrsim 1$, $\varepsilon_o \gtrsim \varepsilon_{cr}$ as well as $\varepsilon_o > 1/n_o$ (or $\varepsilon > n^{-5}$) are the necessary conditions for diffusion, which corresponds to going beyond the perturbation theory and exceeding the quantum border of stability [16]. Thus, for the diffusion to go over a large time interval $\tau^* >> 1$ in the field ε_o (at $\omega_o \approx 1$) the following values are needed

$$n_o >> 1/\varepsilon_o. \tag{7}$$

At $\varepsilon_c \approx 0.03$ this condition is satisfied by $n_o >> 30$. Let us note that the number of levels in one non-linear resonance is $(\Delta n)_r \gtrsim 1$ for $\varepsilon \gtrsim n^{-6}$ [11]. For $\varepsilon_o = 0.04$, $n_o = 45$ the value $\tau^* \sim 6$ agrees well with the results of a numerical experiment (Figure 1). But for τ^* to become comparable with the characteristic diffusion time $\tau_D \sim$ $\sim n_o^2/D_o \sim 1/2\varepsilon_o^2$, we need to satisfy this condition and it is necessary to increase either the number of level n_o or ε_o. At $n_o = 45$, the necessary value $\varepsilon_o \sim (2n_o)^{-1/2} \sim 0.1$. Computer experiments really point to the fact that with the growth of the field, the probabilities of excitation in the classical and quantum cases become comparable. But with the growth of ε the probability of various multiphoton transitions grows also and can lead to a quicker atom excitation (Figure 2e,f). A marked difference of the moments M_1, M_2 from the classical values also points to the quantum nature of excitation in the field with $\varepsilon_o = 0.06$. Thus for $n_o = 45$, $\varepsilon_o = 0.06$, $\tau = 12$, $\omega_o = 1$ the quantum moments are greater than the classical ones ($M_{1,2}^2/$ $/M_{1,2}^{c\ell} = 1.8, 1.7$) and for $\omega_o = 1.025$ smaller than the classical ones ($M_{1,2}^2/M_{1,2}^{c\ell} = 0.41, 0.56$). The given estimates belonged to the case when a one-dimensional approximation can be used, i.e., $n >> n_2 \sim 1$. If $n_{1,2} \sim n_2$ the number of lines in the spectrum $N_L \sim \Delta n_1 \Delta n_2 \sim (\varepsilon_o n_o)$ $)^2\tau$. From which it is clear that the necessary condition of diffusion excitation is $\varepsilon_o \gtrsim 1/n_o$. However when $\varepsilon > \varepsilon_{cr}$ the fulfilment of this condition does not at all denote that there will be no quantum diffusion limitation in the system (although the condition $N_L >$ $> \tau$ is fulfilled). This involves the fact that due to Kepler degeneration N_L contains groups of very close lines, corresponding to one shell, which decreases its efficient value. That is why to find the conditions of diffusion in the range $n_{1,2} \sim n$, $\varepsilon > n^{-5}$, $\varepsilon > \varepsilon_{cr}$ (i.e., for $n \gtrsim 50$) we will need further investigation.

When the frequency of excitation goes up the rate of diffusion slows down [11], which leads to a decrease of $\tau^* \sim D_0(\omega)/\omega$ ($n_2 << n$). The marked growth of the probability of excitation at the decrease of frequency can be qualitatively explained in the following way. At $\varepsilon_0 \approx 0.04$, $\omega_0 \approx 0.7$ the value $n = n_0$ is in resonance, the center of which is at $n_r = n_0\omega_0^{-1/3}$. This leads to an efficient increase of the field $\varepsilon_0 \to \varepsilon_0(n_r/n_0)^4 = \varepsilon_0\omega_0^{-4/3} \approx 1.6\varepsilon_0$ and a more fast classical diffusion. This effect was observed for the classical atom in [20]. This sort of explanation is suitable for the diffusion mechanism of excitation only, since in case of the multi-photon mechanism the probability of excitation should decrease with the lowering of frequency. So, the resonance dependence on ω_0, observed in the experiment (Figure 4) is still being obscure. A further theoretical explanation is also needed for the results on excitation from the states $n_0 = 30,45$ and 66 at $\omega_0 \approx 1$. So, for instance, the theoretical formula, obtained in [21,11], gives value $W_{n\geqslant[1.5n_0]}$ 10 orders less than the experimental one ($n_0 = 30$, Figure 2c).

In case of the two-frequency excitation a substantial spreading of the initial distribution occurs (Figure 6), which corresponds qualitatively to the picture of diffusion excitation. However, in this case as well, a qualitative agreement with classics is observed for the probabilities $W_{n>n_0+4}$ only, while the dependence on frequency for $W_{n\geqslant[1.5n_0]}$ is of a marked resonance nature (Figure 7). Since the quantum diffusion limitation for quasiperiodical excitation shows much weaker than for the periodical one [22], the most optimum case for the diffusion mechanism is the one of two-frequency excitation. But in this case as well, at $\varepsilon_0 \approx 0.02$, $\omega_0,\omega_1 \sim 1$ the values $n_0 \approx 66$ are still in the quantum range (M_1 differs greatly from the classical one). The reason for this difference is evidently connected with the fact that even for $n_0 = 66$ the exceeding of the quantum border of stability $\varepsilon_0 >> \varepsilon_q = 1/n_0$ at $\varepsilon_0 = 0.02$ proves rather small. According to the results obtained for simple models [22], it should be expected that when n_0 increases the time of quantum diffusion limitation will grow exponentially. This question, however, needs further investigation.

For one-frequency excitation quantum corrections lead to a substantial difference of distribution characteristics from the classical ones. This result testifies to the fact that the coincidence of the ionization probability of the classical system [6] with the one obtained in the experiment [1] is evidently accidental. In fact computer experiments have shown that for $n_0 = 66$, $\omega_0 = 0.4327$, $\varepsilon_0 = 0.061$ already at $\tau = 3$ quantum moments are 1.7(M_1), 1.8(M_2) times smaller than the classical ones, and therefore the classical description of ionization is not applicable in this case. The probabilities of excitation are also smaller than the classical ones (for τ = 3 the probabilities $W_{n>n_o+4}$ = 0.45, $W_{n\geqslant[1.5n_o]}$ = 0.23 and corresponding quantum values are equal to 0.29, 0.1).

6. SUMMARY

The investigations carried out have shown that the excitation of atom from states with $n \lesssim 66$ by a monochromatic field with the frequency close to the Kepler one ($\omega \approx n^{-3}$) and amplitude $\varepsilon \sim 0.04n^{-4}$ is of a fundamentally quantum character. The reason for this lies in the fact that on the one hand quantum corrections lead to a quantum diffusion limitation [12,13], and on the other hand there may be strong multiphoton transitions in the atom, that result in many cases in a stronger excitation than the classical diffusion. When $\varepsilon_o \lesssim 0.04$, $\omega_o \approx 1$ the probability of excitation out of the resonance may be one order of magnitude smaller than the classical one. So far as the field grows the probabilities become comparable, but differ 2 - 3 times from each other depending on the frequency ω_o. A more efficient atom excitation is observed in the case when the frequency $\omega_o < 1$. So, when ε_o = 0.03, n_o = 45 the decrease of the frequency from $\omega_o \approx 1$ to $\omega_o \approx 0.7$ involves an increase of the probability $W_{n\geqslant[1.5n_o]}$ by 2 orders of magnitude. At an increase of the there is no such effect.

In the case of the two-frequency excitation in the range of stochasticity there occurs a significant spreading of the packet over unperturbed levels. But, in these conditions as well, when up to 20 levels are excited efficiently, the probability of excitation may have a resonant dependence from the frequency. Quantitative characteristics of quantum distribution differ from the classical ones.

Thus, quantum diffusion limitation leads to the localization of the distribution over a relatively small group of unperturbed levels, from which strong multiphoton transitions start later on. The probability of the latter is significant and cannot be explained by available theoretical estimates. The investigation of the dynamics of excitation in the region where the quantum diffusion limitation takes place can be conducted in experiments similar to [1-3]. Conditions of the one-dimensional approximation will be satisfied if the states with strongly different parabolic quantum numbers ($n_1 \gg\ll n_2$) are excited. Modern experiment methods make it possible to perform this with a high accuracy [23].

The author expresses his sincere gratitude to B. V. Chirikov for his attention to this work and valuable comments, and to N. B. Delone, F. M. Izrailev and V. P. Krainov for helpful discussions.

REFERENCES

1. J. E. Bayfield and P. M. Koch, Phys. Rev. Lett., 33:258 (1974).
2. J. E. Bayfield, L. D. Gardner and P. M. Koch, Phys. Rev. Lett., 39:76 (1977).
3. J. E. Bayfield, L. D. Gardner, Y. Z. Gulkok and S. D. Sharma, Phys. Rev., A24:138 (1981).
4. L. V. Keldysh, Zh. Eksp. Teor. Fiz., 47:1945 (1964).
5. N. B. Delone, B. A. Zon and V. P. Krainov, Zh. Eksp. Teor. Fiz., 75:445 (1978).
6. J. G. Leopold and I. C. Percival, Phys. Rev. Lett., 41:944 (1978); J. Phys., B12:709 (1979).
7. B. I. Meerson, E. A. Oks and P. V. Sasorov, Pis'ma Zh. Eksp. Teor. Fiz., 29:79 (1979).
8. B. I. Meerson, Opt. i Spect., 51:582 (1981).
9. B. V. Chirikov, Phys. Reports, 52:263 (1979).
10. D. L. Shepelyansky, Opt. i Spect., 52:1102 (1982).
11. N. B. Delone, V. P. Krainov and D. L. Shepelyansky, Usp. Fiz. Nauk, 140:355 (1983).
12. G. Casati, B. V. Chirikov, J. Ford and F. M. Izrailev, Lecture Notes in Physics, B., Springer, 93:334 (1979).
13. B. V. Chirikov, F. M. Izrailev and D. L. Shepelyansky, Soviet Scientific Reviews, 2C:209 (1981).
14. D. L. Shepelyansky, Dokl. Akad. Nauk SSSR., 256:586 (1981).
15. S. P. Goreslavsky, N. B. Delone and V. P. Krainov, Preprint Phys. Inst. Acad. Sci., N33, Moscow (1982).
16. E. V. Shyryak, Zh. Eksp. Teor. Fiz., 71:2039 (1976).
17. V. S. Edel'man, Usp. Fiz. Nauk, 130:675 (1980).
18. R. V. Jensen, Phys. Rev. Lett., 49:1365 (1982).
19. L. D. Landau and E. M. Lifschitz, Quantum Mechanics, Moscow, "Nauka", p. 750 (1974).
20. D. A. Jones, J. G. Leopold and I. C. Percival, J. Phys. B., 13:31 (1980).
21. N. B. Delone, M. Yu. Ivanov and V. P. Krainov, Preprint Phys. Inst. Acad. Sci., N42, Moscow (1983).
22. D. L. Shepelyansky, Physica, 8D:208 (1983).
23. P. M. Koch and D. R. Mariani, Phys. Rev. Lett., 46:1275 (1981).

QUANTUM CHAOS AND ANDERSON LOCALIZATION

R.E. Prange, D.R. Grempel and Shmuel Fishman*

Dept. of Physics and Center for Theoretical Physica
University of Maryland, College Park, MD 20742
*Dept. of Physics, Israel Institute of Technology, (Technion)
32000 Haifa, Israel

There is no agreed meaning of the term 'Quantum Chaos'. If 'quantum' has to do with wave equations, and 'chaos' means the generation, in a physical context, of 'random' elements by a relatively simple deterministic equation, then 'Quantum Chaos' ought to concern wave equations which are in some aspect chaotic. Here 'random' means one of the hierarchy of possibilities discussed Joe Ford at this conference.

'Anderson localization'[1] is the subject of the review lecture by Bernard Souillard and of much of the lecture of Jean Bellissard at this conference. It concerns the properties of the Schroedinger equation in the presence of an spatially ergodic potential. Such a potential takes on all possible values from some distribution over and over throughout space. The three most prominent examples are the <u>periodic</u> potential, the <u>incommensurate,</u> or almost periodic potential, and the <u>random</u> potential. [This is not included under the term 'Quantum Chaos' by the above definition since any randomness is externally imposed.] The remarkable fact is, that always in one (and probably also two) dimensions, and sometimes in three dimensions, the wavefunctions of the random potential are <u>localized,</u> whereas the wavefunctions of the periodic potential are <u>extended.</u> The incommensurate case is intermediate, and is sometimes localized, sometimes extended. This kind of localization, which denies the possibility of ordinary conductivity and diffusion, is called <u>Anderson localization</u> and is not at all obvious.

In this lecture we discuss a way in which the two concepts can be connected. The idea is very simple, namely, one adds to the list of potentials giving Anderson localization, the 'chaotic' potential, i.e. a potential arising from a relatively simple deterministic construction, but which selfgenerates sufficient 'randomness' that Anderson localisation results. We shall show that in one of the best studied examples, the periodically kicked rotator, this is the case.[2]

We first review briefly some of the heuristic ideas of Anderson localization, relying on Souillard's lecture for the rigorous results. (It is unnecessary to review 'chaos'.) Next we show the connection between the physics of the electron in a random lattice and that of the rotator. After that, we discuss an important mathematical worry about the claim that any particular 'chaotic' potential gives Anderson localization, just as a truly random one does. The possibility is that the states may not be truly localized, (with a corresponding pure point spectrum), but that the spectrum may be 'singular continuous', and the states 'singularly' or 'sparsely' extended. We will argue that it would be surprising if this case can be distinguished from true localization in real systems. Next, we shall give some arguments, especially one based on the renormalization group method, supporting the idea that the results of this model have some generality. Finally, we discuss quantum chaos more generally.

Considering only one dimension, we recall an argument for localization. Think of a random potential consisting of bumps separated by stretches of zero potential, as in Fig. (1). An intuitive semiclassical-semiquantum picture goes as follows. Consider an 'electron' in the n'th zero stretch which impinges on the bump separating it from the n+1'th stretch. It has a certain probability $|t_n|^2$ of making it to the n+1'th stretch, and the probability $|r_n|^2 = 1 - |t_n|^2$ of being reflected, after which it tries to make it to the n-1'th stretch. This, of course, is perfectly good quantum mechanics. Obviously, the process can be continued, and if the bumps are random, there is a random walk, and the electron thus diffuses away from its initial site.

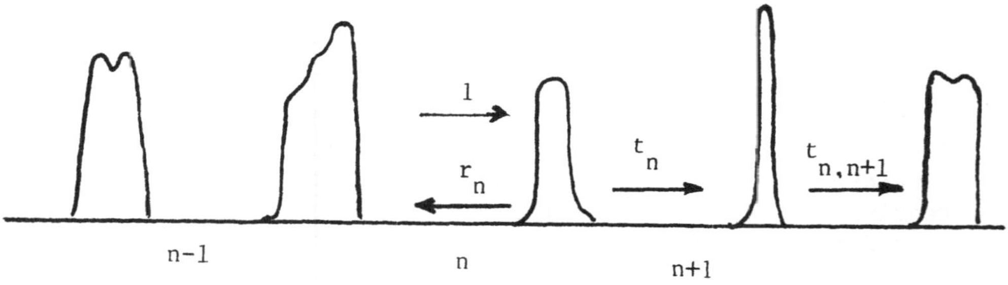

Figure 1

Until Anderson's work in 1958, this kind of argument apparently was convincing. Some classical thinking has been tacitly injected, however, namely it was assumed that the wavefunction 'collapses' after the electron encounters a bump and can be said to be 'either' transmitted 'or' reflected.

To treat the problem without such an assumption, consider the tunnelling probability for transmission between two successive bumps. It is easy to show that this is given by

$$\frac{1}{|t_{n,n+1}|^2} = \frac{1}{|t_n t_{n+1}|^2}\left[1 + |r_n r_{n+1}|^2 + 2\ \mathrm{Re}\ r_n{}^* r_{n+1}\frac{t_n}{t_n{}^*}\right] \qquad (1)$$

If the phases of the t's and r's are random, it is plausible that the cross term in (1) averages out, and thus that $|t_{n,n+1}|^2 < |t_n t_{n+1}|^2$, in some mean sense. This in turn leads to the conclusion that the transmission through a large number of random sites is exponentially small.

Now switch models. Anderson considered the tight binding Schroedinger equation for the wavefunction u_m at site m, and with energy E,

$$T_m u_m + \sum W_r u_{m+r} = E u_m \qquad (2)$$

The hopping coefficient W_r is usually taken to vanish unless $|r|=1$. For this equation there is one eigenstate per site. When $T_m = T_{m+q}$ the states are extended Bloch states. When T_m is random, i.e., chosen independently from some distribution w(T), the states are exponentially localized, and can be labelled by the site ν at which they are centered, that is, $u^\nu{}_m \sim e^{-\gamma|\nu-m|}$ for $|\nu-m|$ large. The localization length, $1/\gamma$, is a function of energy and the other parameters, and is known theoretically in one special case, that of the Lloyd model, [defined by taking $w(T) = 1/\pi(T^2+1)$, a Cauchy distribution]. The eigenstates have more or less random bumps and wriggles around their exponential envelope, but each state is in a statistical sense very similar to all the others: only their centers are systematically different. See Fig. (2) for a picture of such a state.

Other results are that the energy levels E_ν are random, i.e., they have a Poisson distribution, and there is no Wigner level repulsion. For an infinite system the levels form a dense set, so that the density of states is a smooth function. This is easy to understand, as even though there is no symmetry reason to have degeneracy, level crossings between wavefunctions with well separated centers will be scarcely 'avoided'. This is just a consequence of the point made repeatedly by Taylor in this meeting, that matrix elements can be small because the wavefunctions do not overlap, even if there is no small parameter in the Hamiltonian.

This result, by the way, says something interesting about random matrix theory. The Hamiltonian (2) may be regarded as a random matrix of a special form, namely, it has large elements only near the diagonal. The distribution of eigenvalues of large random matrices with this restriction is Poisson. If there is no restriction on the off diagonal elements, the distribution will be that of Wigner. Such a matrix corresponds to a tight binding model with long range hopping, and it is known that such hopping gives extended rather than localised states and thus a continuous instead of a point spectrum.

The local density of states is an important quantity to consider, and can serve to distinguish the two cases, as emphasised by Heller. It is defined by

$$\rho_m(E) = \sum |u^{\nu}_m|^2 \delta(E-E_\nu) \tag{3}$$

and in the case of localization will be discrete, having big contributions at only about $1/\gamma$ energies. These energies, of necessity belonging to nearby sites, of course 'repel', presumably with roughly the Wigner law. The time dependence of wavepackets, given by the Fourier transform of (3) or related objects, consists of an essentially finite Fourier sum of frequencies incommensurate with respect to one another. This is thus an almost periodic dependence in time, and implies recurrences. Recurrences per se are perhaps not so important, since the recurrence time can be unrealistically long, but certainly this means that a wavepacket does not diffuse, as in the semiclassical approximation. If a electron is released at site zero, it will approximately diffuse (if $\gamma \ll 1$) until it has spread out a localization length, or equivalently, until a time has elapsed which is long enough by the uncertainty principle to resolve neighboring energy levels. Only if $\gamma = 0$, and thus the spectrum is continuous, can true diffusion take place for very long times.

Now for the germ of a better argument for localization. Rewrite (1) in matrix form, ($W_{\pm 1} = W$, all other W's = 0.)

$$\begin{pmatrix} u_{m+1} \\ u_m \end{pmatrix} = \begin{pmatrix} (E - T_m)/W & -1 \\ 1 & 0 \end{pmatrix} \begin{pmatrix} u_m \\ u_{m-1} \end{pmatrix} = M_m \begin{pmatrix} u_m \\ u_{m-1} \end{pmatrix} \tag{4}$$

Note that det $M_n = 1$, and thus the eigenvalues of M are of the form $e^{\pm\kappa}$ where κ can be real or pure imaginary. To find the wavefunction far to the right (or left) of two points, say 0,-1, where it is assumed to be known, one multiplies the matrices. Consider such a product. Then it is a theorem (Furstenburg) that for random T_m

$$\lim_{Q\to\infty} (1/Q) \ln \mathrm{Tr}\, M_Q M_{Q-1} \ldots M_2 M_1 = \gamma > 0. \tag{5}$$

Here γ is the Lyapunov exponent and is identical to the inverse localization length. Eq. (5) says that the unimodular matrix which is the product of Q unimodular random matrices will have eigenvalues $e^{\pm\gamma Q}$ if Q is large, and thus that almost any initial starting vector u_0, u_{-1} will give wavefunctions which grow exponentially both to the right and the left. One precise choice of starting vector will give exponential decay to the left, but only for a particular energy can the state be made to decay exponentially in both directions. Actually, this argument is not quite complete, and a positive Lyapunov exponent only implies that the spectrum is singular, that is, lies on a set of measure zero. Thus, ordinary extended states and continuous spectra are excluded, but singular continuous spectra and the corresponding singularly extended states are not. It takes more work to eliminate the latter exotic possibility, and indeed, it cannot always be eliminated. We will have more to say on this matter later.

We were aware of some of this lore of Anderson localization and were struck by the similarity to the behaviour found numerically by Casati, Ford, Chirikov, et. al.[3] in the case of the periodically kicked quantum rotator when we heard it described in a lecture of Ed Ott. This model has the Hamiltonian

$$H = p^2/2I + V(\theta)\sum\delta(t-n) \tag{6}$$

where the kicks are at integer time and usually $V(\theta)$ is taken to be $k \cos \Theta$. Letting p_n, θ_n be the coordinates just after the n'th kick, the classical trajectory is given by

$$\begin{aligned} p_{n+1} &= p_n + k \sin \theta_{n+1} \\ \theta_{n+1} &= \theta_n + K\, p_n/k \end{aligned} \tag{7}$$

where $K = k/I$. If K is large enough, and p_n is not too special, then θ increases significantly between kicks, with the increase having no particular relation to 2π. Thus $\sin \theta_n$ is effectively random, and p_n undergoes a random walk. The mean energy then increases according to $\langle p_n^2 \rangle \sim nk^2/2$. This is chaotic but a more precise concept is developed by the study of the tangent map of (7), giving the evolution of neighboring phase space points;

$$\begin{pmatrix} \delta p_{n+1} \\ \delta\theta_{n+1} \end{pmatrix} = \begin{pmatrix} 1+K\cos \theta_{n+1} & k \cos \Theta_{n+1} \\ K/k & 1 \end{pmatrix} \begin{pmatrix} \delta p_n \\ \delta\theta_n \end{pmatrix} = M_n \begin{pmatrix} \delta p_n \\ \delta\theta_n \end{pmatrix} \tag{8}$$

If $\cos \theta_n$ is random, Furstenburg's theorem again applies, and there is a positive Lyapunov exponent, implying that nearby points in phase space diverge exponentially with increasing time. Even if $\cos \theta_n$ is not strictly random, the exponent may be positive, implying chaos. This is usually established by numerical computation.

When Casati, et. al. numerically solved the quantum rotator, however, there was diffusion for a limited time, but the energy saturated and oscillations were observed. Later, Hogg and Huberman,[4] with different parameters, found many oscillations in the energy and came to the conclusion that the motion was almost periodic, and thus the spectrum pure point.

Seeing that this behavior in angular momentum is just that for an electron moving from site to site in a random lattice, we set out to find a formal relation between the quantum rotator and some lattice model, with angular momentum m corresponding to site. This we were able to do by using Berry's[5] concept of quantum map. In the case of (6), with $p = -i\hbar\partial/\partial\theta$ (= m) the time evolution of a wavefunction ψ from just before one kick to just before the next is given by the 'quantum map',

$$\psi^-(\theta,n+1) = e^{-i\tau p^2}e^{-iV(\Theta)}\psi^-(\theta,n) \tag{9}$$

Here τ is given by $\hbar/2I$, (and we then set $\hbar = 1$). The first exponential gives the time evolution of the free rotator between kicks, the second gets it across the kick. It is convenient to utilise the periodicity in time of (6), and write $\psi = e^{-i\omega t}u(\theta,t)$, where $u(t)=u(t+1)$. Since we consider u only for times just before the kicks, we may suppress the t dependence, and write u^-, the $-$ meaning times just before kicks. The eigenvalue equation for ω is

$$u^-(\theta) = e^{iK(p)}e^{-iV(\Theta)}u^-(\theta) \tag{10}$$

Here $K(p) = \omega-\tau p^2$. Note that the following procedure works for any function of K(p), e.g. a truly random one. Defining W, T by $e^{-iV} = (1+iW)/(1-iW)$, and $e^{iK} = (1+iT)/(1-iT)$, letting $u^- = (1-iW)u$, and multiplying (10) through from the left by $1-iT$ it is found

$$(T + W)u = 0 \tag{11}$$

In its Fourier representation, with $E = -W_o$, this becomes

$$T_m u_m + \sum W_r u_{m+r} = Eu_m \tag{12}$$

where $W(\theta) = \tan(V(\theta)/2)$, and $T_m = \tan(K(m)/2)$. For the rotator model $T_m = \tan((\omega-\tau m^2)/2)$. Note that ω and E have switched roles, from being an eigenvalue to being a parameter of the potential, and vice versa. This however, causes no problems.

This T_m does not fall into any of the catagories, periodic, random or almost periodic, generally discussed. However, we claim that the sequence T_m is pseudorandom (it is somewhat more random than ergodic) and thus chaotic and that the Lyapunov exponent for this case is positive with the same value that it would have for T_m random. In fact, the randomness comes from reducing the argument of the tangent to the interval $(-\pi/2,\pi/2)$, the nonlinear modular arithmetic process so common in random number generators. The distribution of the T_m is Cauchy therefore, and we may thus term this case the pseudoLloyd model, for which the localization length is known theoretically.

The statistics of the correlation functions of T_m look like those for truly (machine generated) random numbers. Further, numerical solution of the model gives localized states which have the theoretically correct exponential decay. Fig. (2) shows such a state. The states in angular momentum space for the rotator have essentially the same structure, and thus we understand why the behavior of the two systems are so similar. To state it again, the time evolution of an initially localized wavepacket is found by expanding this wavepacket in eigenstates. In this case, the eigenstates which enter are all localized with centers no more than a localization length away from the initial wavepacket. Clearly, no matter how long one waits, the system will never go very far from its starting point.

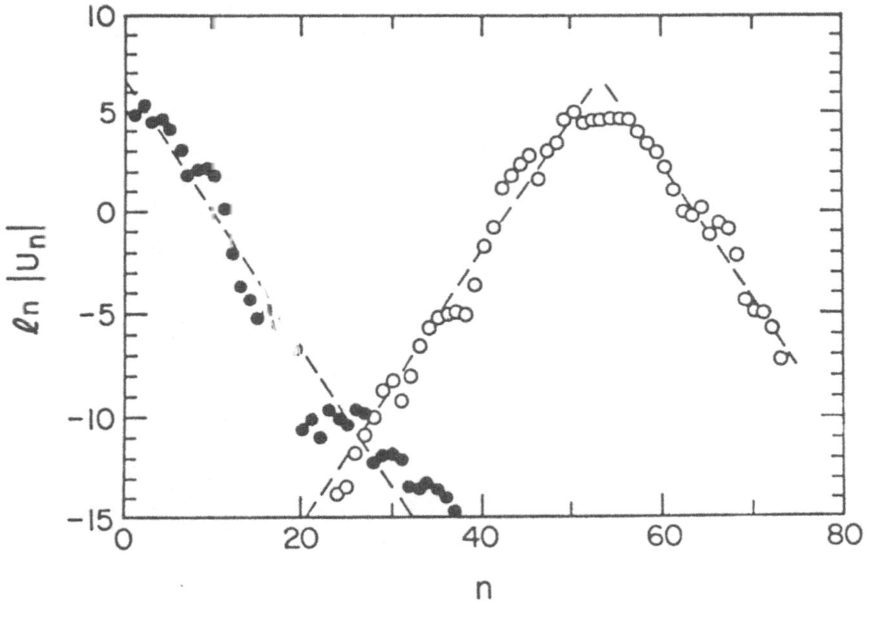

Figure 2.

Two exponentially localized wavefunctions of the pseudo Lloyd model. The dashed lines gives the theoretical exponentially decaying envelope.

To summarise, the rotator in its classical limit was chaotic in its time evolution, this being guaranteed by the positivity of the Lyapunov exponent. The quantum version has an almost periodic time evolution, a much different consequence of a different Lyapunov exponent being positive. However, the energy levels of the quantum version are presumably chaotic, so the chaos has appeared in an entirely different form.

We now turn to the provocative case of the singular continuous spectrum. This unfamiliar kind of spectrum is intermediate between the absolute continuous and pure point cases, and has never been observed in a physical system. These comments are based on what we learned from an exact solution of the rotator-lattice model for the case $K(p) = \omega - \tau p$, but will be phrased here in a more general way.[6] We begin by asking a self-consistency question, namely, why, if one random potential localizes states, some other, equally good, random potential cannot start to delocalize them. Consider two localized states with more or less the same energies whose centers are a long distance L apart in a particular random potential. If we change some site energy T_m systematically, we can cause these levels to cross, except of course the crossing will be 'avoided'. The gap at the crossing will be of order $e^{-\gamma L} \ll 1$. Right at the crossing, the eigenfunctions will consist of two superpositions of the localized functions with each localized function having nearly equal weight. Thus, at the crossing, the eigenfunctions will not be large only over the distance $1/\gamma$ but will be extended in this funny way, a distance L. We call this a <u>sparse</u> extension, since the functions are practically zero for most of the region, except for two bumps a distance L apart. Clearly, by fiddling with more site energies three or more levels could be made to cross simultaneously. Such sparse extension obviously entails a near degeneracy of energies, since it cannot cost much energy to change the relative phase of the bumps.

How probable is it that a random choice of T_m will be just such as to result in sparse extension? Clearly the probability is of order $e^{-\gamma L}$. There are about $4/\gamma$ sites m which appreciably affect the energy of one or the other of the localized states, so, for large L, the probability is very small that a random choice of T's will result in the sparsely extended eigenstates. There may be an occasional pair of states which have such sparse extension to some modest distance L. To change the nature of the spectrum, however, either states with arbitrarily large L must mix, or the T's must be just right so that an infinite number of states have energies lying in a narrow range, both of which have zero probability in the random system.

However, if the T_m are not strictly random, but have some regularity, say they are periodic (or chaotic), then it cannot be easily excluded that there is no systematic near degeneracy of the wavefunctions at long distances, such that the eigenstate dies off at first, as if it were exponentially localized, but rises again and again like a phoenix from the ashes, farther and farther away. If these spatial recurrences are fairly regular, there will be a continuous spectrum, with a band width $e^{-\gamma L}$, where L is the distance between recurrences. In the singular continuous case, the wavefunction has its successive peaks at wider and wider separations, as one goes away from the 'center' of the function. The separations typically increase exponentially or even faster between successive peaks. In either case, an infinite number of repetitions are made, and the eigenstates are not normalisable. The spectrum in the neighborhood of the energy of such a set of sparsely extended states will be an infinite multiplet with splittings of order $e^{-\gamma L_1}$, $e^{-\gamma L_2}$, where L_1, L_2, . . are the successive scales of recurrence which increase rapidly, e.g. $L_n \propto \mu^n$. Clearly, it would take excellent resolution, little dissipation, low temperature, and a large nearly perfect sample to establish, even approximately, that the system had singular continuous states and spectra.

It could even be difficult to observe a sparse extension at the first level if $e^{-\gamma L_1}$ is small enough. The numerical resolution of a computer may not suffice to find such states either. [We did not find any such states in the rotator model, which argues that they do not exist, or that $e^{-\gamma L_1}$ is very small (less than e^{-30})].

One could in principle observe sparse extension by following the motion of an initially localized wavepacket. It first behaves like a particle in a regularly localized situation, but after times of order $e^{\gamma L_1}$, begins to oscillate regularly (with period about $e^{\gamma L_1}$) like an ammonia molecule, between two regions a distance L_1 apart, and with no important probability of finding the particle between the two regions. Then after a time of order $e^{\gamma L_2}$, the oscillations between the first two sites and a third site or sites a distance L_2 from the first ones become apparent, and so on. If an ensemble of many particles are observed, we guess that there might be some sort of apparent anomalous diffusion but this has yet to be worked out in detail.

So, for practical purposes, we think it is usually a good approximation to neglect the singularly extended states, and to approximate the states as localized in the ordinary sense. However, it goes without saying that it would be very exciting to discover a system where the singularly extended case is nicely observable. A further motivation for study of this case is that there is much important and interesting mathematics to be straightened out.

The result found for this model is that the long time behavior of a quantum system, even one whose energy levels are chaotic or random, is different from that of chaotic classical systems, and is simpler and less random. While this result may not be totally generic, it is probably not uncommon. Indeed, it is well known that at long times, the quantum nature of things tends to become important, but we will set down two arguments that give some insight. Both are arguments based on corrections to the eikonal approximation.

In functional integral language, the time evolution of a quantum system is given by Feynman's sum over paths of $e^{iS/\hbar}$, where S is the action, the integral along some path of the Lagrangian, S $= \int L dt$. Most of the sum is contributed by paths in the neighborhood of the one making S stationary and this extremal path is the one followed in classical mechanics. The quantum corrections consist in smearing the path out over an area h. In elementary cases the classical path is smooth, and the corrections cause a additional uncertainty at large times, corresponding to the spread of the wavepacket. A classically chaotic path, however, becomes very convoluted after a long time and smearing out results in simplification rather than more complexity. So the quantum corrections have opposite effects in the two cases.

We next give a conjectural renormalization group argument. Rewrite and iterate Eq. (9) to obtain $\psi(n+N)=U_N\psi(n)$ where $U_N=[e^{-i\tau p^2}e^{-iV}]^N$. Clearly, $U_N = [U_1]^N = [U_2]^{N/2}$. It is a standard renormalization group decimation procedure to try to write U_2 in the same form as U_1 is written, that is, $U_i = f(\lambda_i)$, where i is a power of 2, and λ_i is a set of parameters. Recursion relations, connecting the λ's for successive i's are then sought and studied for many iterations. This procedure was outlined by Mitch Feigenbaum.

In the present case, in the real space representation, the matrix U_1 may be written $U_1(x,y) = \exp[iA_1(x,y)/\hbar]$, where A_1 is the classical action $(x-y)^2/2 + V(x)$. Then

$$U_2(x,y) = \int dz U_1(x,z)U_1(z,y). \tag{13}$$

[The product for U_N can be regarded as a multiple integral. Carrying out the integration of every other variable 'decimates' this integral and results in the expression of U_N in terms of U_2]. We consider the case $h \to 0$. The external point z^*, (assumed unique), will dominate Eq. (13), where $\partial[A_1(x,z^*)+A_1(z^*,y)]/\partial z^* = 0$ and $A_2(x,y)=A_1(x,z^*)+A_1(z^*,y)$. This is the classical path, and has been studied for the rotator by Kadanoff. At some critical parameter in the potential, it can happen that the classical path scales. This means that at this critical value $A_2(x,y) = \alpha\beta[A_1(\alpha x,z^*)+A_1(z^*,\alpha y)]$, where α,β are scaling parameters. Keeping track of the parameter h in U we have, by iteration $U_{2^N}(x,y,\hbar) = U_1(x/\alpha^N,y/\alpha^N,\hbar(\alpha\beta)^N)$. Replacing $2^N = t$, the time, we find

$$U(x,y,\hbar,t) = U(xt^\nu,yt^\nu,\hbar t^\sigma,1) \tag{14}$$

a <u>scaling relation.</u> In known cases, $\sigma > 0$, so as time goes on, the effective value of h grows. In renormalization group lingo, $\hbar$ is a relevant variable to the time evolution. To justify the approximations, we have had to assume that $\hbar$ is small, $\hbar \ll S_0$, where S_0 is some characteristic constant action. Then, at time t* ~ $(\hbar/S_0)^{1/\sigma}$, the classical approximation cannot be correct, the classical scaling will break down, and the finiteness of $\hbar$ will become apparent. The similarity of this result to that of Feigenbaum should be noted. In his case, dissipation was found to be 'relevant', and limited the time for which the classical analysis was valid.

We finally make some more general remarks on the connection between localisztion and quantum chaos. First, we think it likely that if mathematically precise definitions are to be made of quantum chaos, it will be necessary to invoke some limiting process, just as in statistical mechanics, where phase transitions can only be sharply defined in the thermodynamic, infinite volume limit. Of course, at large but finite volumes, we can usually tell pretty well how the limit will turn out. So chaos in a quantum system, regarded as a question of the long time behavior of the system, will not be well defined unless something like an infinite volume limit is taken. In a driven system, as we have seen, the infinite range of possible values of the energy can serve to distinguish sharply between chaos and the alternatives.

Time evolution in quantum theory is closely related to the energy level structure, so it is appropriate to ask about the spectrum, and the type of spectrum in turn is correlated with the localization or extension of the eigenstates. The spectrum, for a spatially bounded time independent system, must be pure point, since the states are localized, being bounded. Only in the infinite volume limit, the spectrum can be continuous and the states extended. It is necessary, in order to have chaos in the time development, that the spectrum be absolutely continuous, for if not, the motion is almost periodic. [We leave aside the possibility of singular continuous spectra.] This means that in a large but finite system, the wavefunctions must be extended over regions with the scale of the whole system, if the spectrum is to become continuous in the infinite size limit.

The large volume limit may often be replaced by some other limit, for example, the limit of small $\hbar$, or high energy, and in fact, the limits may well be identical. For example, in the quantum stadium or Sinai billiard problem, aside from the ratios of lengths characterising the geometry, the only dimensionless parameter is $\varepsilon = m\ell^2 E/\hbar^2$, with m the mass, and ℓ the typical dimension of the system. For $\varepsilon \to \infty$ chaos vs. non-chaos is precisely defined. Achieving this limit by thinking of ℓ large for fixed h appeals to the intuition of the solid state physicist, for whom the problem becomes that of (2-d) Sommerfeld electrons in a sample with a funny boundary. The electronic states are obviously extended, the density of states is independent of the shape of the boundary, and this is often all that one needs to know.

This example makes it clear that for the stronger forms of chaos, the continuity of the spectrum and the extension of the wavefunctions is necessary but not sufficient. Some more global characteristcs at long lengths or times must be considered. It is interesting to try to formulate these characteristics in terms of quantities natural to quantum calculations, made at finite but large ε. An appropriate function to study is the local density of states, since the structure of this function is quite different for localized, and (approximately) extended states. The more global characteristics can perhaps be translated into finer details of the spectrum. An obvious guess is that, at finite size (or $\hbar$), the spectrum must show the Wigner repulsion, if fully developed chaos is to appear in the infinite volume limit. This condition again is probably necessary, but is not likely to be sufficient, except perhaps at the level of ergodicity.

When will Anderson localization be important? In driven systems, and in large systems with complicated potentials, as we have seen. However, in time independent systems of interest to chemists, corresponding to small molecules, there is not likely to be Anderson localization. At high energies, the potential looks smooth on the scale of the wavelength, and does not have many bumps and wiggles globally. Then the states are extended, and there thus may be approximate chaos. At low energy, this concept is inappropriate. In between, there is a broad region where clean distinctions are not possible. We doubt that there is a quantum KAM theory for models like the Henon-Heiles in which some precise and qualitative distinction corresponding to the onset of chaos appears at finite critical values.

ACKNOWLEDGEMENTS

This work was supported in part by NSF Grants DMR 79001172-A02, DMR7908819 and the Center for Theoretical Physics, University of Maryland. S.F. acknowledges support by the Lawrence Deutsch Research Fund. A portion of the work was carried out at the Aspen Center for Physics.

REFERENCES

1. P.W. Anderson, Phys. Rev. 109, 1492 (1958), Rev. Mod. Phys. 50, 191 (1978). For a review see e.g., D.J. Thouless, in Ill-Condensed Matter, Proceedings of Les-Houches Summer School, edited by R. Balian, R. Maynard, and G. Toulouse (North Holland, Amsterdam, 1979).
2. S. Fishman, D.R. Grempel and R.E. Prange, Phys. Rev. Lett. 49, 509 (1982).
3. B.V. Chirikov, F.M. Izrailev and D.L. Shepelyanskii, preprint; G.G. Casati, B.V. Chirikov, F.M. Izraelev, and J.Ford, in Stochastic Behavior in Classical and Quantum Hamiltonian Systems, edited by G. Casati and J. Ford, Lecture notes in Physics Vol. 93 (Springer, Berlin 1979).
4. T. Hogg and B.A. Huberman, Phys. Rev. Lett. 48, 711 (1982) and to be published.
5. M.V. Berry, N.L. Balazs, M. Tabor and A. Voros. Ann. of Phys. 122, 26 (1979).
6. R.E. Prange, D.R. Grempel, and S. Fishman, (to be published).

TRANSITION OF A COHERENT CLASSICAL WAVE TO PHASE INCOHERENCE

Allan N. Kaufman, Steven W. McDonald, and
Eliezer Rosengaus

Lawrence Berkeley Laboratory and Physics Department,
University of California, Berkeley, California 94720

ABSTRACT

A coherent wave may be characterized by a single-valued phase function. As the wave propagates, its rays twist and separate, causing its Lagrangian manifold $\underline{k}(\underline{x})$ to develop pleats. Thereby the phase becomes multivalued, and the wave may be termed incoherent. This process is analyzed by studying the local spectral density, which changes from a line spectrum to a continuous spectrum.

The concept of chaos can be applied to classical waves as well as to the quantum solutions of the Schrödinger equation. In this paper, the term "chaos" will refer to the degree of spatial incoherence of the phase of a linear oscillation of a nonuniform, but nonrandom, classical medium.

The motivation for our study arose in the problem of the controlled heating of magnetically confined plasma by electromagnetic radiation. Although the radiation incident on the plasma is typically coherent (from an antenna structure or a wave guide), in the course of propagation through the nonuniform plasma the wave phase may become multivalued, as we show below. The resulting incoherence has important consequences for the deposition of the radiant energy, i.e., the heating of the plasma. A coherent wave typically traps particles, while an incoherent wave causes diffusion. Nonlinear effects (usually undesirable) have a higher threshold for incoherent waves.

The ideas presented here can of course be applied to other classical media, and (with reinterpretation) to the quantum problem. To a large extent these concepts have been developed previously by M. Berry and co-workers [1]. Our contribution to the study was an outgrowth of the thesis research of S. W. McDonald [2], where several alternative descriptions were explored and developed.

In this presentation, we shall consider the field as a scalar $\phi(x)$ satisfying a linear homogeneous integral equation:

$$\int d^4x' \; D(x,x')\phi(x') = 0 \qquad (1)$$

This is a simplification of the more correct description [2], wherein the kernel D may depend implicitly on ϕ, and where (1) has a (small) inhomogeneous term, representing sources of the field. Further, the scalar $\phi(x)$ should be replaced by the vector electric field, while the kernel becomes a tensor [2].

In (1), x represents position in space-time, while D is (essentially) the two-point linear response function for the field, which is obtained from the underlying dynamics of the system. The description of the latter may be Hamiltonian (e.g., Vlasov or Klimontovich in the plasma case) or dissipative (e.g. a Fokker-Planck kinetic equation for the particle distribution). For present purposes, it is considered known. Because of causality, D is not symmetric in its arguments $(\underline{x},t)$, $(\underline{x}',t')$. However, we may often neglect the antisymmetric part to lowest order, reconsidering it later (not here) as a perturbation.

Thus, if we replace D by its symmetric part (from this point on), the integral equation (1) for $\phi(x)$ is equivalent to the variational principle $\delta S(\phi) = 0$, where

$$S(\phi) = \int d^4x \int d^4x' \; D(x,x')\phi(x')\phi(x). \tag{2}$$

It is convenient to introduce the two-point correlation for the field

$$\phi^2(x',x) = \phi(x')\phi(x); \tag{3}$$

an averaging may be introduced when needed. Thus the action S attains the form of the trace of an inner product:

$$S(\phi) = \int d^4x \int d^4x' \; D(x,x')\phi^2(x',x). \tag{4}$$

For each of the two-point functions, D and ϕ^2, we introduce the local Fourier transforms:

$$D(k,x) = \int d^4s \; \exp(-ik\cdot s) \; D(x+s/2,x-s/2), \tag{5}$$

and obtain (ignoring 2π factors)

$$S(\phi) = \int d^4x \int d^4k \; D(k,x)\phi^2(k,x). \tag{6}$$

(Note that the new functions are denoted by the same symbols as the old; their arguments indicate the representation.)

We now consider fields $\phi(x)$ expressible in eikonal form:

$$\phi(x) = \tilde{\phi}(x)\exp i\theta(x) + c.c, \tag{7}$$

where the amplitude $\tilde{\phi}$ and the gradient of the phase θ are slowly varying. More explicitly, with $x=(\underline{x},t)$ $k=(\underline{k},\omega)$, we may define the local wave-vector and frequency:

$$\underline{k}(\underline{x},t) = \nabla\theta(\underline{x},t); \quad \omega(\underline{x},t) = -\partial\theta/\partial t. \tag{8}$$

The expression (7) is appropriate for a coherent wave. We desire evolution equations for its amplitude and phase.

We use the "slowly-varying" assumption to evaluate the Wigner function (analog of (5)):

$$\phi^2(k,x) = |\tilde{\phi}|^2(\underline{x},t)\ \delta^3(\underline{k} - \nabla\theta(\underline{x},t))\ \delta(\omega + \partial\theta/\partial t). \tag{9}$$

In this approximation, ϕ^2 is singularly concentrated on a "Lagrangian manifold," a 4-dimensional surface embedded in the 8-dimensional phase space $(k,x) = (\underline{k},\omega;\ \underline{x},t)$. Substitution of (9) into (6) now yields

$$S(\tilde{\phi},\theta) = \int dt \int d^3x\ D(\underline{k} = \nabla\theta,\ \omega = -\partial\theta/\partial t;\ \underline{x},t)\ |\tilde{\phi}|^2(\underline{x},t) \tag{10}$$

This has the form of a Lagrangian variational principle. This approach has previously been utilized by Whitham [3] and by Dewar [4].

The Euler-Lagrange equation for the variation of S with respect to the amplitude yields the Hamilton-Jacobi equation for the phase θ:

$$D(\underline{k} = \nabla\theta,\ \omega = -\partial\theta/\partial t;\ \underline{x},t) = 0. \tag{11}$$

The field $J(\underline{x},t)$ conjugate to the phase $\theta(\underline{x},t)$ is defined in the canonical way:

$$J(\underline{x},t) = \frac{\partial D}{\partial\omega}\ |\tilde{\phi}|^2(\underline{x},t), \tag{12}$$

in terms of the Lagrangian density. Variation of S with respect to $\tilde{\phi}$ yields the standard amplitude-transport equation:

$$\partial J/\partial t + \nabla\ (J\ \partial\omega/\partial\underline{k}) = 0. \tag{13}$$

The Poisson structure based on the conjugate fields θ, J has been explored elsewhere [5]. Here we concentrate on the evolution of only the phase θ.

The standard method of solution of (11) is due to Hamilton.

The ray equations:

$$d\underline{x}/d\tau = -\partial D/\partial \underline{k}, \qquad d\underline{k}/d\tau = +\partial D/\partial \underline{x},$$

$$dt/d\tau = +\partial D/\partial \omega, \qquad d\omega/d\tau = -\partial D/\partial t, \tag{14}$$

are to be solved, subject to initial conditions, as discussed further below. The solution is expressed as $\underline{k}(\underline{x},t)$, $\omega(\underline{x},t)$; the phase is then

$$\theta(\underline{x},t) - \theta(\underline{x}_0,t_0) = \int(\underline{k} \ \ d\underline{x} - \omega \ dt). \tag{15}$$

The path of integration is arbitrary, since $k(x)$ is curl-free, i.e., $k = d\theta(x)$ is an exact one-form.

For purposes of illustration of the boundary-value problem, we shall consider a monochromatic wave in a time-independent two-dimensional medium:

$$\phi(\underline{x},t) = e^{-i\omega_0 t} \, \tilde{\phi}(x,y) \, e^{i\theta(x,y)} + c.c; \tag{16a}$$

$$k_x = \partial\theta/\partial x, \qquad k_y = \partial\theta/\partial y; \tag{16b}$$

$$D(k_x,k_y,\omega_0;\ x,y) = 0. \tag{16c}$$

The phase space is 4-dimensional, but we need portray only the three-dimensional (x,y,k_x)-space, since k_y is determined by the dispersion relation (16c). Let the phase be specified on some spatial curve, say $y = 0$; knowing $\theta(x,y=0)$ determines $k_x=\partial\theta/\partial x$ and k_y (from (16c)) on the "boundary" $y=0$. Consider the single-valued curve k_x vs x on the surface $y=0$ (Fig. 1); from each point of that curve, construct the corresponding ray, whose initial conditions are known. The family of rays emanating from the curve generate a smooth surface (the Lagrangian manifold), which represents the desired solution $\underline{k}(\underline{x})$. However, the generic behavior of rays of Hamiltonians with two (or more) degrees of freedom is either twisting about each other ("stable") or exponential separation ("unstable"). In either case, the surface develops pleats, causing the wave-vector field $\underline{k}(\underline{x})$ to become multivalued.

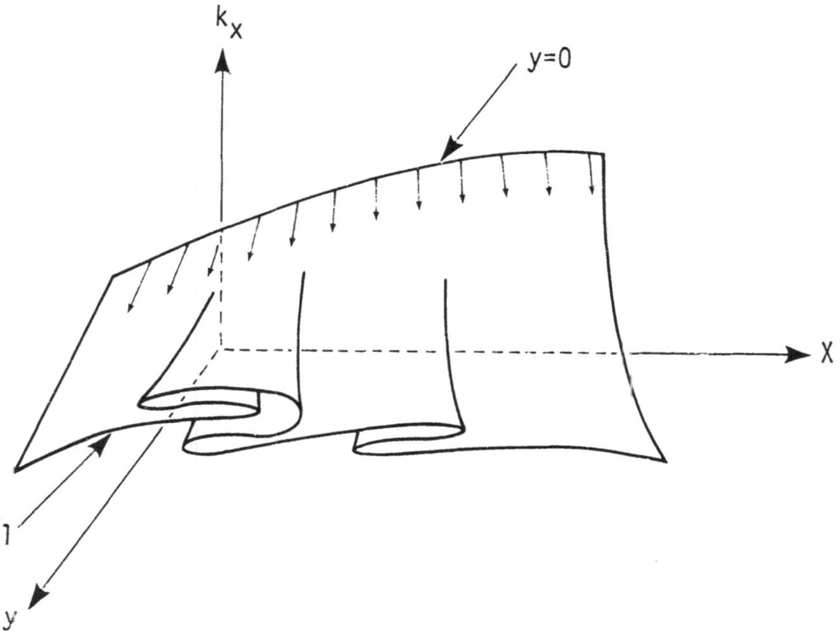

Figure 1

When this occurs, we generalize the eikonal representation (16a) to a sum over the several phases at (x,y):

$$\phi(\underline{x},t) = e^{-i\omega_0 t} \sum_j \phi_j(x,y)e^{i\theta_j(x,y)} + c.c. \qquad (17)$$

We wish to study the local spectral density, going beyond the singular approximation (9). Limiting ourselves (again for simplicity) to a spatial field $\phi(\underline{x})$, not necessarily eikonal, we introduce its local Fourier transform:

$$\phi(\underline{k},\underline{x}) = \int_{\underline{s}} e^{-i\underline{k}\cdot\underline{s}}\, \phi(\underline{x}+\underline{s})\; w(\underline{s}), \qquad (18)$$

utilizing a window function:

$$w(\underline{s}) = \exp - \underline{s}\,\underline{s} : \underset{\sim}{\sigma}^{-2}/2 \quad , \qquad (19)$$

with $\underset{\sim}{\sigma}$ chosen for convenience. We define the local spectral density:

$$I(\underline{k},\underline{x}) = |\phi(\underline{k},\underline{x})|^2; \qquad (20)$$

it is the coarse-grained Wigner function, with Gaussian

averaging of $\phi^2(\underline{k},\underline{x})$ in both $\underline{k}$ and $\underline{x}$. Its Fourier transform:

$$C(\underline{s},\underline{x}) = \int_{\underline{k}} e^{i\underline{k}\cdot\underline{s}}\, I(\underline{k},\underline{x}) \qquad (21)$$

is the coarse-grained field correlation function. When the spectral density changes qualitatively from sharp spectral lines (in $\underline{k}$-space) to a continuous spectrum (Figure 2), then (by Fourier's uncertainty principle) the correlation function changes qualitatively from spatial coherence to spatial incoherence.

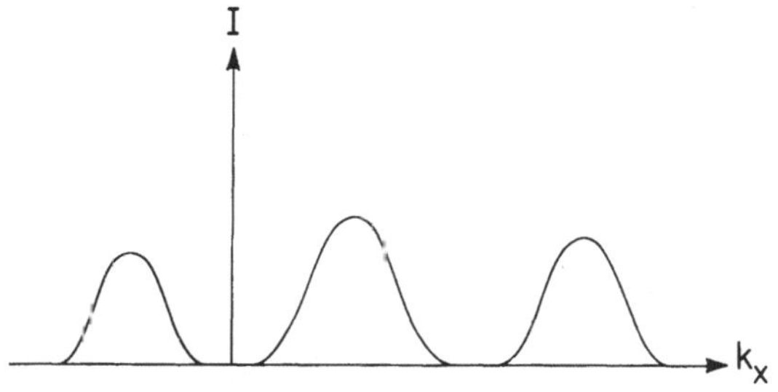

Figure 2a

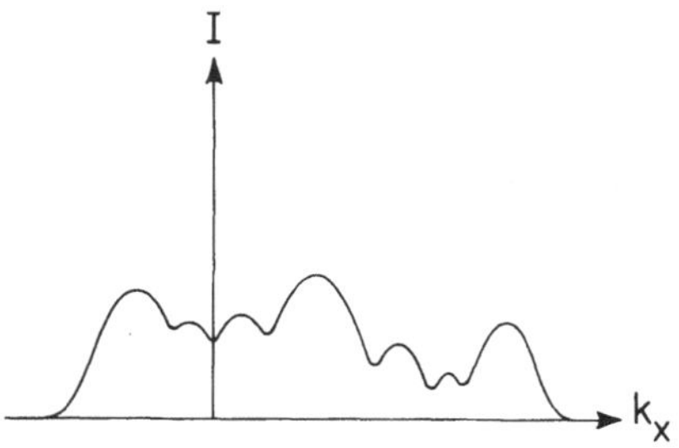

Figure 2b

Limiting ourselves now to a single term in the sum (17), we substitute $\phi(\underline{x}) = \tilde{\phi}(\underline{x}) \exp i\theta(\underline{x})$ into (18), expand $\theta(\underline{x}+\underline{s})$ to second order in $\underline{s}$, and ignore the variation of $\tilde{\phi}$. For (20) we then obtain

$$I(\underline{k},\underline{x}) \sim \exp-[\underline{k}-\underline{k}(\underline{x})][\underline{k}-\underline{k}(\underline{x})]:\mathrm{Re}[\underset{\sim}{\sigma}^{-2}-i\nabla\nabla\theta(\underline{x})]^{-1} \quad . \qquad (22)$$

Thus the spectral density has a Gaussian spread about the Lagrangian manifold $\underline{k}(\underline{x})$. Since the width of the spectral density still depends on $\underset{\sim}{\sigma}^2$, we minimize it with respect to $\underset{\sim}{\sigma}^2$. The minimizing $\underset{\sim}{\sigma}^2$ is diagonal with respect to the principal axes of $\nabla\nabla\theta(\underline{x})$; its components are $\sigma_\mu^2 = |\theta_{\mu\mu}|^{-1}$, in terms of the diagonal elements $\theta_{\mu\mu}$ of $\nabla\nabla\theta$. Thus a spectral line has a width of order $|\nabla\underline{k}(\underline{x})|^{1/2} \sim (k/L)^{1/2}$, where L is the scale-length for variation of $\underline{k}(\underline{x})$. This is indicated in Figure 2a.

As pleating occurs more spectral lines appear. The several values of $\underline{k}(\underline{x})$ must still satisfy the dispersion relation (11), as indicated on Figure 3. Eventually the lines overlap, as in Figure 2b. The spectrum is then broad, the correlation distance is of the order of the wave length, and the wave may be considered incoherent.

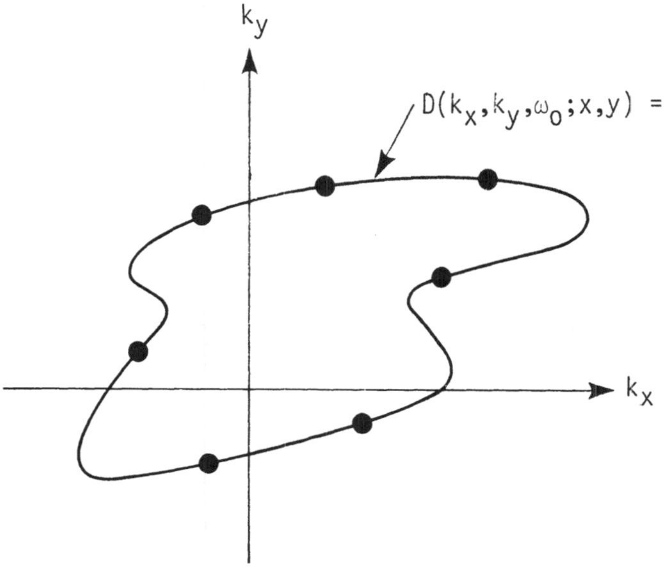

Figure 3

ACKNOWLEDGMENTS

We are grateful for the insightful comments of M. Berry, A. Weinstein, and the participants of the Aspen Center for Physics Workshop on Order and Chaos in Plasma. This work was supported by the Office of Basic Energy Sciences and the Office of Fusion Energy of the U.S. Department of Energy under Contract No. DE-AC03-76SF00098.

REFERENCES

1. M. V. Berry, N. L. Balazs, M. Tabor, and A. Voros, "Quantum Maps," Ann. Phys. 122, 26 (1979).
2. S. W. McDonald, "Wave Dynamics of Regular and Chaotic Rays," Ph.D. thesis, University of California, Berkeley (1983).
3. G. B. Whitham, Linear and Nonlinear Waves (John Wiley and Sons, 1974).
4. R. L. Dewar, "Lagrangian Derivation of the Action Conservation Theorem for Density Waves," Astrophys. J. 174, 301 (1972).
5. A. N. Kaufman, "Poisson Structures of Nonlinear Plasma Dynamics," Physica Scripta T2:2, 517 (1982).

CHAOS IN DISSIPATIVE QUANTUM SYSTEMS

Robert Graham

Universität Essen GHS
Fachbereich Physik
D-4300 Essen - W. Germany

ABSTRACT

We discuss two dissipative quantum systems, which are chaotic in the classical limit. The first system is a single-mode bad-cavity laser. It can be analyzed in the semi-classical limit where it reduces to the complex Lorenz model under the influence of classical external noise. The second system is an exactly solvable dissipative quantum map, which reduces to the Kaplan Yorke map in the classical limit. In both quantum systems the strange attractor of the classical system disappears, but the sensitive dependence on initial conditions remains.

The recent observation of chaos in lasers (see, e.g. [1]) poses new questions: how are features of chaos in dissipative classical systems, like strange attractors and sensitive dependence on initial conditions, modified in quantum systems?

In the present paper we analyze two systems, a laser model, and a map, which may be considered as a simplified version of the Lorenz model. Our results suggest the following answer to the above questions: the strange attractor disappears in the quantized system, but the sensitive dependence of some expectation values on initial conditions remains. Thus, the notion of quantum chaos in dissipative systems seems actually to make sense.

Let us first consider a homogeneously broadened single-mode laser. Its quantum theory was well developed in the sixties, cf. [2]. The basic equation of motion is the master equation

$$\dot{\rho} = L\rho$$

where ρ is the density operator of the laser, and L is a Liouville super-operator acting on ρ . The explicit form of L for a homogeneously broadened single-mode laser is given in ref. [2] eqs. (VI.11.1-6).

In the following we use the Wigner distribution

$$W(\beta,\beta^*,\alpha,\alpha^*,\sigma) = \int d^2\xi \, d^2\eta \, d\zeta \, \exp[i(\xi^*\beta + \xi\beta^* + \eta^*\alpha + \eta\alpha^* + \zeta\sigma)] \times$$

$$\times \mathrm{Tr}\, \rho \exp[-i(\xi^* b + \xi b^+ + \eta^* \sum_\mu S_\mu^- + \eta \sum_\mu S_\mu^+ + \zeta \sum_\mu S_{z\mu})]$$

Here, b, b^+ are the Bose operators of the laser mode, S_μ^+ , S_μ^- and $S_{z\mu}$ are the formal spin - 1/2 operators associated with the 2 energy levels of the atom with index μ in the usual way [2].

Using this quasi-probability distribution, the master equation is transformed exactly into a generalized Fokker-Planck equation [2], which, asymptotically in the classical limit, reduces even further to a classical Fokker-Planck equation. For the quasi-probability distribution, which we employ here, the Fokker-Planck equation was first derived by Risken, Schmid and Weidlich [3], using semi-classical methods from the beginning. This Fokker-Planck equation is stochastically equivalent to the following Langevin equations:

$$\dot{\beta} = -\varkappa\beta - ig\alpha + F_\beta(t)$$

$$\dot{\alpha} = -\gamma_\perp\alpha + ig\beta\sigma + F_\alpha(t)$$

$$\dot{\sigma} = -\gamma_\parallel(\sigma - \sigma_0) + 2ig(\beta^*\alpha - \beta\alpha^*) + F_\sigma$$

β , α , σ are c-number representations of the mode amplitude, the total atomic dipole moment, and the total atomic inversion, respectively. The parameter $\varkappa$ is the inverse life-time of the electromagnetic field in the laser cavity, $\gamma_\perp$ is the inverse life-time of the atomic dipole moment, $\gamma_\parallel$ is the inverse relaxation time of the population difference of the 2 energy levels involved in the laser process. $\gamma_\parallel\sigma_0$ is the external pumping rate of the population difference. g is the dipole-coupling constant, which is proportional to the dipole matrix element. F_β , F_α , F_σ are Langevin forces with the correlation coeffcients

$$\langle F_\beta(t) F_\beta^*(0)\rangle = \frac{1}{2}\varkappa\,\delta(t)$$

$$\langle F_\alpha(t) F_\alpha^*(0)\rangle = \gamma_\perp N\,\delta(t) \quad , \quad \langle F_\sigma(t) F_\sigma(0)\rangle = 2\gamma_\parallel N\,\delta(t)$$

Thermal noise has been neglected in these expressions. The Langevin equations are, formally the equations of motion of the complex Lorenz model under the influence of external classical noise.

For

$$\varkappa > \gamma_\perp + \gamma_\parallel \quad , \qquad \sigma_0 > \frac{\varkappa\gamma_\perp}{g^2} \cdot \frac{\varkappa + 3\gamma_\perp + \gamma_\parallel}{\varkappa - \gamma_\perp - \gamma_\parallel}$$

the corresponding system without noise has a strange attractor [4]. The condition on $\varkappa$ defines a bad-cavity laser. We note that despite its classical appearance, the noise has its origin in quantum effects, namely the vacuum fluctuations of the electromagnetic field and the spontaneous emission of the atoms. At the same time this noise is proportional to the dissipative decay rates $\varkappa$, $\gamma_\perp$, $\gamma_\parallel$ and is therefore typical of a dissipative quantum system. Here is an important general difference to conservative quantum systems, which are noiseless in the classical limit, if described in terms of the Wigner function [5].

The classical noise which remains in a dissipative quantum system asymptotically in the classical limit eliminates the finestructure of the classical strange attractor in a way which is now well understood. E.g. the real Lorenz model with external classical noise has been analyzed in [6]. An exactly solvable map with external classical noise was considered in [7]. On the other hand, sufficiently small classical noise cannot eliminate the sensitive dependence on initial conditions, if the classical strange attractor of the noiseless system has a basin of attraction of non-zero measure. Thus, at least asymptotically in the classical limit, chaos in dissipative quantum systems is precisely defined by the usual criterion that the largest Lyapnuov characteristic exponent is positive.

Let us now turn to a second system, which can be analyzed exactly also in the deep quantum region. It is defined by the master equation [8]

$$\langle q|\rho_{n+1}|q'\rangle = \frac{1}{2}\exp\left[-\frac{2i}{\hbar}\left(g(q/2)-g(q'/2)\right)\right]\times$$
$$\times\left(\left\langle \frac{q}{2}\right|\rho_n\left|\frac{q'}{2}\right\rangle + \left\langle \frac{q+1}{2}\right|\rho_n\left|\frac{q'+1}{2}\right\rangle\right)$$

Here, n is a discrete, integer time-parameter, i.e. we are considering a discrete map. ρ_n is the density operator of the map at time n. The variable q is a phase-like coordinate in the unit interval $0 \leq q < 1$. It is extended to the entire real axis by requiring periodicity in q with period 1. The matrix $\langle q|\rho_n|q'\rangle$ is the density matrix in q-representation. The function g (q) is like a potential. We assume that it is periodic in q with period 1/2. A possible choice would be $g(q) = g_0 \cos 4\pi q$.

In order to interprete the quantum map, we consider the equations of motion for the expectation values of q and its canonically conjugate momentum

$$p = \frac{\hbar}{i}\frac{\partial}{\partial q} \quad , \quad \langle q\rangle_n = \mathrm{Tr}\,\rho_n q$$

We obtain from the diagonal elements of the master-equation

$$\langle q\rangle_{n+1} = \langle (2q)(\mathrm{mod}\,1)\rangle_n$$

Making use of the relation

$$\langle p\rangle_n = -\frac{\hbar}{i}\int dq\,\left[\frac{\partial}{\partial q'}\langle q|\rho_n|q'\rangle\right]_{q'=q}$$

we obtain from the off-diagonal elements of the master-equation

$$\langle p\rangle_{n+1} = \frac{1}{2}\langle p\rangle_n - \langle g'(q)\rangle_n$$

Therefore, by Ehrenfest's theorem the classical map corresponding to the master-equation is given by

$$q_{n+1} = (2q_n)(\mathrm{mod}\,1)$$
$$p_{n+1} = \frac{1}{2}p_n - g'(q_n)$$

This map arises as a qualitative approximation to a return map of the Lorenz model. There, q would be a coordinate along the expanding direction of a suitably chosen cross-section transverse to the attractor, while p would be a coordinate along an attracting direction transverse to the attractor. The cross-section in the Lorenz model could e.g. be chosen like in [6]. We note that the classical map was first studied by Kaplan and Yorke [9] numerically and was solved analytically by Mayer and Roepstorff [10]. An exact solution of the classical map with classical noise was given in [7]. The quantum map was constructed and solved exactly in [8]. If we introduce again the Wigner function by

$$W_n(q,p) = \int_{-\infty}^{+\infty} \frac{dx}{2\pi}\, e^{-ipx} \langle q + \tfrac{\hbar x}{2} | \rho_n | q - \tfrac{\hbar x}{2} \rangle$$

where p corresponds again to the canonical momentum, the steady state ensemble is obtained as [8]

$$W_\infty(q,p) = \lim_{n\to\infty} \frac{1}{2^n} \sum_{m=0}^{2^n-1} F_m^{(n)}(q,p,\hbar)$$

with

$$F_m^{(n)}(q,p,\hbar) = \int_{-\infty}^{+\infty} \frac{dx}{2\pi} \exp\Big[-ipx - \frac{2i}{\hbar}\sum_{\ell=1}^{n}\Big(g\Big(\frac{(q+\frac{\hbar x}{2})(\mathrm{mod}\,1)+m}{2^\ell}\Big) - g\Big(\frac{(q-\frac{\hbar x}{2})(\mathrm{mod}\,1)+m}{2^\ell}\Big)\Big)\Big]$$

The most interesting features of this solution are the following:

(i) In the classical limit $\hbar \to 0$ the integral over x can be carried out by the method of stationary phases and we obtain

$$\lim_{\hbar\to 0} F_m^{(n)}(q,p,\hbar) = \delta(p - \tilde{F}_m^{(n)}(q))$$

with

$$\tilde{F}_m^{(n)}(q) = -\sum_{\ell=0}^{n} 2^{-\ell} g'\Big(\frac{q+m}{2^{\ell+1}}\Big)$$

Therefore, the Wigner distribution in the classical limit reduces to an uncountable sum over δ-functions of p where p is concentrated on the branches $F_m^{(\infty)}(q)$. These branches and their closure make up the strange attractor. This distribution agrees with the result of

Mayer and Roepstorff [10]. There remains no classical noise in the limit $\hbar \to 0$, but this is consistent with the general statements made before, since the classical map, in the present case, is locally area preserving. Still, it is globally dissipative, since it is contracting phase space by a factor 1/2 at each step.

(ii) Asymptotically for small $\hbar$, the integral over x can still be carried out and gives Airy functions. Ignoring any complications due to caustics we obtain

$$F_m^{(n)}(q,p,\hbar) \simeq |\Delta_m(q)|^{-1/3} \, Ai\left(\operatorname{sgn} \Delta_m(q) \, \frac{p - \tilde{F}_m^{(n)}}{|\Delta_m(q)|^{1/3}}\right)$$

with

$$\Delta_m(q) = \hbar^2 \sum_{k=1}^{\infty} 2^{-3k-2} \, g'''\left(\frac{q+m}{2^k}\right)$$

which shows how the strange attractor is gradually smeared out due to quantum interference effects.

In the exact expression for $F_m^{(n)}(q, p, \hbar)$ the strange attractor has completely disappeared. The exact quantum distribution of the momentum p is obtained by integrating $W_\infty(q, p)$ over q. We find [8]

$$W_\infty(p) = \sum_{\ell=-\infty}^{+\infty} \delta(p - 2\pi\hbar\ell) \, W_\ell$$

i.e. the momentum is quantized in units of $2\pi\hbar$. The distribution over the quantum states is given by W_ℓ, which we obtain as

$$W_\ell = \lim_{n\to\infty} \frac{1}{2^n} \sum_{m=0}^{2^n-1} \left| \int_0^1 dq \, e^{-2\pi i q \ell - \frac{2i}{\hbar} \sum_{k=1}^{n} g\left(\frac{q+m}{2^k}\right)} \right|^2$$

In contrast to the full Wigner distribution, the distributions $W_\infty(p)$ and W_ℓ are genuine positive probability distributions, which can, in principle, be measured.

Even though the strange attractor has disappeared in the quantum map, it is obvious that the sensitive dependence on initial conditions remains. This follows from the fact that the master-equation for the diagonal elements of the density matrix is given by

$$\langle q|\rho_{n+1}|q\rangle = \frac{1}{2}\left(\langle \tfrac{q}{2}|\rho_n|\tfrac{q}{2}\rangle + \langle \tfrac{q+1}{2}|\rho_n|\tfrac{q+1}{2}\rangle\right)$$

and describes the evolution of the probability density $\langle q|\rho_n|q\rangle$ by a Bernoulli shift. A positive Lyapunov characteristic exponent $\lambda = \ln 2$ is therefore still present in the exact quantum map.

In conclusion we summarize that the notion of attractors in phase space is a classical concept for dissipative dynamical systems, which looses its sharp meaning in quantum theory at least in the two examples we considered. On the other hand the notion of sensitive dependence on initial conditions seems still to make sense in the quantum domain. This opens the possibility that a precise notion of quantum chaos in dissipative dynamical systems can be defined by the property of sensitive dependence on initial conditions.

REFERENCES

1 R.S. Gioggia, N.B. Abraham, Routes to Chaotic Output from a Single-Mode, dc-Excited Laser, preprint 1983

2 H. Haken, Laser Theory, Encyclopedia of Physics XXV/2c, ed. S. Flügge, Springer, New York 1970

3 H. Risken, C. Schmid, W. Weidlich, Z. Physik 194, 337 (1966)

4 H. Haken, Phys. Lett. 53A, 77 (1975)

5 E.J. Heller, J. Chem. Phys. 65, 1289 (1976)

6 M. Dörfle, R. Graham, Phys. Rev. A27, 1096 (1983)

7 R. Graham, Phys. Rev. A, to appear

8 R. Graham, Quantum Noise and Strange Attractors, preprint 1983

9 J.L. Kaplan, J.A. Yorke, Lecture Notes in Math. 730, 228 (1979)

10 D. Mayer, G. Roepstorff, J. Stat. Phys. 31, 309 (1983)

SEMICLASSICAL QUANTIZATION ON FRAGMENTED TORI

William P. Reinhardt

Molecular Spectroscopy Division,[a] National Bureau of Standards, Washington, DC 20234

ABSTRACT

In many cases the method of Einstein-Brillouin-Keller (or EBK quantization on tori) gives excellent semiclassical quantum levels when the classical motion is integrable. Analysis of the primitive semiclassical quantum energy levels suggests a Poisson distribution of nearest neighbor level spacings. Lack of integrability -- classical chaos -- is then associated with (i) failure of the EBK method and (ii) level repulsions, and conjectures as to the form of P(S) the normalized level spacing distribution, as $S \to 0$. The expectation that classical chaos leads to robust avoided crossings (strong level repulsion) seems to have been verified by numerical experiment: however, an expected result does not always verify the initial premise. In this lecture it is argued that even in chaotic volumes of phase space nonintegrability sometimes does not completely destroy the underlying time independent manifold structure of classical phase space: Fragments of the invariant tori remain and may be used as a basis for EBK quantization. This is illustrated for the Hénon-Heiles problem, and for the truncated $\pi/4$-right triangular rational billiard -- both nonintegrable systems. In both cases the underlying quantum level structure follows from

---

[a]Visiting Scientist 1982-83; Permanent address: Department of Chemistry, University of Colorado, and Joint Institute for Laboratory Astrophysics, National Bureau of Standards and University of Colorado, Boulder, Colorado, 80309 USA.

integrable approximations to the dynamics, and avoided crossings are easily rationalized via the primitive (as opposed to uniform) quantization used -- leading to the conjecture that classical chaos may have little, per se, to do with the results of currently available numerical experiments.

> Chaos. . . It has also some other significations among the alchemists.
>
> Ephraim Chambers, Cyclopaedia (Supplement of 1753)

1. INTRODUCTION: QUANTIZATION ON TORI

One-dimensional bound classical motion may be (approximately) quantized by use of the Bohr condition

$$\frac{I}{2\pi} = \frac{1}{2\pi} \oint p\,dq = (m + \frac{\alpha}{4})\hbar \quad . \tag{1}$$

Here, p,q are the conjugate momentum and position, m the (integer) quantum number, α the Keller-Maslov index. The cyclic integral is about one classical period and determines the value of the action I. Einstein first suggested, in 1917, that the appropriate generalization of (1) to the nonseparable n-dimensional case was what is now referred to as the Einstein-Brillouin-Keller (EBK)[1] quantum condition: an approximation to a quantized energy is obtained at an energy E if the n independent quantum conditions

$$\frac{I_i}{2\pi} = \frac{1}{2\pi} \oint_{C_i} \bar{p}\cdot d\bar{q} = \left(m_i + \frac{\alpha_i}{4}\right) \hbar \qquad i=1,2...n \tag{2}$$

are simultaneously satisfied. In Eq. (2),

$$\bar{p}\cdot d\bar{q} \equiv \sum_{i=1}^{n} p_i dq_i$$

is the invariant 1-form of Poincaré. The quantization condition has an elegant geometric interpretation in that the n paths C_i are the n independent cycles on the surface of an invariant (i.e., time independent) torus of dimension n, embedded in the 2n dimensional phase space.[1,2] The existence of such tori and the quantization rule are illustrated in Figs. 1, 2 and 3 for the case n = 2. Quantization via Eq. (2) has been carried out in many ways,[2a,d] one of which is to use numerically generated sections (e.g., Fig. 2) to determine the tori.

An alternative, and completely algebraic, method follows from the fact that for a Hamiltonian of form

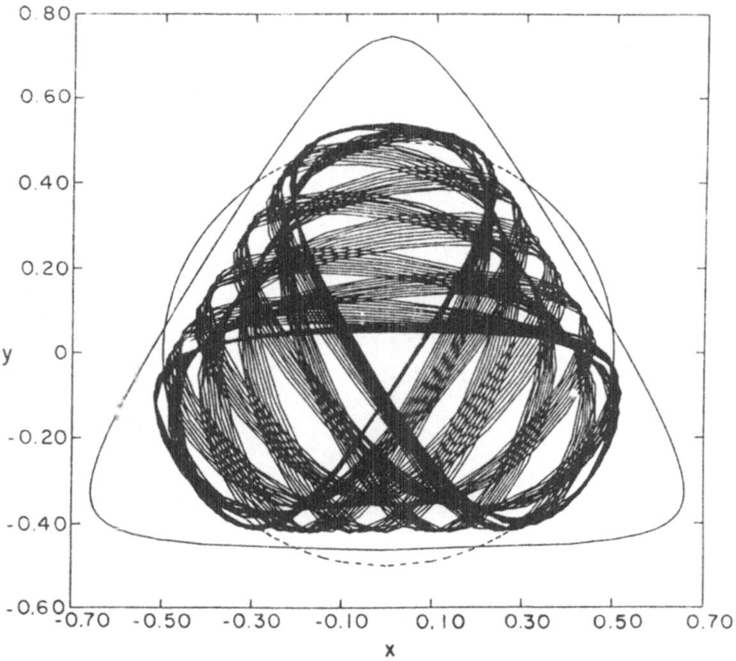

Fig. 1. A quasi- periodic, or regular, trajectory for two-dimensional motion on the Hénon-Heiles potential $1/2\ x^2 + 1/2\ y^2 + x^2y - y^3/3$, Ref. 9.

$$H(\bar{p},\bar{q}) = H(p_1,\ldots p_n;\ q_1,\ldots q_n) \quad , \tag{3}$$

existence of invariant tori allows the Hamiltonian to be written as a function of the n-actions, I_i, of Eq. (2).[3] That is, a canonical transformation may be found such that

$$H_{new} = \tilde{H}(\bar{I}) = \tilde{H}(I_1,\ldots I_n) \quad , \tag{4}$$

the I_i being generalized momenta conjugate to the angles θ_i. Such a system is said to be integrable if the I_i's are independent and have vanishing Poisson brackets with one another, and are analytic and single valued functions of $(\bar{p},\bar{q})$. The fact that the θ_i are ignorable (i.e., do not appear explicitly in $\tilde{H}$) implies that, as

$$\dot{I}_i = -\frac{\partial\tilde{H}(\bar{I})}{\partial\theta_i} = 0 \quad , \tag{5}$$

the actions are constants of motion, as befits their association with the time independent torus. The angles θ_i evolve at constant

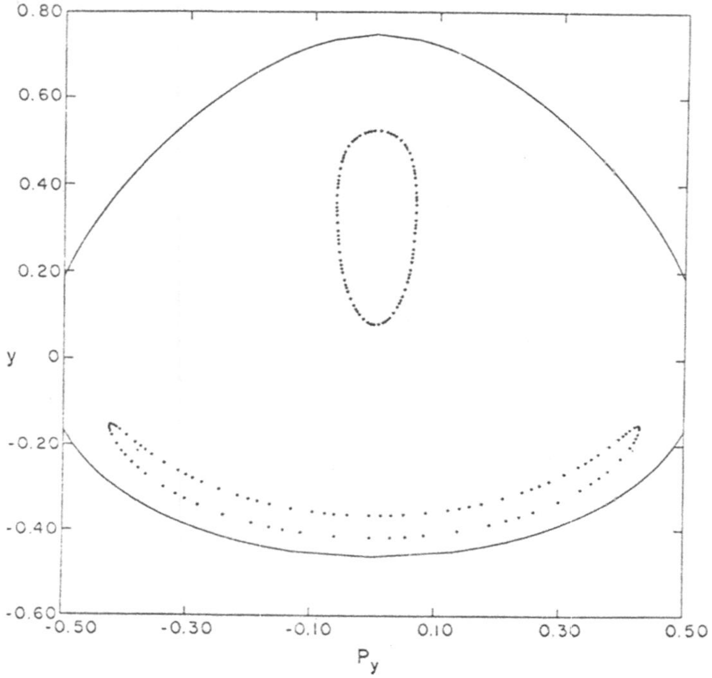

Fig. 2. Poincaré section of the trajectory of Fig. 1. Plotted are the intersections of the trajectory with the surface x = 0, projected onto the (y, p_y) plane. Values with both signs of p_x are included. Examination of the figure suggests that this trajectory is confined to a two-dimensional surface in the four-dimensional phase space (p_x, x, p_y, y), and is consistent with motion on a torus, which has been cut twice by the x = 0 section.

frequency

$$\omega_i(\bar{I}) = \frac{\partial \tilde{H}(I)}{\partial I_i} \tag{6}$$

and provide a coordinatization of a trajectory as it evolves, confined to the torus. The quantization condition, in terms of the new Hamiltonian, $\tilde{H}$, in the new variables, I_i, is

$$E_{m_1, m_2 \ldots m_n} = \tilde{H}\left((m_1 + \frac{\alpha_1}{4})\hbar,\ (m_2 + \frac{\alpha_2}{4})\hbar, \ldots (m_n + \frac{\alpha_n}{4})\hbar \right) \tag{7}$$

implying that if the transformation to the form $\tilde{H}(\bar{I})$ can be analytically carried out, quantization follows at once. In the case of motion near an equilibrium point, where $H(\bar{p},\bar{q})$ is of the form

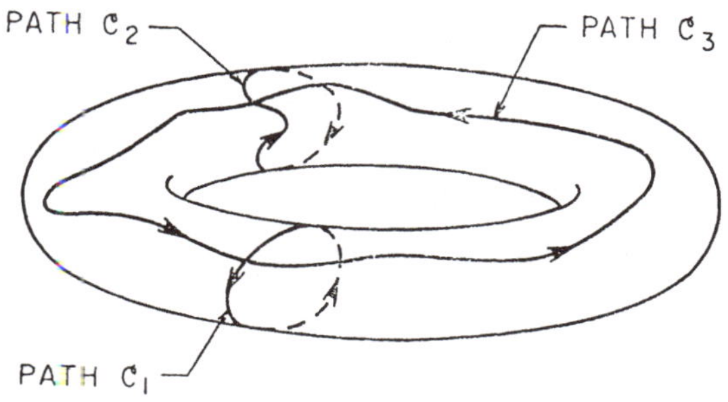

Fig. 3. Quantization on a torus for a system of two degrees of freedom. The two fundamental 1-cycles are shown. Note that as the torus is a Lagrangian manifold (Ref. 3) continuous path distortion does not change the value of cyclic integrals of the invariant 1-form $\bar{p}\cdot d\bar{q}$. Thus, for example there is no difference between quantization on paths C_1 and C_2. The path C_3 gives an independent quantization as it cannot be continuously deformed into a path of the type C_1 or C_2 unless it leaves the surface of the torus.

$$H(\bar{p},\bar{q}) = \sum_{i=1}^{n} \tfrac{1}{2}(p_i^2+\omega_i^2 q_i^2) + \varepsilon V(\bar{q}) \quad , \tag{8}$$

the method of Birkhoff[4] may be used to implement the transformation to the form of Eq. (4), as a formal expansion in "ε", if the frequencies ω_i are not rationally related. This has been implemented by Swimm and Delos[5] for the $n = 2$ case. If ω_i are rationally related a modification of Birkhoff's method, due to Gustavson,[6] can be used to find a formal[7,8] series for the transformation to the form ($n = 2$)

$$H(\bar{p},\bar{q}) \rightarrow \tilde{H}(I_1,I_2,\theta_2) \quad . \tag{9}$$

As I_1 is conserved [see Eq. (5)] two constants, $\tilde{H}(\bar{I},\theta_2)$ and I_1 have been found, and a torus is thus defined, allowing quantization. This has been implemented, also by Swimm and Delos[5] in the case $\omega_1 = \omega_2 = 1$ for the Hénon-Heiles[9] problem, in those regions of phase space where the dynamics is dominated by the presence of tori.

All might be thought to be well, but for the existence of chaos!

2. CLASSICAL CHAOS: POSSIBLE IMPLICATIONS FOR LEVEL SPACINGS

Figure 4 shows a fixed energy composite Poincaré surface for the classical dynamics related to the quadratic Zeeman effect in atomic hydrogen. In the regularized coordinates used[10]

$$H(\bar{p},\bar{q}) = \frac{1}{2}(p_1^2+q_1^2) + \frac{1}{2}(p_2^2+q_2^2) + \varepsilon(q_1^6+q_1^4q_2^2+q_1^2q_2^4+q_2^6) \quad . \tag{10}$$

In the central region of the Figure motion appears to be on invariant tori -- allowing quantization via the EBK scheme. However, outside this region a single trajectory apparently fills an area on the sectioning surface, and thus a volume in the three-dimensional energy conserving submanifold of the four-dimensional phase space. The trajectory is apparently not on a torus -- if this is the case the system is nonintegrable, or chaotic. A composite surface of section for the more familiar Hénon-Heiles problem (see Figs. 1, 2) at energies above $E \approx 1/10$ would similarly show motion of both types.[9,11] The celebrated KAM theorem[7] indicates that a perturbed integrable system will in general still have motion on tori in

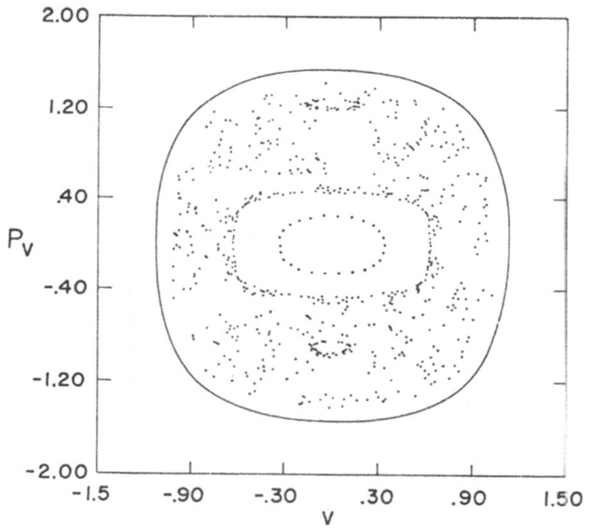

Fig. 4. Composite Poincaré surface for the quadratic Zeeman Hamiltonian See Ref. 10. The central region contains invariant tori, the "chaotic" outer region is generated by a single trajectory. Comparison with Fig. 2 suggests that the motion in the outer region is not on a torus, or perhaps that the torus is now a surface of a higher, perhaps fractal, dimension (see B. Mandelbrot, The Fractal Geometry of Nature W. H. Freeman, San Francisco, 1983). In any case, a dramatic change has occurred.

large measure provided that the perturbation is weak. Figure 4 is a clear illustration of the coexistence of tori with complementary volumes of chaos, both in large measure.

Acknowledging the existence of chaos suggests the following line of reasoning: chaos → no tori → no EBK quantization → "quantum" chaos. In this instance, all that is meant by "quantum" chaos is a change in the nearest neighbor level distribution, P(S). Percival[12] has thus suggested existence of an irregular quantum spectrum. Marcus[13] has noted a possible analogy between classical overlapping nonlinear resonances (which in Chirikov[14] theory can give rise to classical chaos) and multiple avoided crossings, although he is careful to note that analogy does not necessarily imply a one-to-one correspondence. At least part of the intuitive origin of the expectations that there should be some quantum ramification of classical chaos is summarized by Berry[15]:

> In particular, any transition of classical orbits from regular to random ought to be reflected in the form of the wave functions for quantum states, and in the distribution of quantum energy levels, especially in the semiclassical limit where these are densely populated.

More specifically, Berry and Tabor[16] have analyzed the level spacings expected for classically integrable systems by assuming that all quantum states follow from EBK quantization.* They find a Possion distribution

$$P(S) = e^{-S} \tag{11}$$

where P(S) is the nearest neighbor level distribution normalized to unit level spacing. As $S \to 0$, $P(S) \to 1$, and is maximized. This has been verified in many instances, see for example Ref. 17 and Fig. 8. Berry[2c,18] has suggested that, in the nonintegrable or chaotic case, rather than degeneracy (S = 0) being the most probable level spacing, $P(S) \sim S^{j-1}$ as $S \to 0$, j being the dimension of the parameter space needed to ensure crossing of levels. This, for the generic case j = 2, is consistent with the small S limit of the the Wigner distribution,

$$P(S) = \frac{1}{2}\pi S\, e^{-\pi S^2/4} \quad , \tag{12}$$

often empirically found to give an excellent fit to "generic" quantum level distributions — i.e., those arising from ensemble averages of eigenvalues of random matrices. However, it is of

*That is, all tunneling is neglected.

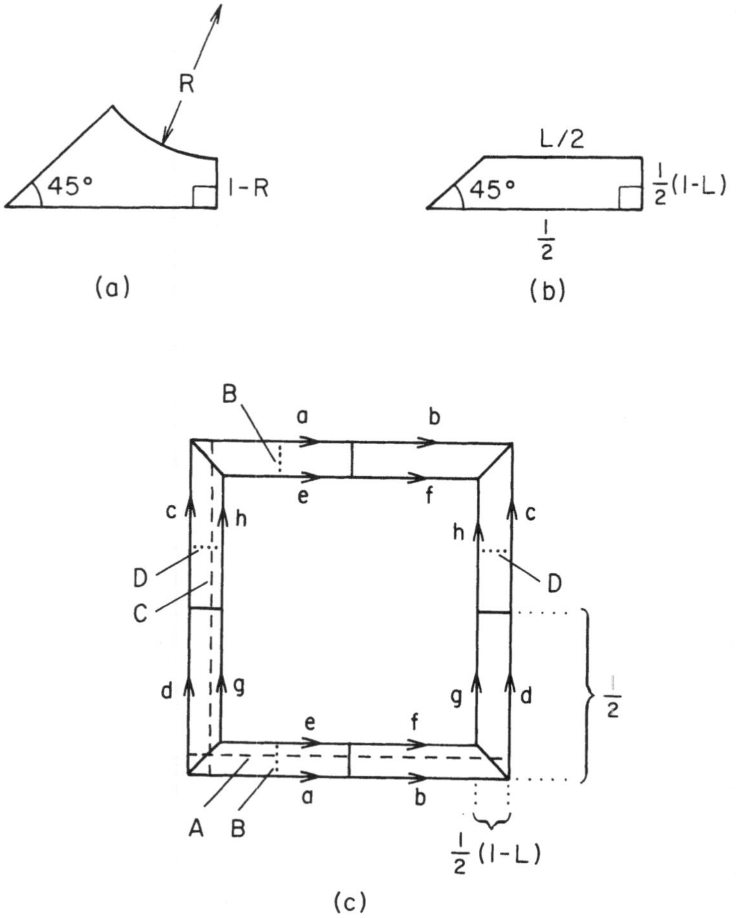

Fig. 5. (a) Desymmetrized Sinai billiard as quantized by M. Berry, Ref. 18. Free two-dimensional motion takes place inside a hard enclosure which is a 45° right triangle, truncated by an arc of radius R. Classical motion is chaotic. Quantum levels are illustrated in Figs. 6 and 7. See Ref. 18 for further discussion. (b) Desymmetrized "square torus" billiard discussed by Richens and Berry, Ref. 32. In this case the 45° right triangle has been truncated by a line, rather than an arc. Exact quantum levels are shown in Fig. 11 where they are compared with those of an integrable approximation. (c) Phase space topology for the rational billiard of Fig. 5b. The surface is of genus 2 ("2 holes") as may be seen by noting that the labelled (lower case Roman) directed lines are to be identified. (See Ref. 32 for a more detailed discussion.) Quantization on the (fragmented) torus defined by paths A and B gives the dotted curves of Fig. 11.

exceptional importance to note that Berry has only derived a limiting (i.e., $S \to 0$) behavior, and this contains no estimate of the domain of validity. In particular Berry's arguments do not imply Eq. (12), except in the small S limit. For example, if exponentially small tunnelling splittings are included, the Poisson law of Eq. (11) would suddenly turn over in the $S \to 0$ limit, but on a scale exponentially small compared to the visible e^{-S} part of the distribution, and thus be unobservable on the same scale.

Berry[18] has carried out quantization of the "desymmetrized" Sinai billiard (see Fig. 5a) and obtained the quantum levels shown in Fig. 6, with the empirically obtained level spacing distribution of Fig. 7. Similar results have been obtained by McDonald and Kaufman[17] for the "stadium" billiard problem. In both cases the classical dynamics of these billiard systems is highly chaotic, and the level repulsion histograms seem to bear out the expectation of a relationship between classical chaos and strong level repulsions. In fact, the results seem to imply more than has been rigorously shown: The empirically derived P(S) histograms appear to peak quite near the average level spacing: that is, not only does $P(S) \sim S$, as $S \to 0$ but the distributions have the robust anticrossings of the type characterized by the Wigner distribution. In fact a more extensive analysis of the Sinai problem[19] of Figs. 6 and 7 gives a fine scale histogram in remarkable agreement with the Wigner distribution.

An alternative situation is shown by the Hamiltonian of Eq. (10), the M = 0 quadratic Zeeman problem. Even parity energy levels[20] as a function of external field, B, are shown in Fig. 8, with the level spacing distribution corresponding to the chaotic part of phase space shown in Fig. 9. The dashed line in Fig. 8 indicates the boundary between integrable dynamics, as quantized by Delos, Knudson and Noid,[21] and those regions where chaos is of substantial measure (see Fig. 4). Visually the levels of Fig. 8 do not seem to show any strong correlation with the onset of classical chaos in large measure. The histogram of Fig. 9 shows a Poisson-like distribution (on the scale of $\hbar^2$) in the chaotic region of phase space. As the apparent crossings of eigenvalues in Fig. 8 are actually exponentially small (and thus invisible in the figure) anticrossings, the $S \to 0$ limit would presumably vanish linearly in S, as per the Berry analysis, but only on a scale small [i.e. $\sim \exp(-\text{const}/\hbar)$] compared to the mean spacing. The avoided crossings are thus not robust; i.e., they do not dominate the large scale behavior. This example immediately suggests that:

(1) there is no necessity of a relationship between robust avoided crossings and classical chaos;
(2) as the quantum structure is well accounted for using integrable dynamics in the regular region, it is reasonable to

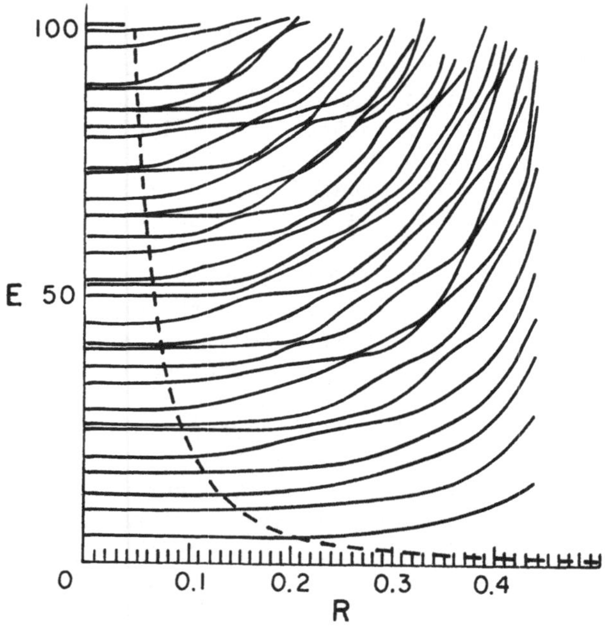

Fig. 6. Quantum levels for the Sinai billiard of Ref. 18. Large-scale avoided crossings set in once a perturbative region (dashed line) is left behind. Figure from Ref. 18, reproduced with permission.

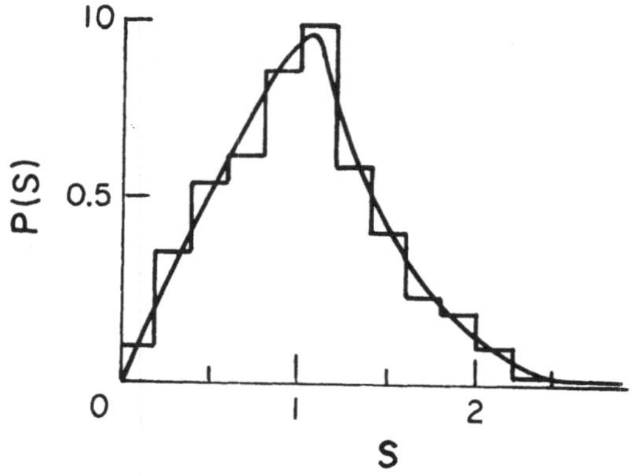

Fig. 7. Level spacing distribution, P(S), for the levels of Fig. 6. Note that the distribution peaks near the average spacing: i.e., the robust repulsions in a sense "produce" the average spacing. The smooth curve has been drawn by eye, and is not the result of a particular theoretical model. Redrawn from Ref. 18, with permission.

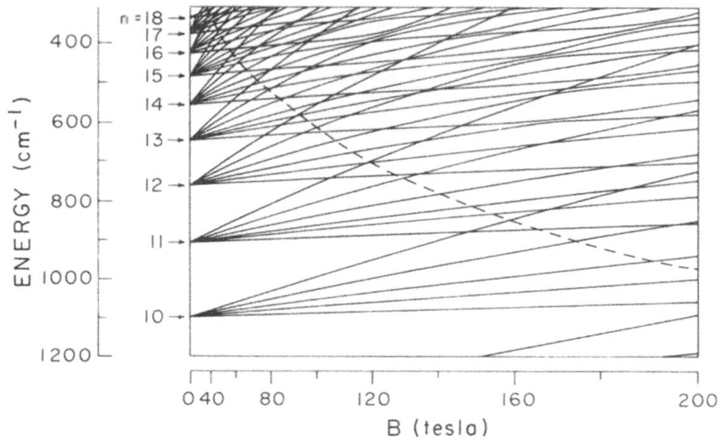

Fig. 8. Quantum levels (see Ref. 20) for the quantum Zeeman problem, as a function of external field B. All of the levels shown are of the same symmetry. The dashed line indicates the onset of classical chaos, as seen in the Poincaré surface (e.g., Fig. 4). In contrast to Fig. 6, there is no wide spread onset of avoided crossings. Figure from D. Farrelly and W. P. Reinhardt, J. Chem. Phys. 78, 606 (1983).

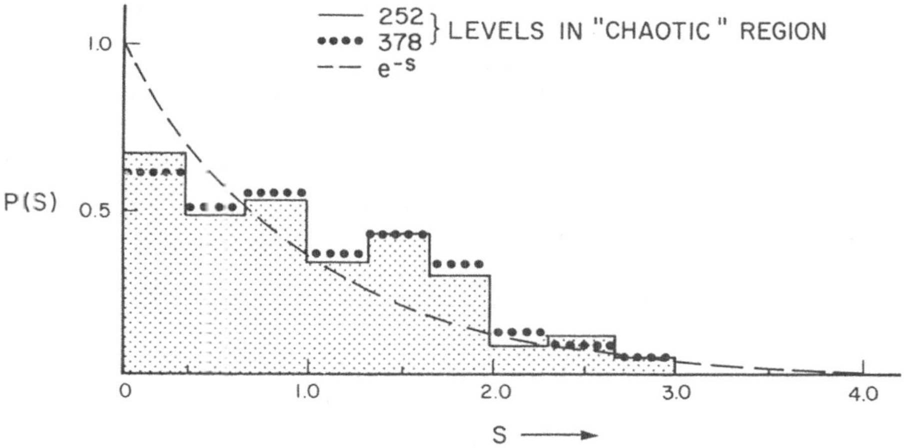

Fig. 9. Level spacing distribution for the levels above the dashed curve of Fig. 8. The histogram was obtained from the data of Fig. 8, and is a composite of data from ~20 fields. The dashed line is $P(S) = e^{-S}$, the distribution "expected" for integrable dynamics. It is evident that the underlying classical chaos is not detected in this quantum level distribution.

assume that an integrable approximation to the chaotic dynamics might well give the entire spectrum of Fig. 8.

But how can this be, as any integrable approximation will confine trajectories to tori, and thus entirely fail to reproduce a surface of section such as that of Fig. 4? In the next sections it is suggested that by focusing on trajectories, rather than the geometric structure of phase space, the extent of chaos is greatly over-emphasized, and that many of the systems currently being studied are, in a global and structural sense, "almost" integrable. This point of view is elaborated and discussed in the next two sections.

3. QUANTIZATION OF CLASSICAL CHAOS: FRAGMENTED TORI, I

The subject of this section is quantization of the classically chaotic classical phase space on the Hénon-Heiles surface. The Hamiltonian

$$H(\bar{p},\bar{q}) = \frac{1}{2}\,(p_1{}^2+q_1{}^2) + \frac{1}{2}\,(p_2{}^2+q_2{}^2) + \alpha(q_1^2 q_2 - \frac{1}{3}\,q_2^3) \qquad (13)$$

has been the subject of numerous inquiries.[9,22-24] For the value $\alpha = (80)^{-1/2}$ used in the studies discussed here ($m = \hbar = 1$) chaotic dynamics is present in appreciable measure over the energy range from $E \cong 9$ to 13.2, dissociation being at $E \cong 13.2$. In the 1978 paper by Swimm and Delos,[5] mentioned earlier, not only were the quantum levels obtained in the integrable "low energy" regime ($0 < E < 9$), but, astonishingly, the algebraic technique used by these authors gave good estimates a modest fraction of levels in the chaotic parts of phase space. This suggested that, even though the surface of section indicated classical chaos, integrable approximations to the dynamics could allow implementation of the EBK scheme to quantize chaotic volumes of phase space. These results, combined with more qualitative ideas based on a local stability analysis (Toda,[25] Brumer and Duff,[26] Cerjan and Reinhardt,[27] Kosloff and Rice[28]) suggested that the algebraic methods be further developed.

The Birkhoff-Gustavson[5,6,24] procedure may be used to systematically transform the Hamiltonian of Eq. (13) into the form $H(I_1 I_2 \theta_2)$. As θ_1 is ignorable, $\dot{I}_1 = 0$, and $\{H,I_1\} = 0$. I_1 is thus a global second constant of motion, with a formal power series representation

$$I_1(\bar{p},\bar{q}) = \sum_{i=3}^{k} \varepsilon^i I_1^{(i)}(\bar{p},\bar{q}) \qquad (14)$$

where the $I_1^{(i)}(\bar{p},\bar{q})$ are analytically calculable homogeneous polynomials in $\bar{p},\bar{q}$. The series of Eq. (14) cannot converge[29] to an analytic function as $k \to \infty$, otherwise the Hénon-Heiles problem would be integrable -- a contradiction, as the existence of classical chaos precludes this. However, Shirts and Reinhardt[24] undertook an extensive study of the convergence properties of this series. The $I_1^{(i)}(\bar{p},\bar{q})$ were analytically calculated for n = 3 through 13, and the partial sums, and corresponding Padé table, were generated. Results of this investigation are discussed in great detail in Ref. 24. The general conclusion is illustrated in Fig. 10. The figure

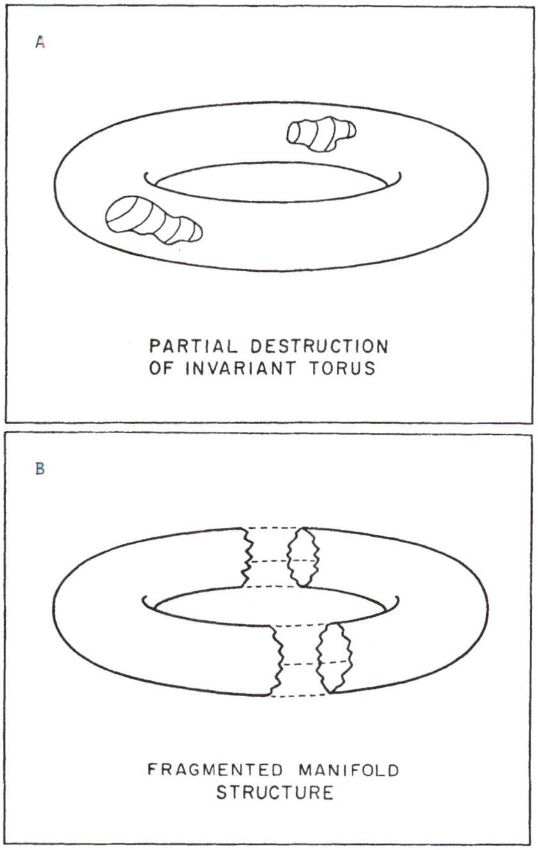

Fig. 10. Partially destroyed tori. These heuristic drawings are an attempt to indicate the conclusions of Refs. 23 and 24, which contain extensive discussion. (a) indicates a partially destroyed torus which might be detected in a short time surface of section -- i.e., a trajectory might be confined to the remaining manifold for many periods before entering a region of complete destruction. EBK quantization can be attempted on this fragmented manifold. (b) indicates a fragmentation which would be difficult to observe classically by integration of (individual) trajectories. Nevertheless it suggests applicability of EBK quantization.

illustrates (heuristically) the fact that the partial sums and the Padé approximations seemed to be convergent in large volumes of the chaotic phase space. However in other, quite localized, parts of the chaotic phase space strong local divergence of the series was noted by the occurrence of dense clusters of zeros and poles, which remained approximately invariant in different members of the Padé table. The qualitative conclusion, (consistent with the ideas of Refs. 25-28) is that even in those volumes of phase space where trajectories appear to be chaotic, much of the manifold structure remains intact. The topology is no longer that of a torus, as the tori have (partially) fragmented -- perhaps developing seams or patches of strong nonanalyticity. However, if enough of the manifold surface remains, use of EBK quantization is suggested. Table I indicates a sampling of quantum levels obtained in a complete quantization of the chaotic phase space by Jaffé and Reinhardt.[23] The fact that excellent semiclassical estimates of quantum eigenvalues were obtained from an integrable approximation to the chaotic dynamics on the Hénon-Heiles surface calls into question any definite relationship between integrability (and lack thereof) to level spacing statistics. Reference 24 contains examples of individual trajectories which seem to bear out, in another way, the idea of Fig. 10a. However, as noted in the caption to Fig. 10, remnants of manifold structure are not necessarily easily found by running trajectories.

In summary, empirical evidence has been presented which indicates that much of the manifold structure -- formerly associated only with complete tori -- may well remain intact in the chaotic phase space. It is then this time-independent underlying structure -- rather than the dynamics of orbits -- which provides a basis for the semiclassical quantization.

4. FRAGMENTED TORI, II: RATIONAL BILLIARDS

The rational billiards, recently discussed by Keller and Rubinow,[30] Zemlyakov and Katok,[31] Berry,[18] Richens and Berry,[32] and Eckhardt and Ford[33] consist of free motion in two-dimensional configuration space, bounded by hard walls consisting of straight lines intersecting at rational multiplies of π. These differ from the Sinai and Stadium problems discussed in Section 2, in that there, one, or two sides respectively, of the restraining boundary consisted of arcs of a circle. The remarkable fact about the rational billiards is that although the motion is analytically soluble (in terms of Fourier expansions) the motion need not be integrable, i.e. the phase space motion may define a manifold but if the genus is not 1, a torus is not defined.[32] The test for integrability is simple.[30-33] At each collision of the moving body with a wall the subsequent specular trajectory is found by reflecting the boundary about the side in question. If the collection of

Table I. High Lying Quantum Levels for the Hénon-Heiles problem, as obtained by numerical solution of the Schrödinger equation (QM),[b] uniform semiclassical quantization (JR)[c] and primitive semiclassical quantization (SD).[d] All of these levels correspond to classically chaotic phase space.

Quantum numbers[a]	Symmetry of quantum states[a]	QM[b]	JR[c]	SD[d]
12 0	A	11.966	12.011	11.864
12 ± 2	E	11.968	12.017	---
12 ± 4	E	12.206	12.217	---
12 ± 6	A*	12.277 12.334	12.274 12.332	12.310
12 ± 8	E	12.480	12.490	12.491
12 ± 10	E	12.712	12.749	12.750
12 ± 12	A*	13.077 13.087	13.0975 13.0976	13.097
13 ± 1	E	12.762	12.824	---
13 ± 3	A*	12.748 13.032	12.827 13.060	---
13 ± 5	E	13.081	13.104	---
13 ± 7	E	13.233	13.238	---
13 ± 9	A*	--- ---	13.4397 13.4474	---
13 ± 11	E	---	13.7351	

[a]D. W. Noid and R. A. Marcus, J. Chem. Phys. 67, 559 (1977).
[b]D. Noid, as reported in Ref. 5.
[c]C. Jaffé and W. P. Reinhardt, Ref. 23.
[d]R. Swimm and J. Delos, Ref. 5.

all such reflections tessellates the plane uniquely and exactly the system is integrable. This is the case, for example, for the square, rectangle, equilateral triangle, and for 45°-45°-90° triangles. It is not the case for the $\pi/3$ rhombus, nor for the truncated 45°-90° triangluar billiard of Fig. 5b. The connectedness of the phase space of this latter, truncated, system is illustrated in Fig. 5c. Phase space motion is on a surface of genus 2 (a sphere with two handles and thus is not integrable.[32]) Richens and Berry[32] have obtained the exact quantum levels for the truncated 45°-90° triangle billiard as a function of L, which defines the extent of truncation. They find large scale avoided crossings, again seemingly consistent with the original idea that nonintegrability →

robust avoided crossings. Can we analyze this system of semiclassically?

Keller and Rubinow[30] (repeated with more detail in Ref. 32) give concise discussions of semiclassical quantization of the rectangle and equilateral triangle, where the manifold topology is that of a torus. In these cases the semiclassical results are exact. Richens and Berry comment that, as a phase space manifold of genus 2 implies four independent "actions" (i.e. four independent minimal cycles), EBK quantization is ambiguous, and only gives quantum levels in a relatively few special cases. However, Richens and Berry are attempting to find an "exact" semiclassical scheme. If we relax this a bit we can ask, can we give a sensible approximate semiclassical quantization (as a function of L) for the truncated 45°-90° triangular billiard? Can we then see how the approximate quantization can be patched up?

The answer is yes to both questions. Rather than regarding motion of the 45°-90° truncated triangular billiard as being on a "sphere with two handles," we regard the motion as being on a torus with one extra handle, and ask, when can we ignore the extra handle? This is in the same spirit of the quantization via the remnants of tori of Fig. 10. To illustrate the workability of this idea, consider motion in the billiard of Fig. 5b, in the case when $L \approx$ "$1-\varepsilon$" thus $1/2(1-L)$ is small, and the enclosure approximates a box of dimension 1/2 by $\varepsilon/2$. In the case that $p_y \gg p_x$ the motion will chatter in the "short" of direction for a long period of time, suggesting that the motion is described by extended residence on a torus with occasional hopping to another torus.[34,35] This motion can be quantized using the two independent paths A and B indicated in Fig. 5c, giving the quantum levels

$$E_{n_x,n_y} = \frac{4n_y^2}{(1-L)^2} + 4n_x^2 \quad ; \quad n_x > n_y \geq 1 \tag{15}$$

as a first approximation. This quantization simply ignores the "holes" in the torus, which would allow change of sheet. The condition $n_x > n_y$, leads to the levels of Eq. (15) being exact in the $L \to 0$ limit. Figure 11 shows a comparison of the quantum levels obtained via Eq. (15) with the exact levels of Richens and Berry.[32] The overall correspondence is excellent, the quantization on the tori giving crossings, where the exact levels avoid. However, this is as expected. EBK quantization on tori always is a primitive quantization, in that torus-torus tunneling is ignored. If two distinct tori have the same quantum energy the actual quantum states will split, given an appropriate uniform approximation.[13,23,36] Our expectation is that the splittings shown in the exact quantum results are not a fundamental consequence of the genus 2 nature of the manifold, and that the semiclassical dynamics will be well

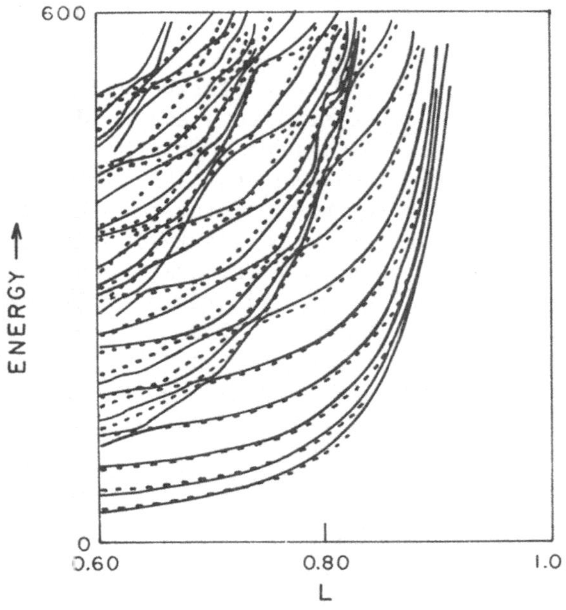

Fig. 11. Quantization of the rational billiard of Fig. 5b. The exact quantum levels (solid curves) are those of Richens and Berry, Ref. 32. The dotted curves are the result of an EBK quantization on the fragmented torus obtained by ignoring paths "C" and "D" in Fig. 5c [see Eq. (15)]. The result of this integrable approximation has been shifted horizontally ~0.014 units towards larger L: no rescaling was necessary. There is a one to one correspondence between crossings in the integrable EBK results and avoided crossings in the exact results. This suggests that a uniform quantization be carried out. We conjecture that such a uniform quantization would indicate that classical chaos has little to do with the magnitude of the avoided crossings.

well approximated using a nearby integrable approximation, with an appropriate uniform approximation. As the desymmetrized Sinai and Stadium billiards are well approximated by rational billiards, one might expect that, as far as semiclassical quantization is concerned, that these systems are nearly integrable, and given use of an appropriate uniform approximation, may also be quite accurately quantized using integrable approximations to the dynamics.

5. DISCUSSION AND CONCLUSIONS

Standard techniques for quantization on tori have been briefly reviewed, and the basic idea of quantization using the time-independent underlying phase space structure extended to the case where only part of that structure is intact. This has been done in an intuitive, rather than rigorous manner. The author is, as yet, unable to give a criterion for when an integrable approximation will be an adequate approximation to the full dynamics, at a given value of $\hbar$. Formulation of such a criterion would seem to have little to do with the dynamics of individual trajectories. Were such a criterion available it would allow routine primitive quantization on the tori of the approximation. Such a primitive quantization will, in general, lead to the type of approximation shown in Fig. 11, with its attendant level crossings. A multiple-torus uniform approximation is then needed to allow estimation of the magnitude of the crossings. Only when this has been systematically carried out will there be any indication of the role of classical chaos in determining the magnitude of the splittings: I conjecture it to be quite minor in the types of systems discussed here. If this is, indeed, the case a proper interpretation of the results of, say, Figs. 6 and 7 and their possible connection with the underlying classical chaos will be to note only that the results are classically and quantum mechanically generic: Outside the regime of validity of low order perturbation theory the generic classical dynamical system is nonintegrable, and the generic quantum system has a Wignerian-like level distribution.

ACKNOWLEDGMENTS

This research has been supported, in part, by National Science Foundation Grants CHE80-11442 and PHY82-00805 to the University of Colorado. This support is gratefully acknowledged. The author is particularly grateful to the National Bureau of Standards for the hospitality shown him during his tenure as Visiting Scientist; and, to his several coworkers, in particular C. Cerjan, C. Jaffé, R. Shirts and D. Farrelly. Finally, Messrs Casati and Ford, those modern sons of A. Volta, are to be congratulated for their enthusiastic organization of an excellent meeting.

REFERENCES

1. A. Einstein, Verh. Dtsch. Phys. Ges (Berlin) 19, 82 (1917); L. Brillouin, J. Phys. Radium 7, 353 (1926); J. B. Keller, Ann. Phys. (N.Y.) 4, 180 (1958).
2. Recent reviews and extended expositions include (a) I. C. Percival, Adv. Chem. Phys. 36, 1 (1977); (b) M. V. Berry in S. Jorna (Ed.) AIP Conference Proceedings #46 (Am. Inst. Phys.,

New York. 1978), p. 16; (c) M. V. Berry, Lectures Notes for the 1981 Les Houches Summer School; (d) D. W. Noid, M. L. Koszykowski, and R. A. Marcus, Ann. Rev. Phy. Chem. 32, 267 (1981); (e) V. P. Maslov and M. V. Fedoriuk Semiclassical Approximation in Quantum Mechanics (Reidel, Boston, 1981).
3. For example, V. I. Arnold, Mathematical Methods of Classical Mechanics (Springer, New York, 1978).
4. G. D. Birkhoff, Dynamical Systems, Am. Math. Soc. Colloq. #9 (Providence, RI, 1979); J. Moser, Stable and Random Motion in Dynamical Systems (Princeton University Press, Princeton NJ, 1973).
5. R. T. Swimm and J. B. Delos, J. Chem. Phys. 71, 1706 (1979).
6. F. G. Gustavson, Astron. J. 71, 670 (1966).
7. See for example, J. Moser, in AIP Conference Proceedings #46 (Am. Inst. Phys., New York, 1978), p. 1.
8. C. Siegel, Ann. Math. 42, 806 (1941).
9. M. Hénon and C. Heiles, Astron. J. 69, 73 (1964).
10. W. P. Reinhardt and D. Farrelly, J. Phys. (Paris), Colloq. Suppl. 11, 29 (1982).
11. See for example, Fig. 1 of Ref. 24.
12. I. C. Percival, J. Phys. B 6, L229 (1973); see also Ref. 2a-d.
13. R. A. Marcus, in Horizons of Quantum Chemistry, Eds. K. Fukui and B. Pullman (Reidel, Boston, 1980), p. 107; D. W. Noid, M. C. Koszykowski and R. A. Marcus, J. Chem. Phys. 78, 4018 (1983), and references therein; see also Ref. 2d.
14. B. V. Chirikov, Phys. Rep. 52, 263 (1979).
15. M. V. Berry, in S. Jorna (Ed.) in AIP Conference Proceedings #46 (Am. Inst. Phys., New York, 1978), p. 16. Quote from p. 18.
16. M. V. Berry and M. Tabor, Proc. Roy. Soc. Lond. A 356, 375 (1977).
17. S. W. McDonald and A. N. Kaufman, Phys. Rev. Lett. 42, 1189 (1979).
18. M. V. Berry, Ann. Phys. (N.Y.) 131, 163 (1981).
19. M. J. Giannoni, unpublished remarks at the Como Meeting, 1983.
20. M. L. Zimmerman, M. M. Kash and D. Kleppner, Phys. Rev. Lett. 45, 1092 (198).
21. J. B. Delos, S. K. Knudson, and D. W. Noid, Phys. Rev. Lett. 50, 579 (1983).
22. M. Hénon and C. Heiles, Ref. 9; and; R. L. Churchill, G. Pecelli and D. L. Rod, in Stochastic Behavior in Classical and Quantum Hamiltonian Systems, G. Casati and J. Ford (Eds.) Springer Lecture Notes in Physics 93, p. 76, Springer Verlag (New York, 1979); R. H. G. Helleman and T. Bountis, ibid. p. 376. See also Ref. 2d.
23. C. Jaffé and W. P. Reinhardt, J. Chem. Phys. 77, 5191 (1982).
24. R. B. Shirts and W. P. Reinhardt, J. Chem. Phys. 77, 5204 (1982).
25. M. Toda, Phys. Lett. A 48, 335 (1974).
26. P. Brumer and J. W. Duff, J. Chem. Phys. 65, 3566 (1976).

27. C. Cerjan and W. P. Reinhardt, J. Chem. Phys. 71, 1819 (1979).
28. R. Kosloff and S. A. Rice, 74, 1947 (1981).
29. See Refs. 7 and 8.
30. J. B. Keller and S. I. Rubinow, Ann. Phys. (N.Y.) 9, 24 (1960).
31. A. N. Zemlyakov, A. B. Katok, Math. Zametki 18, 291 (1975), English Trans: Math. Notes 18, 760 (1976).
32. P. J. Richens and M. V. Berry, Physica 2D, 495 (1981).
33. B. Eckhardt, M.S. Thesis, Georgia Institute of Technology, 1982, unpublished.
34. This was confirmed, in conversations at Como, by P. J. Richens, private communication, 1983.
35. See also the discussion in N. Saito, H. Hirooka, J. Ford, F. Vivaldi, and G. H. Walker, Physics 5D, 273 (1982); and, S. J. Shenker and L. P. Kadanoff, J. Phys. A 14, L23 (1981), where hopping between approximate tori (the vague tori of Ref. 24) is discussed.
36. M. J. Davis and E. J. Heller, J. Chem. Phys. 75, 246 (1981).

QUANTUM ERGODICITY AND INTENSITY FLUCTUATIONS

E. J. Heller and R. L. Sundberg

Los Alamos National Laboratory

Los Alamos, New Mexico 87545

I. INTRODUCTION

Flow in phase space is intimately related to the concept of ergodicity. An ergodic system will sample all regions of phase space democratically in the time average, subject only to known a priori constants of the motion. For convenience, we shall refer to the accessible phase space as the "energy shell." This shell may have some thickness if a distribution of energies are initially populated. An energy envelope is defined by the initial distribution of energy and the envelope plays a role both in classical and quantum phase space flow, as discussed before[1-3] and as will become evident in the theory and examples to follow.

Flow in phase space is of paramount interest in chemistry because all intramolecular energy transfer following some nonequilibrium preparation (collisions, photoabsorption, etc.) can be formulated in terms of phase space flow.[1-5]

Perhaps we should try to define what "phase space flow" is in a quantum system, given that conjugate positions and momenta cannot be simultaneously known in quantum mechanics, and that most standard quantum texts speak of Hilbert space rather than phase space. There are two widely known and rigorous ways of viewing quantum mechanics in phase space. The first is the Wigner formulation, which takes wavefunctions and density matrices into classical-like phase space distributions.[6] The second is closely related to the first and uses the overlap with phase space localized functions to determine the phase space disposition of wavefunctions and density matrices.[1-4,6-8] Most often, these localized functions are the coherent states, or equivalently, gaussian wavepackets.[9] In any case, once one becomes familiar with these pictures, it is easy to visualize Hilbert space in terms of phase space. Also, it is never a very big jump from the rigorous quantum phase space description to the analogous classical one.

Let us consider an example. Suppose we want to know about the propensity of energy to flow from local mode "a" to local mode "b." The relevant quantity is[1,3]

$$P(a|b) = \lim_{T\to\infty} \frac{1}{T} \int_0^T |\langle b|a(t)\rangle|^2 \, dt \tag{1}$$

where $|a\rangle$ and $|b\rangle$ are the non-stationary local mode states and $|a(t)\rangle$ is the time evolved $|a\rangle$. This is a purely quantum mechanical problem, but it has a close classical analog: we can prepare a classical phase space distribution which is ergodic on the N-dimensional torus and corresponds to an energy of E_n in the local mode state. This torus is defined (as in the state $|a\rangle$) as the stationary motion of some zero-order Hamiltonian for which the local modes are exact eigenfunctions (or eigentori). The b-torus is defined in a similar way, and the phase space flow between them can be classically monitored. A calculation which is similar to our example has been described by Stechel and Heller[2] for HCN where the classical and quantum flow in phase space for local mode C-H stretch initial conditions were monitored. More commonly, we[3,6,10] and others[7,8] have used gaussian wavepackets (which have a direct Wigner transform and classical analog as gaussian phase space distributions centered at arbitrary coordinates and momenta) as elementary units in phase space to measure the flow. Far from being arbitrary, such gaussian initial distributions are often very nearly approximated in the laboratory in electronic absorption, emission, and Raman experiments.[11] Indeed, since any state Ψ can be expressed as an expansion in terms of simple gaussians with various phase space positions and momenta, and since these gaussians have classical-like evolution (at least for short times)[9], it makes even more sense to study their evolution.

This paper continues the study of these gaussians and their phase space flow for a new highly anharmonic potential that has extensive regions of classically chaotic motion without the presence of a dissociative continuum. The possibility of dissociation, together with finite computer budgets (which restricts the number of states below dissociation) has tended to muddle the issue of the quantum response to classical chaos, something we wish to study here. The problem has been muddled because the "standard" potentials, such as Henon-Heiles[12] and Barbanis[13], reach classically allowed dissociation just as classical chaos is dominating nearly all of phase space. This is not a problem in the "demonic" potential defined in Section III and used exclusively in this paper.

In several previous publications we have demonstrated, both on rigorous theoretical grounds (reviewed in Section II) and as numerical examples, that there is a close connection between spectral intensities and phase space flow.[1-3,10,14] Loosely speaking, let D(E) be the density of symmetry allowed states in a spectral transition and d(E) be the sub-density of such states with significant intensity. Then the fraction of the available phase space actually accessed by the spectroscopically prepared nonstationary state, whose spectral envelope is centered at E_n, is just $d(E_n)/D(E_n)$. This simple prescription is extended and quantified considerably in Section II, but the basic notion is: a higher fraction of spectral lines will be present for systems which are more ergodic (quantum mechanically). If there is classical-quantum correspondence regarding ergodicity, then a classically ergodic system should show a much more dense spectrum for a phase space localized nonstationary state than should a classically quasiperiodic system. At this loose level of discussion, the classical-quantum correspondence has been verified repeatedly in numerical experiments.[2,3,10,15]

The purpose of this paper is threefold. First, we want to present results, for a new Hamiltonian, which is probably the most extensive quantum mechanical numerical study to date on a highly classically chaotic system. We find several very interesting features classical-quantum correspondence when the classical motion is chaotic. Second, we want to investigate the presence of (probably gaussian-) random fluctuations in spectral intensities which are expected in a quantum system that is "as ergodic as it can get." Such fluctuations have significant consequences for the self-overlap $P(a|a)$, without necessarily affecting the cross-overlaps $P(a|b)$, $b\neq a$. (Stechel and Sundberg[16] have made a theoretical study of the effects of the fluctuations on $P(a|a)$, and Stechel[17] has has given a rigorous account of quantum ergodicity and its relevance to spectra.) Third, we wish to make contact with the statistical theories of nuclear reactions[18], which have long considered related questions regarding "spectral intensities," which crop up as resonance widths in nuclear (and chemical!) reactions. In the spirit of the nuclear reaction literature we also investigate the eigenvalue spacing distributions, which have been the subject of more attention than spectral intensities. The spacing distributions provide clues to the presence or absence of chaos, but the spectral intensities give <u>quantitative</u> information about phase space flow.

The paper is organized as follows. In Section II we review the spectral intensity and spacing conditions for quantum ergodicity. Section III gives a brief description of the statistics of nuclear reactions and their connection to spectral fluctuations. In Section IV we present our model Hamiltonian and give some numerical results in terms of spectra, Porter-Thomas distributions,[19] wavefunctions, energy spacings, and the classical mechanics of the system. Section V gives some interpretation and discusses some interesting reluctance of the quantum system to become as chaotic as the classical. This reluctance goes beyond the random fluctuations in spectral intensities mentioned above and has to do with the downright recalcitrance of some eigenstates to become chaotic. Possible explanations in terms of the short-time effect[1,2], vague tori[20] and quantization around unstable orbits[21] are given. This section also contains a short conclusion.

II. SPECTRA AND ERGODICITY

A. Review of Fundamental Principles

Ergodicity is the weakest form of chaos in classical mechanics[22,23]; yet it is a very strong statement about the dynamics from a chemist's (or nuclear physicist's) point of view. It says that the molecule covers its whole allowed phase space evenly on the time average. Certainly ergodicity is therefore sufficient for RRKM theory to hold for molecular dynamics, such as in slow unimolecular reactions.

Consider a localized nonstationary state $|a\rangle$ (which for example may be produced spectroscopically in the laboratory). We may ask: what does ergodicity imply about $|a(t)\rangle$, and what are the consequences for the high resolution spectrum of $|a\rangle$, which can be determined from $|a(t)\rangle$ as[24]

$$\begin{aligned}\varepsilon(\omega) &= \omega\kappa \int_{-\infty}^{\infty} e^{i\omega t} \langle a|a(t)\rangle\, dt \\ &= 2\pi\omega\kappa \sum_n p_n{}^a\, \delta(\omega - E_n/h)\end{aligned} \tag{2}$$

where κ is a constant,

$$\begin{aligned}p_n{}^a &= |\langle a|n\rangle|^2, \\ \langle a|a\rangle &= 1,\end{aligned} \tag{3}$$

and $|n\rangle$ is an eigenstate of the Hamiltonian with eigenenergy E_n. The time averaged probability of starting in $|a\rangle$ and being found in $|a\rangle$ at some later time is just Eq. (1) with a=b, i.e. $P(a|a)$, for a nondegenerate spectrum, is[1,3,25]

$$P(a|a) = \sum_n (p_n{}^a)^2 \tag{4}$$

Also, $P(a|b)$ is calculated separately from the spectra of $|a\rangle$ and $|b\rangle$ and is just

$$P(a|b) = \sum_n p_n^a \, p_n^b \tag{5}$$

Already in Equations (4) and (5) we see a direct connection between spectra (the measurables p_n^a, p_n^b,...) and phase space flow ($P(a|a)$, $P(a|b)$,...).

First we prove a trivial ergodicity property, one shared by quantum as well as classical systems: wherever $|a(t)\rangle$ does go in phase space, be it only a small fraction of the available space or nearly all of it, it goes ergodically. Let $|a'\rangle = |a(\tau)\rangle$ for arbitrary τ. Then,

$$P(a|a) = P(a|a'), \tag{6}$$

for $|a\rangle$ and $|a'\rangle = e^{-iH\tau}|a\rangle$ have the same spectral intensities, p_n^a. Property (6) is analogous to the classical result that if the motion is confined to torus and is quasiperiodic with irrational ratios of the classical frequencies, then the classical motion is ergodic on the torus. If rational ratios of frequencies hold, the motion is ergodic in a space of even lower dimension.

We have heretofore dodged the question of just what is the "available" phase space. On hindsight, all systems are trivially ergodic, for the reason just proved above: once we have executed the full dynamics, the "available" space is just where $|a(t)\rangle$ actually went. But in foresight, the situation is different and quite analogous to the usual experimental questions. Starting in $|a\rangle$, does the dynamics access the state $|b\rangle$, both of which we define at the begining? This is where the concept of elligible or accessible phase space must come in, both in classical and quantum mechancial systems. "Accessible" means all of the phase space (all states $|b\rangle$) consistent with the known prior constants of the motion. These constants include: total energy E (or really $\langle E\rangle$ for distributions), $\langle E^2\rangle$,..., known conserved symmetries (e.g. a A_1 state cannot evolve into B_2), J (angular momentum) conservation, if present, and other constants of the motion which may be known a priori. The spectral moments are conveniently summarized in the concept of a spectral envelope or strength function, $S(\omega)$, which is simply the smoothed version of the spectrum. Some ambiguity may exist as to how to smooth the spectrum, but a brief Ansatz is to

1. Determine $\langle a|a(t)\rangle$ by inverse Fourier transform of the spectrum.
2. Back transform to the frequency domain, setting $\langle a|a(t)\rangle$ to zero for all times after its initial fall of to zero or nearly zero.

The resulting $S(\omega)$ is the envelope. (Ambiguities arise when $\langle a|a(t)\rangle$ refuses to go nearly to zero, which can happen when only a very few lines are present in $\varepsilon(\omega)$). $S(\omega)$ is a constant of the motion, and any definition of ergodicity must account for it. Thus, we define ergodicity with respect to $S(\omega)$ as $P(a|a) = P(a|b) = P(b|b)$ etc. for all localized states $|a\rangle$, $|b\rangle$ which have the same envelopes $S(\omega)$. (Methods for dealing with non-equal envelopes are easily developed, see Ref. (1)). Stechel[17] gives a rigorous quantum ergodic theory based on these concepts.

The envelope $S(\omega)$ defines the energy shell mentioned earlier. While $S(\omega)$ is a necessary concept in the quantum systems because $|a\rangle$ is nonstationary and therefore has energy dispersion, $S(\omega)$ is also a rigorous and necessary concept for classical phase space distributions, which by the Wigner phase space picture, are the correct classical analog to wavefunctions.

The envelope gives the energy distribution of the localized state $|a\rangle$. The "first decay" of the autocorrelation function, function, as just described, gives the envelope via inverse Fourier transform. Now, if we investigate the spectrum at somewhat higher resolution, or, what is the same thing, carry the inverse Fourier transform out to longer time, we may see finer spectral structure or bands, composed of even finer structure, etc. If these bands can be resolved down to the baseline, then each band corresponds to a well-defined nonstationary state $|b_\ell\rangle$, which has a narrower energy dispersion than the original state $|a\rangle$. (An example of this is: $|a\rangle$ is a coherent state or gaussian wavepacket, and $|b_\ell\rangle$ are zero order vibrational states of a single mode such that

$$|a\rangle = \sum_\ell a_\ell |b_\ell\rangle$$

Then if we consider a particular $|b_\ell\rangle$ to be our localized state, the appropriate envelope is given by the energy dispersion corresponding to $|b_\ell\rangle$, which is again obtained by Fourier transform of the "first decay" of $\langle b_\ell | b_\ell(t)\rangle$. To restate: for each localized state $|\alpha\rangle$ spanning sufficiently many eigenstates, the appropriate envelope $S(\omega)$ is given by the Fourier transform of $\langle\alpha|\alpha(t)\rangle$ over the first decay region, i.e.,

$$S(\omega) = \int_{-\tau}^{\tau} e^{i\omega t} \langle\alpha|\alpha(t)\rangle \, dt \tag{7}$$

where τ is the first decay of $\langle\alpha|\alpha(t)\rangle$; $\langle\alpha|\alpha(\tau)\rangle \approx 0$. Careful study of this concept wil help to remove possible ambiguities concerning appropriate envelopes. These ideas are further discussed in Ref. (1) and (3).

For reasons which might all be lumped into the category of in terference effects, quantum systems (two or more dimensional) can never be mathematically as ergodic as classical ones, as long as we are dealing with quantum pure states. The reasons for this will become evident in this paper and have been discussed in the context of a symmetry related effect earlier[26] and a more general context more recently.[16] One can get around the interference problem with deliberate coarse graining (use of incoherent density matrices, as done by Kay[22]), but we feel this is throwing out the baby with the bathwater since most isolated molecule experiments, for example experiments by Parmenter[27] and Smalley,[28] deal with, or would like to deal with, pure states.

One key word in the above definition of quantum ergodicity is "localized." Confusion over this point has arisen in the past, and we would like to discuss it further. If we were free to define <u>any</u> state $|b\rangle$ with the envelope $S(\omega)$, then we could easily define $|b_1\rangle$ and $|b_2\rangle$ such that $P(b_1|b_2) = 0$ merely by <u>defining</u> spectral intensities $p_n^{b_1}$, $p_n^{b_2}$ which are mutually exclusive, i.e. $p_n^{b_1} \cdot p_n^{b_2} = 0$, for all n. This is quite possible to do keeping the same $S(\omega)$, because of the smoothing of the spectrum involved in forming the envelope. It is also clearly possible to arrange this independent of the intrinsic dynamics. This seems paradoxical, but the problem stems from "tampering" with the spectral intensities. By such tampering we lose sight of the physical nature of $|a\rangle$. In the example above, both $|b_1\rangle$ and $|b_2\rangle$ would prove to be delocalized states. By delocalized, we mean spread randomly over the allowed energy shell. Clearly such random functions cannot be expected to be sensitive probes of the intrinsic dynamics. By excluding such functions from consideration, we at once limit ourselves to states in which, in molecular terms, the energy is localized in some mode or combination of modes, and we remove the paradox of $P(b_1|b_2) = 0$.

Such difficulties (i.e. the necessity to exclude delocalized states) can be traced to the fact that interference effects never go away for pure states even as $\hbar \to 0$. For example, wavefunctions have nodes i.e. $\Psi(x) = 0$, therefore $|\Psi(x)|^2 = 0$ and the system has a zero probability of being found at a position where it would be found in a classically ergodic system at the same energy.

In any case, the requirement $P(a|a) = P(a|b)$ for all localized states $|a\rangle, |b\rangle$, with the same envelope $S(\omega)$ immediately results in[1,3]

$$p_n^a = p_n^b = \dots \tag{8}$$

i.e. all spectra the same, independent of the localized state! Since it can be shown that the <u>average</u> spectral intensities $\langle p_n^b\rangle$, averaged over all localized $|b\rangle$, smoothly fills in the envelope[1], we have the <u>ergodic ideal</u> (never to be achieved in two or more dimensional quantum pure state systems because of interference effects), where

$$p_n{}^a = \frac{S(E_n)}{D(E_n)}, \tag{9}$$

and $D(E_n)$ is the local density of states. The local density of states must come in if the $p_n{}^a$'s are to re-generate the energy envelope, or strength function $S(\omega)$.

This ergodic ideal corresponds to <u>maximal spectral congestion</u>: every symmetry allowed eigenstate has intensity, none are "missing" from the spectrum. Since, e.g.

$$P(a|a) = \sum_n (p_n{}^a)^2$$

and

$$\sum_n p_n{}^a = 1,$$

the reader can easily be convinced that maximal spectral congestion corresponds to minimal $P(a|a)$, i.e., an ergodic system spends a minimum of time "at home" consistent with the notions 1) the system spends equal time everywhere it goes and 2) it goes more places if it is ergodic. Indeed, from knowledge of $P(a|a)$ and the total molecular density of states, the fraction of available phase space covered by $|a(t)\rangle$ may be computed.[2,3b]

The spectral congestion is expected from the correspondence principle: chaotic motion in classical mechanics is known to give rise to a continuous frequency spectrum. A bounded quantum system never can have a continuous spectrum, but a bounded "ergodic" quantum system tries as hard as it can to have one, i.e. every possible eigenfrequency has intensity. The discrete spectrum guarantees that certain classically chaotic properties, such as mixing, can never be attained in quantum pure state systems[29], but a kind of imperfect ergodicity, which might be termed "ergodicity with interference effects," can be achieved. We now discuss the imperfection in more detail.

B. Ideal Versus Real Quantum Ergodicity

The requirement $P(a|a) = P(a|b) =$ etc. leads mathematically to $p_n{}^a = p_n{}^b = \ldots$etc. for all n and all a, b... having the envelope $S(E)$. Two important things are settled in this section. One, <u>we must expect this ideal never to be achieved for pure state quantum systems, no matter how small h is or how classically ergodic the system.</u> Two, <u>the practical implications of this imperfect ergodicity in chemistry are normally nil.</u>

The imperfections manifest themselves as random fluctuations in the $p_n{}^a$'s no matter how ergodic the system is classically. The need to consider the fluctuations was noted by Heller and Davis[3a], but much progress on the nature and effect of the fluctuations has been made by Stechel and Sundberg.[16]

Why should the p_n^a's fluctuate about the ergodic ideal, Eq.(9)? In the quantum ergodic case, we are speaking of random fluctuations, not the qualitatively very different fluctuations seen in the quasiperiodic domain. In the classically quasiperiodic regions as $h \to 0$ most p_n^a's will tend to vanish (or nearly vanish), if $|a\rangle$ is localized in phase space. Many of those p_n^a's that do not vanish are very large compared to Eq. (9). This is because localized in phase space implies localized in action-angle space (for this is also phase space by a cannonical transformation), and only those quantum states whose actions $(j_1, j_2 \ldots j_n)$ are near the average actions $(j_1^a, j_2^a \ldots j_n^a)$ of $|a\rangle$ can have significant intensity. The remaining states, by far the vast majority for large n at high energy, have negligible intensity.

When there are no good action variables, the above argument loses significance, and we expect far more states to have intensity in the spectrum of a localized state $|a\rangle$. Viewed from the point of view of the eigenstates, if $\Psi_n(\vec{x})$ is globally or quasi-microcannically distributed over phase space, then it will be found in $|a\rangle$'s region of phase space too, provided $|a\rangle$'s energy envelope overlaps E_n. Thus, a significant p_n^a is to be expected, subject to envelope considerations, for every $|n\rangle$. However, the proximity of $|n\rangle$'s to $|a\rangle$'s phase space locale does not guarantee large overlap. We have to consider the nature of the integral

$$\langle n|a\rangle = \int \Psi_n^*(\vec{x})\, \Psi_a(\vec{x})\, d\vec{x} \qquad (10)$$

If $\Psi_n(\vec{x})$ is a quasiperiodic state, then $\Psi_n(\vec{x})$ might be essentially zero everywhere $\Psi_a(\vec{x})$ is large. If $\Psi_n(\vec{x})$ is a quantum ergodic state, then the integrand in Equation (10) will be nonvanishing for most values of x. However, the integrand is still expected to fluctuate between positive and negative values. Indeed, something with a gaussian distribution peaked about zero might be expected.

We motivate the fluctuations in the ergodic case by appealing to a very plausible, yet unproven, conjecture by Berry[30]: As functions of x, chaotic $\Psi_n(\vec{x})$ are gaussian random. At a given $\vec{x}$, there is a unique value of kinetic energy available, namely $E-V(\vec{x}) = T$. Locally, $\Psi_n(\vec{x})$ is to be thought of as a superposition of many plane waves, with random phases

$$\Psi_n(\vec{x}) \propto \mathrm{Re}\,\Big(\sum_k e^{i/\hbar\, \vec{p}_k\cdot\vec{x} + i\phi_k}\Big), \qquad (11)$$

where

$$T = \vec{p}_k^{\,2}/2m$$

Crudely and semiclassically speaking, trajectories keep returning to the same place with random direction but more importantly with random phase ϕ_k.

Now, suppose we consider the overlap of $\Psi_n(\vec{x})$ with a coherent state $|\alpha\rangle$:

$$\langle\vec{x}|\alpha\rangle = \eta \exp[-\vec{x}\cdot\overleftrightarrow{A}\cdot\vec{x} + i\vec{B}\cdot\vec{x}] \tag{12}$$

where η is a normalization constant. Then, it is easy to show that $\langle\alpha|\Psi_n\rangle$ is complex gaussian random, i.e. both the real and imaginary parts of the overlap are separately gaussian random with zero mean.

If this is the case, then P_n^α is a χ^2 distribution of two random variables (separately the real and the imaginary parts), i.e. χ_2^2.

$$P_n^\alpha = p_n^\alpha/p_n^{\alpha,STO} = |\langle\alpha|\Psi\rangle|^2/p_n^{\alpha,STO} \tag{13a}$$

$$= \{[\mathrm{Re}(\langle\alpha|\Psi_n\rangle)]^2 + [\mathrm{Im}(\langle\alpha|\Psi_n\rangle)]^2\}/p_n^{\alpha,STO} \tag{13b}$$

where $p_n^{\alpha,STO}$ is the ergodic ideal for p_n^α, see Eq. (9). Since for nondegenerate systems $\Psi_n(\vec{x})$ is real, if B=0 in Eq. (12) we have $\langle\alpha|\Psi_n\rangle$ = real, and P_n^α would be a χ_1^2 distribution. Dividing by $p_n^{\alpha,STO}$ in Eq. (13a) locally adjusts the intensities for the fact that the p_n^a's are expected to be small when $p_n^{\alpha,STO}$ is small. While we have a separate section on the relevance of the present ergodic theory to the statistical theory of nuclear reactions, we cannot help but mention that the χ_1^2 distribution is the famous "Porter-Thomas" distribution[19], which seems to be well obeyed by neutron resonance width fluctuations.

If we define, as in Ref. (1),

$$P^{STO}(a|a) = \sum_n (p_n^{a,STO})^2, \tag{14}$$

with $p_n^{a,STO}$ given by Eq. (9), and "STO" refers to "stochastic" or the ergodic ideal, then with the χ_1^2, distribution Stechel and Sundberg show that because of the fluctuation the <u>expected</u> value of the quantum ergodic P(a|a) is

$$P^{QE}(a|a) = 3\,P^{STO}(a|a) \tag{15}$$

and the same equation holds for χ_2^2, with a factor of 2 instead of 3.

The factor of 3 (or 2) is certainly a significant deviation from the ergodic ideal, especially since $P(a|a)$ can be interpreted as

$$P(a|a) = h^{\ell}/V \tag{16}$$

where V is the total phase space volume explored by the dynamics of $|a(t)\rangle$ and ℓ is the number of degrees of freedom. The factor of three means that the most ergodic quantum system misses 2/3 of the total phase space volume! This seems to be a disaster for quantum ergodicity, but really it is not. We finish this section with a discussion showing why the factor of 2/3 (or 1/2) is not very serious.

We state the "bottom line" of our argument first, then motivate it. The 2/3 (or 1/2) of missing phase space volume is quite real; it is really missing but is <u>filamentary</u>. That is, the missing volume that $|a(t)\rangle$ does not explore in a QE system is distributed throughout phase space in a delocalized, filamentary manner, as in the magnesium wire in a flash bulb. Thus, other localized states $|b\rangle$, $|b'\rangle$, etc. will not see the defects in the time averaged $|a(t)\rangle$, and the $P(a|b)$'s in a QE system will generally be much closer to P^{STO}, than $P(a|a)$ is to $P^{STO}(a|a)$.

The reason for this is mathematically rather simple. Suppose we have the set $\{p_n{}^a\}$, which has random fluctuation about $p_n{}^{a,STO}$, and likewise $\{p_n{}^b\}$. If $|a\rangle$ and $|b\rangle$ are far from each other in phase space, there is no reason to assume that for a QE system the fluctuations will be correlated. Then

$$P(a|b) = \sum_n p_n{}^a \, p_n{}^b \approx P^{STO}(a|b).$$

Of course, there will be some fluctuation of $P(a|b)$ about $P^{STO}(a|b)$, but their fluctuations are small if many terms contribute to the sum (i.e. many lines in the spectrum). Stechel and Sundberg[16] have shown that the expected value of $P(a|b)$ is $P^{STO}(a|b)$, and have given the variance also.

The 1/2 or 2/3 of <u>missing</u> phase space in the time evolution of $|a(t)\rangle$ (assuming totally QE motion) can be found by using those "illegal", highly delocalized probe states defined earlier. These states $|b'\rangle$ are all <u>defined</u> to have large $p_n{}^{b'}$ where $p_n{}^a$ is small and vise-versa. Since the most likely $p_n{}^a$ is zero for $\chi_1{}^2$, there are many such $|b'\rangle$ states with small $P(a|b')$. But again, the $|b'\rangle$'s are delocalized states.

C. Summary of Quantum Ergodicity

The considerations given above, together with the references cited, give our view of quantum ergodicty and its manifestations for spectra. Here we avoid the subtleties and summarize our conclusions, for clarity and for a point of departure for the remainder of the manuscript.

1. Quantum Ergodicity is defined in terms of phase space flow between localized, nonstationary states.

2. Ergodicity must be defined with respect to prior constraints; in the quantum case (and in the classical case for the analagous phase space distribution) these constraints include the strength function, or energy envelope S(E), together with any rigorous symmetries.

3. Ideal ergodicty is defined as equal time averaged flow to any other state satisfying the constraints.

4. Because of interference effects, quantum fluctuations in the amount of phase space flow to various states satisfying the constraints always exist. But if the deviations from ideal ergodicity are random, we say that the system is quantum ergodic, (QE), i.e. as ergodic as quantum mechanics allows.

5. An ideally ergodic system will have smooth spectral intensities p_n^a. A quantum ergodic system has local adjusted intensities p_n^a taken from a χ^2 distribution of one degree of freedom for real amplitudes and two degrees of freedom for complex amplitudes.

6. Since the spectral intensities are measurable, it is possible to test for quantum ergodicity by taking a spectrum and obtaining the p_n^a's. Locally adjusting them to $P_n^a = p_n^a/p_n^{a,STO}$ a χ^2 distribution is obtained for QE systems. Computing $P(a|a)$ from the p_n^a's, we obtain the volume of phase space covered by $|a(t)\rangle$. Computing $P^{STO}(a|a)$ from the $p_n^{a,STO}$ values obtained from Eq. (9) we have the total available volume.

Comparing the sample volume to the total volume, a measure of ergodicity is obtained. A QE system of one (two) degree(s) of freedom is expected to sample 1/3 (1/2) of the total volume.

7. Explicit measures of phase space flow can be made by taking the spectrum of $|a\rangle$ and $|b\rangle$ separately. Then $P(a|b) = \sum p_n^a p_n^b$. But also $P^{STO}(a|b) = \sum p_n^{a,STO} p_n^{b,STO}$. This measures is still useful even if $|a\rangle$ and $|b\rangle$ have different envelopes, since each has a $p_n^{a \text{ or } b,STO}$ determined by its own envelope, so that it is meaningful to compare the actual $P(a|b)$ with $P^{STO}(a|b)$.

8. The hallmark of non-QE systems will be more zero or near zero p_n^a's and correlations between p_n^a and p_n^b such that $P(a|b)$ can be much larger or much smaller than QE allows.

The numerical results presented below show that the $P(a|b)$'s are rather more sensitive to failure to achieve full QE than are the $P(a|a)$'s. However, $P(a|a)$ is still a useful indication of ergodicity.

D. Properties of Individual Eigenstates

We have concentrated on the statistical properties of the spectral intensities p_n^a for given localized states $|a\rangle$, with n varying over the eigenstates of the system. Within certain limitations, the statistics are the same if we keep $|n\rangle$ fixed and vary $|a\rangle$. That is, if we examine various localized states $|a\rangle$ for the same eigenstate $|n\rangle$. we find $p_n^a = |\langle a|n\rangle|^2$ is usually zero or near zero for most $|a\rangle$ when $|n\rangle$ is quasiperiodic and randomly fluctuating p_n^a (coming from a χ^2 distribution) if the state $|n\rangle$ is chaotic.

Indeed, when we examine a picture of an eigenstate (plotted as contours of $|\Psi|^2$) we are really examining it with a localized state $|a\rangle = |\vec{x}\rangle$ (position eigenstates). Berry et al[31] used the measure

$$K_n = \int |\Psi_n(\vec{x})|^4 \, d\vec{x} \tag{17}$$

and noted that, for quasiperiodic dynamics, the various caustics that arise as a function of position give rise to very different K_n behavior as $h \to 0$ than for chaotic dynamics, where no such caustics (except anti-caustics) can exist. Indeed, the various types of caustic (e.g. swallowtail, fold, etc.) were shown to give rise to different h dependencies.

To see the relation of Eq. (17) with the present work, we write

$$\begin{aligned} K_n &= \int |\Psi_n(\vec{x})|^2 \, |\Psi_n(\vec{x})|^2 \, d\vec{x} \\ &= \int d\vec{x} \;\; p_n^{\vec{x}} \, p_n^{\vec{x}} \end{aligned} \tag{18}$$

In this last form we see explicitly that K_n is closely related to the spectral mesures we have been using; it is like "$P(a|a)$" value except that $a = \vec{x}$ is being summed over, not the eigenstates $|n\rangle$. K_n is definitely a measure of the locality of $|n\rangle$. Since

$$\int d\vec{x} \; p_n^{\vec{x}} = 1, \tag{19}$$

K_n is smallest if $p_n^{\vec{x}}$ is evenly distributed over $\vec{x}$, and larger if $p_n^{\vec{x}}$ occurs in localized clumps.

III. THE STATISTICAL THEORY OF NUCLEAR REACTIONS

A. General Considerations

The statistical theory of nuclear reactions is perhaps the first practical theory of "quantum chaos." The theory has enjoyed development over a period of more than 30 years.[18] Strangely, it has never been embraced by chemists. Its assumptions are surely compatible with the RRKM theory of unimolecular decay, for example, but it is much more explict about the statistics of individual resonance level fluctuations (widths and spacings). However, some of the motivation for adoption of a statistical theory of scattering is lost in chemistry for two reasons. First, the Hamiltonian is very well known in chemistry so random matrix assumptions and their associated statistics are less natural. Second, individual channel resonances are more difficult to measure in chemistry, and very often the non-resonant, direct scattering processes dominate. Still indirect unimolecular decay processes, which are really by definition resonance controlled, are ideal subjects for the statistical point of view. A future publication will deal, in part, with this case.[33]

Here, we are dealing with bound states, not unbound or scattering cases. In practice, many different phase space probes can be devised for bound state systems in the laboratory. We have already discussed the spectroscopically accessible states, of which a whole host are in principle available for any given molecule.

Dealing with bound states and general phase space probes (S-matrix of scattering theory, which treat the internal scattering dynamics as a "black box," uses "channel functions" as a probe) gives rise to a somewhat different formulation than in the statistical nuclear reaction theory. Also, the chemical (specifically molecular spectroscopic) motivation for the theory put forth here has a known Hamiltonian at its foundation. This means that we will want to investigate many of the statistical quantities (energy level distribution, strength fluctuations, etc.) for a smooth, many body potential, rather than for random matrices. The study of random matrices forms the extremely useful limiting case of quantum chaos but leads to chaos or random behavior by construction.[34] Smooth Hamiltonians have to give spontaneous birth to chaos or randomness in a more subtle way.

B. Phase Space Flow, $p_n{}^a$ and Γ_{nc}

The concept of flow in phase space, $P(a|a)$, $P(a|b)$, etc. does not seem to be a part of the language of the statistical theories in nuclear physics. There are, however, analogous concepts. The only "flow" of real importance to scattering theory is contained in the cross sections, σ_{ab}, between channel a and channel b.[35] If this cross section is dominated by narrow (i.e. long lived) resonance, then the concepts of statistical properties of the resonance states Ψ_n and resonance energeis E_n are natural. If the Hamiltonian is strongly coupled, random statistics (e.g. gaussian random matrix elements) might be expected.

The resonance partial widths Γ_{na} play an analogous role to the spectral intensities $p_n{}^a$. Here n again refers to the eigenstate in the case of $p_n{}^a$, and the resonance state in the case of Γ_{na}. The label "a" refers to a phase space state for $p_n{}^a$, and a reaction channel state for Γ_{na}. Both $p_n{}^a$ and Γ_{na} are essentially squared overlaps of eigenstates with localized states. Defining

$$\tilde{\Gamma}_{na} = \Gamma_{na}/(\Gamma_n)^{1/2} \tag{20}$$

where

$$\tilde{\Gamma}_n = \sum_n \Gamma_{nc}$$

is the total resonance width, we obtain the cross section $\langle\sigma_{ab}\rangle$ averaged over several or many resonances as

$$\langle\sigma_{ab}\rangle = \sum_n \tilde{\Gamma}_{na}\tilde{\Gamma}_{nb} \tag{21}$$

For the decay of a single nuclear resonance ("unimolecular decay" in chemistry) the partial width Γ_{nc} gives the decay into various channels c. The statistics of the Γ_{nc} have been given much attention, and under the standard assumption that the amplitudes γ_{nc} are gaussian random with zero mean, then $\Gamma_{nc} = 2P_c\gamma_{nc}^2$, where P_c is a kinematic factor, fits a one-degree-of-freedom chi-squared distribution, i.e. χ_1^2. We have already discovered why the $P_n{}^a$'s might likewise be expected to be χ^2, for real valued $|a\rangle$, and χ_2^2 for complex-valued $|a\rangle$. This is checked in Section IV on a classically chaotic, smooth potential.

Seen in the light of these analogies, the spectral theory of quantum ergodicity put forward here is a natural extension of the much older statistical theories of nuclear reaction. It has the same foundation and should eventually prove a very useful probe of molecular chaos for bound states, and indeed for nuclear chaos in the case of giant dipole resonances, for example, if the spectral features can be resolved (i.e. if the $p_n{}^a$'s can be measured).

C. Phase Space Explored and Effective Number of Channels

One of the concepts emerging from the $P(a|a)$ analysis was the volume V accessed in phase space, given by

$$P(a|a) = h^{\ell}/V = \sum_n (p_n{}^a)^2 \tag{22}$$

where ℓ is the number of degrees of freedom. A more meaningful quantity is the fraction F of available volume, i.e.

$$F = V/V^* \tag{23}$$

where

$$P^{STO}(a|a) = \sum_n (p_n^{a,STO})^2 = h^\ell/V^*$$

with $p_n^{a,STO}$ given by Eq. (9).

Brody et al.[34] have reviewed the results of Draayer et al.[36] for the effective number of open channels ν. This number is low when the system exhibits "strong collective behavior" and only a few states share in the strength ("singular spectrum"), i.e. simple quasiperiodic motion in our current language, and is high when the system exhibits a"rich spectrum," i.e. quantum ergodic. Thus ν is the analog of our number $N = V/h^\ell$ of phase space cells accessed in the dynamics of $|a\rangle$.

IV. NUMERICAL RESULTS

In the previous sections we have said a great deal about what is expected of spectra, $P(a|a)$'s, $P(a|b)$'s, and wavefunctions for a QE system. Here we will present results of extensive calculations on what is probably the most nearly QE system with a smooth potential studied to date. As mentioned earlier, the "demonic" potential investigated here was designed to have extensive chaotic regions without the annoyance of imminent dissociation. The demonic potential has this property, along with regions of co-existing quasiperiodicity and choas at lower energies, and nearly pure quasiperiodicity still lower in energy.

It is probably best to define the Hamiltonian and study its classical dynamics first, which we do with 1) the usual surface of section approach and 2) a check of classical $P(a|a)$ values compared to $P^{STO}(a|a)$, to see if the classical mechanics is indeed erogdic at higher energies.

Next, we study the eigenvalue spacing distributions in various energy ranges and compare with the Wigner and exponential distributions. These distributions are not our main point here, but it is becoming standard to examine them as they are a qualitative indicator of chaos.

The quantitative indicators of quantum ergodic behavior are examined next: spectral intensities, $P(a|a)$'s, $P(a|b)$'s, and wavefunctions. It is found that the demonic potential is not quite QE even at the highest energies we were able to examine, but that all the indicators are close to their QE limits. The reaons for the interesting reluctance of this system to be fully QE in spite of a fairly small h (high density of states) and extensive classical chaos are discussed in the conclusion.

A. Model Hamiltonian and Classical Dynamics

The two-dimensional Hamiltonian considered is

$$H(u,s) = 1/2\ (P_u^2 + P_s^2) + 1/2\ (\omega_u^2 u^2 + \omega_s^2 s^2) + \lambda u^2 s + \beta(u^4 + s^4) \qquad (24)$$

with $\omega_u = 1.1$, $\omega_s = 1.0$, $\lambda = -3.24$ and $\beta = .324$. This Hamiltonian is similar to the Hamiltonian studied by Barbanis[13], but we have added a quartic term to keep the potential bound. With the parameters used in this study, the potential bears little resemblance to the Barbanis Hamiltonian. It is a highly anharmonic potential with two deep minima separated by a saddle point region near u=0, s=0 and E≈0.02. A contour plot of the "demonic" potential is shown in Fig. 1,

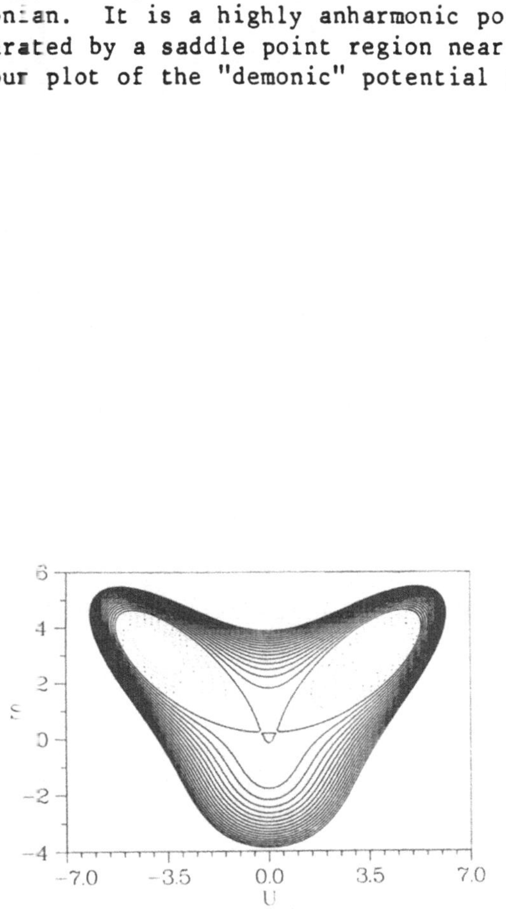

Fig. 1. The demonic potential discussed in the text, see Eq. (24). Contours range from E=80.0 to E=-25.0 in increments of 5.0; contours for E<0.0 are plotted with a broken line.

where the contours are plotted in increments of 5 energy units and contours for $E<0$ are plotted with broken lines.

As mentioned above the classical dynamics of the demonic Hamiltonian was studied by calculating Poincare surface of sections (SOS). The SOS's are obtained for the conjugate variables (δ, P_δ) where $\delta = 1/\sqrt{2}(u-s)$ and $P_\delta = 1/\sqrt{2}(P_u - P_\delta)$. Using the standard technique, a point in the (δ, P_δ) plane is mapped out whenever $P_\xi = 1/\sqrt{2}(P_u + P_\delta) > 0.0$ and $\xi = 1/\sqrt{2}(u + s) = 0.0$. This leads to maps in which quasiperiodic motion forms closed curves and chaotic motion creates a "random" pattern of points.

The classical dynamics of the demonic system is predominantly quasiperiodic in the deep potential wells. The (δ, P_δ) SOS for $E=-10.0$ is shown in Fig. (2) (top), where the quasiperiodic motion does indeed form closed curves and segments of closed curves. The first chaotic trajectories are found when trajectories have sufficient energy to surmount the saddle point near $u=0$ and $s=0$. As higher energies are considered, a larger fraction of phase space becomes chaotic. This coexistence region is show in Fig. (2) (bottom), where the $E=10.0$ SOS is plotted. We see one large island of quasiperiodic space surrounded by a sea of chaos; there are actually two quasiperiodic islands, one above each potential minima, but only one island is mapped into the (δ, P_δ) SOS as we have defined it ($P_\xi > 0.0$ and $\xi = 0.0$).

At even higher classical energies, chaos eventually totally dominates the phase space. The $E=60.0$ SOS is shown in Fig. (2) (right), and it is clear that there are no significant regions of quasiperiodic phase space. To verify that the dynamics is indeed stochastic at this energy, classical $P(a|a)$'s were calculated and compared to estimates for $P^{STO}(a|a)$. In Table I we summarize several calculations of classical $P(a|a)$ and $P^{STO}(a|a)$. In all cases the agreement is with in the error bounds for the Monte Carlo technique used to calculate the classical $P(a|a)$ values. The details of classical $P(a|a)$ and $P^{STO}(a|a)$ calculations are discussed in greater detail in Refs. (2,3).

To summarize, the classical dynamics is quasiperiodic at low energies ($E<0$), chaos and quasiperiodicity coexist at intermediate energies ($0<E<40$) and chaotic dynamics dominates at higher energies ($E>40$). The dynamics should again become predominantly quasiperiodic at sufficiently high energies when the potential is dominated by the quardic terms, but our numerical evidence suggests that this must occur at energies several orders of magnitude above the highest energy we will consider.

At this point we would like to briefly discuss the quantum calculation used to obtain eigenfunctions and eigenvalues. The quantum states were calculated variationally using basis functions of the form:

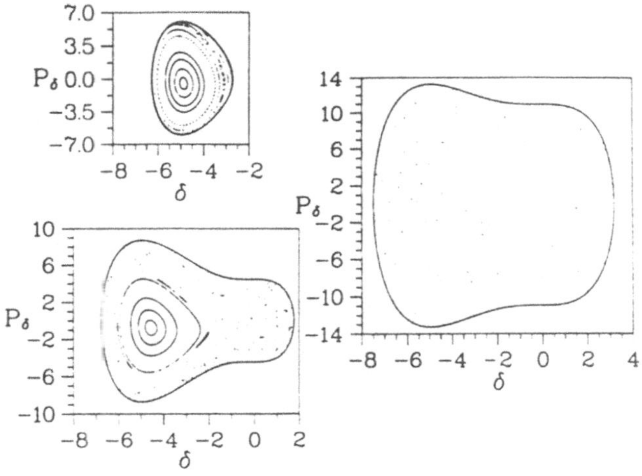

Fig. 2. Classical surface of sections for E=-10.0 (top), E=10.0 (middle) and E=60.0 (bottom). See text for more details.

$$G_k(q_i,p_i) = [g(q_i,p_i) + c_k g(q_i,p_i)]$$

$$+ c_s[g(-q_i,p_i) + c_k g(-q_i,p_i)] \qquad (25)$$

where

$$g(q_i,p_i) = \left(\frac{2\alpha_i}{\pi}\right)^{1/2} \exp[-\alpha_i(q-q_i)^2 + ip_i q] \qquad (26)$$

The parameter c_k allows for the combination of complex functions to obtain the real basis functions $G_k(q_i,p_i)$. The coefficient $c_s$enables the basis to take advantage of any reflection symmetry in the potential. For the Hamiltonian in Eq. (24), c_s= 1 for the u coordinate and c_s = 0.0 for the s coordinate. The coherent state basis in Eq.(26) was introduced by Davis and Heller.[37] The functions are typically placed on a grid in phase space and multi-dimensional basis functions are built up as a product of one-dimensional functions.

TABLE I

Classical $P(a|a)$'s and $P^{STO}(a|a)$'s

Initial state parameters [a]

alpha	s_0	P_{s0}	u_0	P_{u0}	$P^{Cl}(b\|a)$ [b]	$P^{STO}(a\|a)$ [b]
2.0	3.0	0.0	5.0	10.453	.0011	.0010
2.0	-3.55	0.0	.5	0.0	.0016	.0014
2.0	-3.0	0.0	.5	7.301	.0019	.0017
.5	3.0	9.607	3.5	9.607	.0016	.0017
.5	3.0	9.222	3.0	9.222	.0023	.0019

a) Initial states have the form

$$\left(\frac{2\alpha}{\pi}\right)^{1/2} \exp\{-\alpha[(s-s_0)^2 + (u-u_0)^2 + i[sp_{s_0} + up_{u_0}]\}$$

b) The techniques used to calculate $P^{Cl}(a|a)$ and $P^{STO}(a|a)$ are detailed in Ref.(2).

For the Hamiltonian in Eq. (24) a total of 1750 basis functions were used to variationally calculate the eigenvalues and eigenvectors. For the highest energy considered in our study of the dynamics ($E \approx 120$), 575 even states and 559 odd states were converged through the first decimal place ($\approx .2\%$).

As an interesting side light, we speculate that the same accuracy would be much more difficult to achieve using harmonic, adiabatic[38] or any other conventional basis.

B. Nearest-neighbor spacing distributions

The distribution of nearest-neighbor spacings of eigenlevels in a regular system, regular meaning that the corresponding classical dynamics in integrable[39], has been shown by Berry and Tabor[40] to usually be exponential. In fact, these authors demonstrate that the nearest-neighbor spacing distribution for any regular system which has curved energy contours in action space fits an exponential distribution. (The energy contours of uncoupled Harmonic oscillators are straight lines in action space, and the distribution of nearest-neighbor spacings is non-exponential, as well as non-Wigner.) The distribution of nearest-neighbor spacings for the quasiperiodic portion of the demonic potential ($-30.0<E<0.0$) is shown in Fig. (3a). We have also plotted an exponential (dotted line) and Wigner distribution[18] (dashed line) which has the form

$$P(s) = \frac{\pi s}{2} e^{-\pi s^2/4} \tag{27}$$

where s in the nearest neighbor spacing divided by the average spacing. All three distributions are normalized equivalently. Since there are only 21 eignestates in this energy range, the statistics are too poor to make any firm conclusions about the measured distribution.

The distribution of nearest-neighbor spacings for any irregular or QE system is believed to follow a Wigner distribution.[34,41] In Fig.(3b) the distribution of nearest neighbor spacings for the energy range $40.0<E<120.0$ is plotted along with an exponential (dotted) and Wigner distribution (dashed). The agreement between the calculated distribution and the Wigner distribution is excellent.

C. Wavefunctions, Spectra, and $P(a|b)$'s

The break up of nodal patterns in wavefunctions has been correlated with the one of classical chaos.[42] This idea seems reasonable since a QE wavefunction is expected to have an irregular nodal pattern,[31] and very simple regular systems, such as box-type wavefunctions, usually have very regular nodal patterns. However,

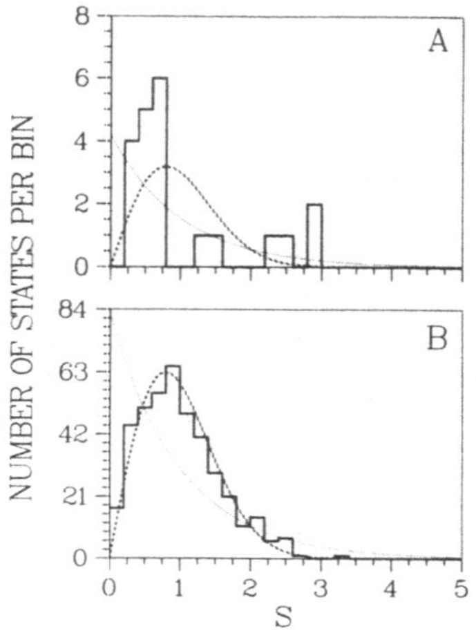

Fig. 3. Distribution of nearest-neighbor eigenlevel spacing for the demonic system (solid line) between a) E=30.0 to E=0.0 and b) E=40.0 to E=120.0. Also shown are the Wigner (dashed) and exponential (dotted) distributions.

these are purely integrable classical phenomena, such as the classical resonance studied by DeLeon, Davis and Heller,[43] which can also complicate the nodal patterns. Therefore, nodal patterns provide only a piece of the answer and other dynamical tests must be performed in order to quantify the nature of quantum dynamics. With these ideas in mind, we will first examine the qualitative features of wavefunctions near E=-10.0, 10.0 and 60.0 and then use spectra and P(a|b)'s to further test the quantum dynamics.

Figure (4) shows contour plots for four consecutive even wavefunctions near E=-10.0 where the classical dynamics is quasi-periodic [see Fig.(2) (top)]. The wavefunctions have two "good" quantum numbers, just as the classical dynamics must have two

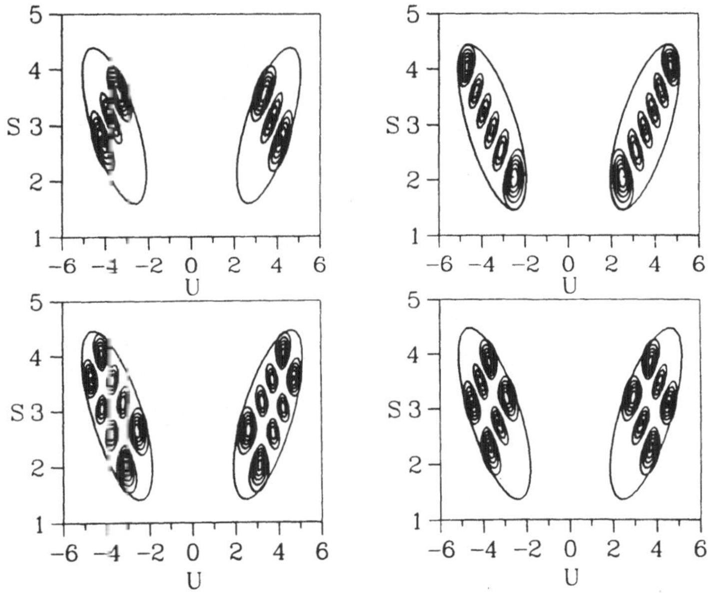

Fig. 4. Contour plots of four consecutive even wavefunctions near E=-10.0 and the eigenlevel potential energy contour for the potential shown in Fig. 1.

constants of the motion. There are no surprises here, the quantum dynamics in this energy range is regular.

The next set of wavefunctions shown in Fig. (5) have energies near E=10.0. The classical dynamics in this energy range has coexisting quasiperiodic and chaotic regions. The upper right wavefunction in Fig. (5) is confined to the two quasiperiodic phase space islands. Classically the two islands are isolated since they are separated by chaotic phase space, but quantum mechanically one can tunnel from one island to the other, so the eigenstate has amplitude in both islands. This type of tunneling has been termed dynamical tunneling[44], since the tunneling is not through an actual potential barrier, but through a classical dynamical barrier. The other wave-

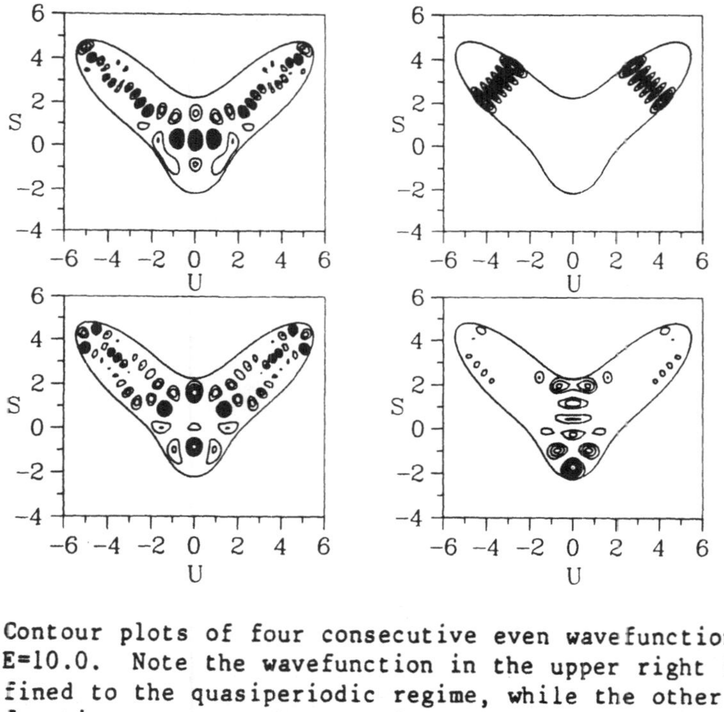

Fig. 5. Contour plots of four consecutive even wavefunctions near E=10.0. Note the wavefunction in the upper right is confined to the quasiperiodic regime, while the other wavefunctions cover much more of the available position space.

functions at this energy cover more of their available phase space, but the coverage is by no means uniform or even terribly "irregular."

The last energy range we will consider is near E=60.0 where the classical dynamics is fully stochastic. Four consecutive even wavefunctions are shown in Fig. (6) and the states generally cover a large part of their available space with fairly irregular nodal patterns. It would, for example, be difficult to assign two good quantum numbers to either of the wavefunctions on the right side of Fig. (6). However, there is not total uniform coverage of phase space, in fact the upper left wavefunction avoids a large part of its phase space. This particular state seems to have built up amplitutde along the short time history of a classical trajectory. The state and the short time evolution of a trajectory are plotted

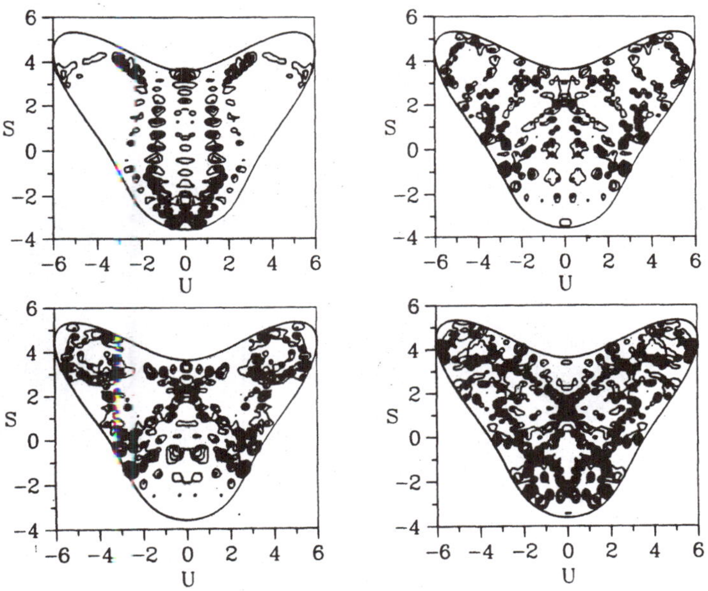

Fig. 6. Contour plots of four consecutive even wavefunctions near E=60.0.

together in Fig. (7). The trajectory eventually "becomes" chaotic (see Fig. (7)), but for a short time it appears quasiperiodic. A second wavefunction and trajectory are also shown in Fig. (7), where the wavefunction again avoids a large part of its available phase space and seems to have amplitude only in the regions sampled by the early dynamics of a trajectory. The influence of short term quasi-periodic-like classical motions will be further discussed in the following section.

Now we have established the general trends in the classical dynamics and have examined the qualitative features of wavefunctions at various energies, we intend to examine the quantum mechanical spectra arising from several different choices for initial states.

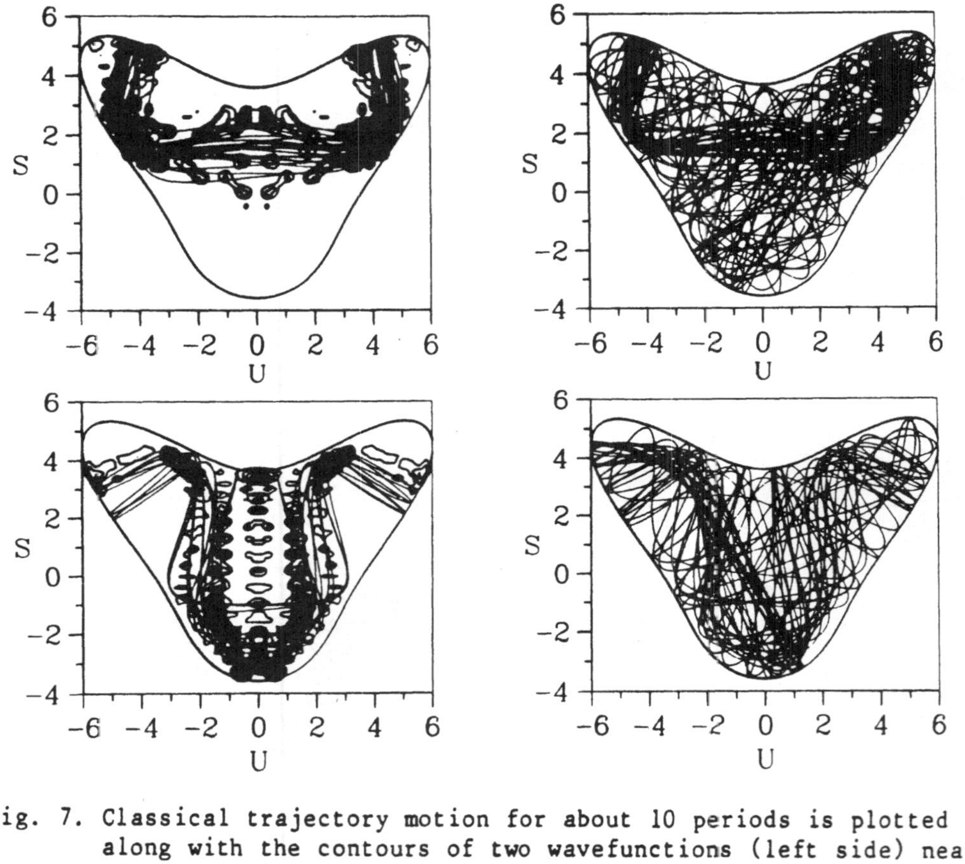

Fig. 7. Classical trajectory motion for about 10 periods is plotted along with the contours of two wavefunctions (left side) near E=60.0. The longer time evolution of the classical trajectories are shown on the right side.

We will be discussing initial states which have energy envelopes peaked near E=-10.0, 10.0 and 60.0. We will use gaussian coherent states of the form given in Eq. (26) for the spectroscopic initial states. The parameters for the specific initial states discussed are summarized in Table II.

Initial state A is placed in quasiperiodic phase space and has a peak envelope energy of approximately -10.0. The FWHM contour of the initial state is plotted along with the E=-10.0 SOS in Fig. (8). The spectra for state A is shown below the SOS, and it is quite sparse with only a single progression of intense p_n^a's. The density of states is shown in the enlarged portion of the spectrum, and even though the density is fairly low there are several p_n^a's which have negligible intensity. This last statement is quantified by examining the histogram plot in Fig. (8), which shows the distribution of $X = P_n^a/\langle P_n^a\rangle_{avr}$ compared to the Porter-Thomas distribution (dotted line). There are extra states with small values of X, as expected for regular (non-stochastic) quantum systems.

Figure (9) includes spectra and X distribution plots for two initial states with energy envelopes peaked near E=10.0. Again the FWHM contours of the initial states are plotted along with the appropriate SOS. Initial state B was placed in a quasiperiodic island, and the resulting spectrum is very sparse, with many, in fact most, p_n^a's having negligible intensity. These are the "missing "lines" that one expects when the spectrum is a result of quasiperiodic dynamics. On the other hand, the spectrum for initial state C, which was placed in chaotic phase space, is very congested

TABLE II

Spectroscopic Initial States

State[a]	alpha	s_o	Ps_o	u_o	Pu_o
A	2.0	+1.923	0.0	-1.923	0.0
B	1.0	- .425	7.02	.425	5.46
C	1.0	.425	8.45	.425	-2.83
D	2.0	-3.55	0.0	.5	0.0

a) All initial states have the form

$$\left(\frac{2\alpha}{\pi}\right)^{1/2} \exp\left\{-\alpha\left[(s-s_o)^2 \quad (u-u_o)^2\right] + i\left[s\, p_{s_o} + u\, p_{u_o}\right]\right\}$$

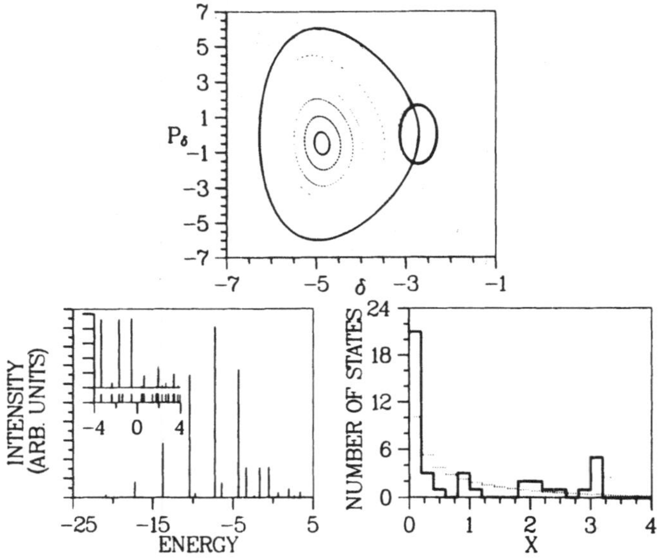

Fig. 8. Classical surface of section for E=-10.0 and FWHM contour of initial state A (top), see Table II for initial state parameters The spectrum for initial state A (lower right) and the distribution of spectral intensities compared to Wigner (dashed) and exponential (dotted) distributions (lower left). The inset shows a magnified portion of the spectrum along with the density of states. See text for more details.

and from the enlarged portion of the spectrum it is clear that nearly every p_n^a is present with healthy intensity. The missing line in spectrum C (see arrow in enlarged spectrum) is by no coincidence the very intense line found in spectrum B. This line corresponds to an eigenstate, or in fact a pair of nearly degenerate eigenstates, which are localized in the quasiperiodic region. Such a state was already shown in Fig. (5). Since there is no classically allowed phase space flow between the chaotic region and the quasiperiodic one, and since

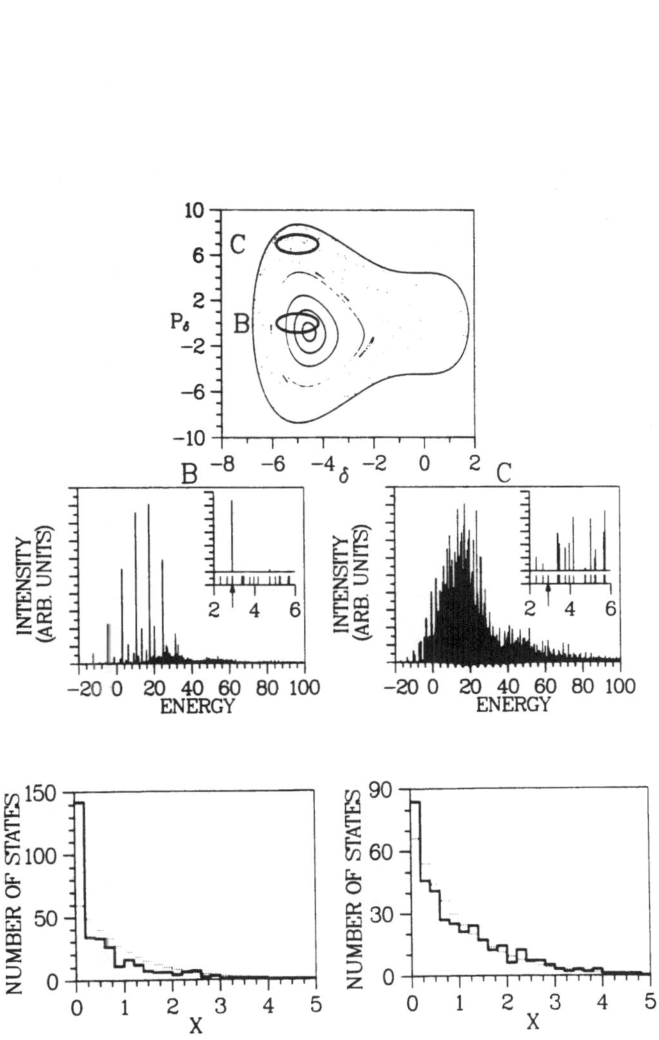

Fig. 9. Classical surface of section for E=10.0 and the FWHM contours of initial states B and C. The spectrum and distribution of spectral intensities for states B (left) and C (right). See text for more details.

$P(a|b) = \sum p_n^a p_n^b$, the spectra should indeed be mutually exclusive, except for the possibility of quantum tunneling.

Initial states B and C are complex since they have a "kick" in momentum (see Table I.) Consequenty, their X distributions should be compared to the two-degree of freedom chi-squared distribution. The X distributions for both initial states are shown in the bottom panel of Fig. (9). Initial state B has many more states with small X values than the chi-squared distribution predicts, i.e. many negligible p_n^a's. On the other hand, the χ-distribution for state C nearly matches the chi-squared distribution, with only a

TABLE III.

Fraction of Sampled Phase Space

Real Initial State Parameters[a]

α	s_o	o	F^b
2.0	-3.55	.5	.37
2.0	-2.62	2.0	.28
2.0	2.05	5.0	.25
2.0	.625	4.0	.30
2.0	5.1	4.0	.24

Complex initial state parameters[c]

α	s_o	p_{s_o}	u_o	p_{u_o}	F
2.0	-3.0	0.0	.5	7.301	.48
2.0	3.0	7.939	.5	0.0	.40
2.0	3.0	-8.372	3.0	10.0	.42
2.0	3.0	0.0	5.0	10.453	.54
2.0	-1.2	-6.63	2.4	0.0	.61

a) Real initial states have the form $\left(\frac{2\alpha}{\pi}\right)^{1/2} \exp\{-\alpha[(s-s_o)^2 + (u-u_o)^2]\}$.

b) Fraction of sampled phase space $F \equiv P^{STO}(a|a)/P(a|a)$

c) Complex initial states have the form:

$$\left(\frac{2\alpha}{\pi}\right)^{1/2} \exp\{-\alpha\,[(u-u_o)^2 + (s-s_o)^2] + i\,[sp_{s_o} + up_{u_o}]\}$$

few extra small X states, since states localized in the quasiperiodic region will have negligible overlap with initial state C.

The classical dynamics near E=60.0 is stochastic, and hence any localized initial state with approximately the same energy envelope would have qualitatively the same spectrum. Individual p_n^a intensities would vary for states |a>, |a'>, etc., but the spectrum would look essentially the same for all such states. The FWHM contour for a possible state is shown along with the E=60.0 SOS in Fig. (10). The quantum spectrum for state D is quite congested and from the enlarged portion of the spectrum, which also shows the

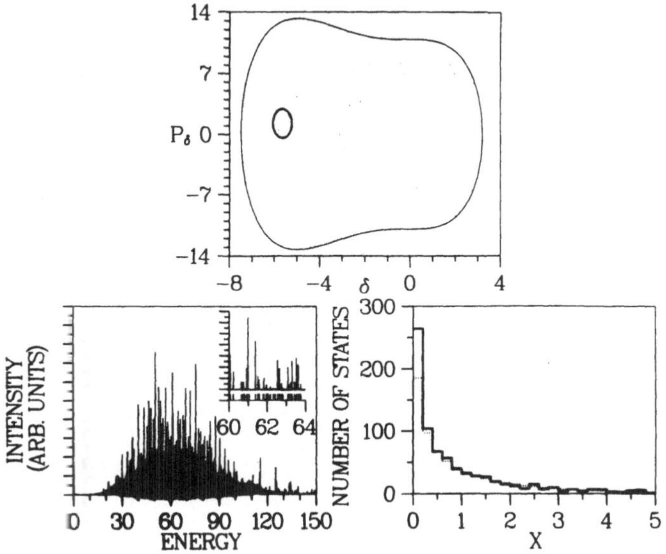

Fig. 10. Classical surface of section for E=60.0 and FWHM contour of an initial state with the same width as state D (top). The spectrum (lower left), distribution of spectral intensities and Porter-Thomas distribution (lower right.) The inset shows a magnified portion of the spectrum, and the tick marks show the eigenvalues of the system.

density of states, it is evident that most p_n^{a}'s are present with significant intensity. There are, however, a few missing lines. Indeed, low intensity is the most probable in a χ^2 distribution, so there will be a few missing lines. However, the χ^2 distribution in Fig. (10) shows a few extra states with small p_n^{a}'s when compared to the Porter-Thomas χ^2 distribution (dotted line). This indicates that the demonic system is not a fully QE system even when the classical dynamics is fully stochastic. This reluctance to be fully QE is consistent with the localized wavefunctions discussed [see Fig. (7)] and will be further discussed in the next section.

A final test of the system is provided by examining the fraction of phase space sampled by various initial states and the fraction that would be sampled in a QE system. The $P(a|a)$ for a QE system is expected to be $2\,P^{STO}(a|a)$ for a complex initial state and $3P^{STO}(a|a)$ for a real initial state. The fraction of phase space sample is defined as $F = P^{STO}(a|a)/P(a|a)$ and values of F for 5 real and 5 complex initial states are summarized in Table III. The fraction of real states ranged from .23 to .37, and for complex states .40 to .61. These values are in good agreement with the expected QE values of 1/3 and 1/2 for real and complex states, respectively. As noted earlier $P(a|b)$'s provide an even more sensitive and direct measure of phase space flow from one region of phase $|a\rangle$ to another $|b\rangle$. Table IV summarizes ten $P(a|b)$ calculations in which all states a and b have energy envelopes peaked near

TABLE IV.

$P(a|b)$'s

State a[a]		State b[a]		$P(a\|b)$	$P^{STO}(a\|b)$
s_0	u_0	s_0	u_0		
-3.55	.5	-2.62	2.0	.81(-3)	.14(-2)
-3.55	.5	2.05	5.0	.10(-2)	.11(-2)
-3.55	.5	6.25	4.0	.12(-2)	.12(-2)
-3.55	.5	5.1	4.0	.57(-3)	.11(-2)
-2.62	2.0	2.05	5.0	.40(-3)	.11(-2)
-2.62	2.0	6.25	4.0	.67(-3)	.13(-2)
-2.62	2.0	5.1	4.0	.38(-3)	.11(-2)
2.05	5.0	0.625	4.0	.13(-2)	.10(-3)
2.05	5.0	5.1	4.0	.52(-3)	.84(-3)
.625	4.0	5.1	4.0	.52(-3)	.94(-3)

a) All initial states have the form

$$\left(\frac{2\alpha}{\pi}\right)^{1/2} \exp\{-\alpha[(s-s_0)^2 + (u-u_0)^2]\} \text{ with } \alpha=2.0.$$

E=60.0. The values of P(a|b) is yet another indication that the demonic system is not fully QE. If the fluctuation in P_n^a, p_n^b were truly random and χ^2 distributed, then the variance and χ^2distributed, then the variance in P(a|b) as the state b is changed would be much smaller than our numbers seem to indicate.This variance is discussed in Stechel and Sundberg.[16]

V. DISCUSSION

In this paper we have discussed a theory of quantum ergodicity which includes quantum interference effects (e.g. nodal structure). We have also examined the consequences of quantum ergodicity on spectra, P(a|a)'s, P(a|b)'s and wavefunctions. Our general results can be summarized as follows.

1) Quantum ergodicity is defined in terms of phase space flow between phase space localized nonstationary states.
2) Ergodicity must be defined with respect to prior constants of the motion, such as the energy envelope S(ω).
3) A Quantum Ergodic (QE) system must deviate from ideal ergodicity (the direct quantum analog of classical ergodicity) because of quantum fluctuations in the phase space flow caused by quantum interference effects.
4) In pure state quantum systems the interference effects are always present, even as $\hbar \to 0$.
5) The spectral intensities p_n^a's of an ideal ergodic system smoothly fill the energy envelope. The p_n^a fluctuations present in a QE system follow a χ^2 distribution of one degree of freedom for real initial states and two degrees of freedom for complex initial states.
6) A QE system of one (two) degree(s) of freedom is expected to sample 1/3 (1/2) of the total energetically allowable phase space. The unsampled volume is spread throughout the phase space in a nar row ribbon so that localized probe states never fall entirely in the unsampled volume. Consequently, the missing phase space has little effect in chemistry.

The numerical results presented in this paper indicate that the demonic system is not quite fully QE. The fluctuations of the spectral intensities showed more "missing states" and the values of P(a|b) had larger fluctuations than is expected from a QE system. In addition, several wavefunctions were shown to be localized to small regions of the available phase space, even though the underlying classical dynamics was fully stochastic. These localized eigenstates seemed to build up their amplitude along shortlived quasiperiodic-like trajectories which at long times cover all of the allowable phase space. An explaination for these localized eigenstates is suggested by the Time-Energy Uncertainly Principle. Loosely speaking, there is no "new" quantum dynamics after a time $T>h/\Delta E$, where ΔE is the minimum energy level spacing.[1] Stated in a slightly different way, the quantum dynamics is "frozen in" after a length of time sufficient to resolve neighboring eigenstates. This implies that when classical mechanics is slow to become fully chaotic, i.e. a small Lyopanov exponent[45], the quantum mechanics may freeze in before the classical dynamics has sampled more than just a small part of its available phase space. This could lead to an eigenstate which would be confined to only a small part of its energetically accessible phase space. The example given in Ref. (1) was the stippled oscillator. In two or more dimensions, narrow stipples on an otherwise smooth integrable potential can wreak havoc with the classical dynamics on a long time scale. But on a short time scale the classical dynamics is nearly unaffected, and the quantum eigenstates will remain almost unaffected by narrow stipples, retaining their basic nodal structure.[46] Chaotic trajectories which exhibit short term quasiperiodic-like motion have been recently described by Shirts and Reinhardt.[20] These authors found trajectories which jumped from one short term quasiperiodic-like motion to another, such that the overall motion was chaotic, but for a short time dynamics appeared to be quasiperiodic. They discussed the concept of a vague torus to describe the stability of the classical dynamics and were able to calculate approximate constants of the motion. Using the language of Shirts and Reinhardt, if a trajectory is confined to a vague torus long enough to resolve neighboring eigenstates, the corresponding quantum dynamics will only sense the short term quasiperiodic motion. This leads to quantum eigenstates that are confined, like a fully integrable system, to only a small region of the allowable phase space. Clearly there is some overlap between the concepts to short time dynamics and the vague torus.

The recent use of closed orbits by Gutzwiller[21] and Tabor[47] to semiclassically quantize ergodic systems suggests that the dense but measure zero set of unstable orbits in the demonic potential may also contribute to the quasiperiodic-like wavefunctions. The techniques used by these authors have been used to obtain only the eigenvalues, never the eigenfunctions, and successful use of the technique depends upon enumerating all the closed orbits of a system. This has been done for only a few select systems as of yet, such as the anisotropic Kepler problem[21,48] and for a class of area-preserving maps.[47] The use of closed orbits to quantize chaotic systems is a potentially powerful technique which still requires more work.

Acknowledgements

We are grateful to M. P. Sundberg for help in preparation of this manuscript. Conversations with E. B. Stechel and M. V. Berry are also gratefully acknowledged.

References

1. E. J. Heller, Quantum intramolecular dynamics: Criteria for stochastic and nonstochastic flow, J. Chem. Phys. 72:1337 (1980).

2. E. B. Stechel and E. J. Heller, Quantum-classical correspondence for vibrational motion of HCN, submitted.

3. a) E. J. Heller and M. J. Davis, Criteria for quantum chaos, J. Phys. Chem. 86:2118 (1982).
b) M. J. Davis and E. J. Heller, Comparisons of classical and quantum dynamics for initially localized states, submitted.

4. a) J. Brickmann and P. C. Schmidt, Quantum Dynamics of Wave-packets on Phase Space in Nonlinearity Coupled Oscillators, Int. J. Quant. Chem. XXIII:147 (1983).
b) J. Brickmann, Quantum dynamics of gaussian wave packets in anharmonic vibrational systems, in "Intramolecular Dynamics," edited by J. Jortner and B. Pullman. Reidel, Boston (1983), and references therein.

5. J. Stone, E. Thiele, and M. F. Goodman, An analytic approach to time averages and chaotic behavior in quantum mechanics, in "Intramolecular Dynamics," edited by J. Jortner and B. Pullman, Reidel, Boston (1983).

6. E. J. Heller, Wigner phase space method: Analysis for semi-classical applications, J. Chem. Phys. 65:1289 (1976).

7. E. B. Stechel and R. N. Schwartz, Spreading and recurrence in anharmonic quantum-mechanical systems, Chem. Phys. Lett. 83:350 (1981), and references therein.

8. Y. Weissman and J. Jortner, What are the quantum manifestations of classically stochasticity in a discrete level structure?, Chem. Phys. Lett. 78:224 (1981).

9. See for example, E. J. Heller, Time-dependent approach to semiclassical dynamics, J. Chem. Phys. 62:1544 (1975) and references therein.

10. a) M. J. Davis, E. B. Stechel and E. J. Heller, Quantum dynamics in classically integrable and non-integrable regions, Chem. Phys. Lett. 76:21 (1980) and b) E. J. Heller and E. B. Stechel, Quantum correspondence to classical chaos, Chem. Phys. Lett. 90:484 (1982).

11. E. J. Heller, The semiclassical way to molecular spectroscopy, Acc. Chem. Res. 14:368 (1981), and references therein.

12. M Henon and C. Heiles, The applicability of the third integral of motion: some numerical experiments, Astron. J. 69:73 (1964).

13. B. Barbanis, On the isolating character of the "third" integral in a resonance case. Astron. J. 71:415 (1966).

14. E. J. Heller, The correspondence principle and intramolecular dynamics, Faraday Discuss. Chem. Soc., to be published.

15. E. J. Heller, E. B. Stechel and M. J. Davis, Molecular spectra, Fermi resonances, and classical motion, J. Chem. Phys. 73:4720 (1980).

16. E. B. Stechel and R. L. Sundberg, in preparation.

17. E. B. Stechel, in preparation.

18. J. E. Lynn, "The Theory of Neutron Resonance Reactions," Clarendon Press, Oxford (1968).

19. C. E. Porter and R. G. Thomas, Fluctuations of nuclear reaction widths, Phys. Rev. 104:483 (1956).

20. R. B. Shirts and W. P. Reinhart, Approximate constants of mo-tion for classically chaotic vibrational dynamics: Vague tori, semiclassical quantization, and classical intramolecular energy flow, J. Chem. Phys. 77:5204 (1982).

21. M. C. Gutzwiller, The quantization of classically ergodic system, Physica 5D:183 (1982).

22. K. G. Kay, Toward a comprehensive semiclassical ergodic theory, J. Chem. Phys., to be published.

23. A. Voros, Semi-classical ergodicity of quantum eigenstates in the Wigner representation, in: "Stochastic Behavior in Classical and Quantum Hamiltonian Systems," G. Casati and J. Ford, ed., Springer, Berlin (1979).

24. E. J. Heller, Quantum corrections to classical photodissociation models, J. Chem. Phys. 68:2066 (1978)).

25. K.S.J. Nordholm and S. A. Rice, Quantum ergodicity and vibrational relaxation in isolated molecules, J. Chem. Phys. 61:2036 (1976); Quantum ergodicity and vibrational relaxation in isolated molecules. I. λ-independent effects and relaxation to the asymptotic limit, J. Chem. Phys. 61:768 (1974).

26. E. J. Heller, Quantum effects in intramolecular energy transfer, Chem. Phys. Lett. 60:338 (1979).

27. R. A. Coveleskie, D. A. Dolson and C. S. Parameter, A direct view of intramolecular vibrational redistribution in S_1 p-difluorobenzene, J. Chem. Phys. 72:5774 (1980).

28. J. B. Hopkins, D. E. Powers and R. E. Smalley, Vibrational relaxation in jet-cooled alkyl benzenes. III. Nanosecond time evolution, J. Chem. Phys. 73:683 (1980).

29. R. Kosloff and S. A. Rice, The influence of quantization on the onset of chaos in Hamiltonian systems: The Kolmogorov entropy interpretation. J. Chem. Phys. 74:1340 (1981).

30. M. V. Berry, Regular and irregular semiclassical wavefunctions, J. Phys. A10:2083 (1977).

31. M. V. Berry, J. H. Hannay and A. M. Ozorio De Almeida, Intensity moments of semiclassical wavefunctions, Physica D forthcoming.

32. M. G. Kendall and A. Stuart, "The Advanced Theory of Statistics," Charles Griffin and Company, Ltd., London (1963).

33. R. L. Sundberg and E. J. Heller, Predissociation rates as a probe of intramolecular dynamics, submitted.

34. T. A. Brody, J. Flores, J. B. French, P. A. Mello, A. Pandey and S.S.M. Wong, Random-matrix physics: spectrum and strength fluctuations, Rev. Mod. Phys. 53:385 (1981).

35. J. R. Taylor, "Scattering Theory," John Wiley and Sons, Inc., New York (1972).

36. J. D. Draayer, J. B. French and S.S.M. Wong, Strength distributions and statistical spectroscopy. I. General theory, Ann. Phys. 106:472 (1977).

37. M. J. Davis and E. J. Heller, Semiclassical Gaussian basis set method for molecular vibrational wavefunctions, J. Chem. Phys. 71:3383 (1979).

38. R. M. Roth, R. B. Gerber and M. A. Ratner, Vibrational levels in the self-consistent-field approximation with local and normal modes. Results for water and carbon dioxide, J. Phys. Chem. 87:2376 (1983) and references therein.

39. I. C. Percival, Regular and irregular spectra, J. Phys. B6:L229 (1973).

40. M. V. Berry and M. Tabor, Closed orbits and the regular bound spectrum, Proc. Roy. Soc. Lond. A356:375 (1977).

41. P. Pechukas, Distribution of energy eigenvalues in the irregular spectrum, Phys. Rev. Lett., submitted.

42. R. M. Stratt, N. C. Handy and W. H. Miller, On the quantum mechanical implications of classical ergodicity, J. Chem. Phys. 71:3311 (1980).

43. N. DeLeon, M. J. Davis and E. J. Heller, Quantum manifesta tions of classical resonance zones, J. Chem. Phys., submitted.

44. M. J. Davis and E. J. Heller, Quantum dynamical tunneling in bound states, J. Chem. Phys. 75:246 (1981).

45. G. Benettin, L. Galgani and J. M. Strelcyn, Kolmogorov entropy and nmerical experiments, Phys. Rev. A14:2338 (1976); G. Benettin and L. Galgani, Transition to stochasticity in a one-dimensional model of a radiant cavity, J. Stat. Phys. 27:153 (1982).

46. P. Pechukas, Semiclassical approximation of multidimensional bound states, J. Chem. Phys. 57:5577 (1972).

47. M. Tabor, A semiclassical quantization of area-preserving maps, Physica 6D:195 (1983), and references therein.

48. M. C. Gutzwiller, Classical quantization of a Hamiltonian with ergodic behavior, Phys. Rev. Lett. 45:150 (1980).

ASPECTS OF INTRAMOLECULAR DYNAMICS IN CHEMISTRY

R. A. Marcus

Noyes Laboratory of Chemical Physics
California Institute of Technology
Pasadena, CA 91125

ABSTRACT

The role of chaotic and quasiperiodic behavior of isolated molecules is discussed. The quantum analog of classical quasiperiodic states appears to be relatively well understood. A possible quantum analog of classical chaotic behavior is described, using overlapping quantum mechanical Fermi resonances as a way of generating irregularities. The relation of these Fermi resonances to avoided crossings in eigenvalue plots and to various treatments of intramolecular dynamics in the chemical literature is outlined.

INTRODUCTION

In this lecture we review several aspects of the role of chaos in the behavior of isolated molecules:

(1) interests of chemists in the properties of isolated molecules

(2) possible quantum analogs of classical chaotic and quasiperiodic motion, and the relation to quantum mechanical 'overlapping Fermi resonances' and overlapping avoided crossings,

(3) relation of 'overlapping Fermi resonances' to theories of various experimental results.

Since we have discussed some of this material elsewhere[1-4] we shall abbreviate this written version of the lecture.

RELATION TO CHEMISTRY

The relation of chaotic and quasiperiodic nonlinear behavior to the properties of isolated molecules is twofold: (a) chemical reactive behavior, (b) spectral properties. As an example of (a), we note that the rate of dissociation of a molecule, such as

$$C_2H_6 \rightarrow 2CH_3 \quad , \tag{1}$$

and isomerization of a molecule, such as

$$CH_3NC \rightarrow CH_3CN \quad , \tag{2}$$

depends on the vibrational energy of the molecule and on whether the motion is quasiperiodic or chaotic. (More precisely it depends on the quantum mechanical analogs of these properties.) Spectral properties, absorption and fluorescence, will also depend on these properties.

A review of some of the experimental methods of excitation of the vibrations of molecules, the experimental results, and the relevant statistical theory of unimolecular dissociations and isomerizations are given in ref. 1-4. The statistical theory ("RRKM" theory) frequently used in interpretation of the experiments is in agreement with most of these data.[1-4]

Recently, the possible ability of a heavy metal atom to partially isolate dynamically the parts of a molecule that are attached to it has been discussed.[2, 5, 6] If this isolation can be accomplished it will offer a possibility of intramolecular laser selective chemistry, in which one selectively excites one part of a molecule and causes mainly that part to dissociate or isomerize. Normally, the vibrational energy of excitation would, at least in the absence of the heavy atom, rapidly redistribute between the various parts of the molecule (in ~ 1 picosecond). A discussion has been given elsewhere of the heavy metal atom studies, both experiment[5] and theory.[2, 6]

ON THE QUANTUM ANALOG OF QUASIPERIODIC AND CHAOTIC BEHAVIOR

We first recall that nonlinear classical resonances provide an effective way of exchanging energy between vibrational modes, a point stressed by many authors (e. g., Chirikov, Ford). It offers a route to classical chaos. The role of these nonlinear resonances in coupling vibrational modes is very evident in refs. 2 and 6.

The semiclassical connection between classical quasiperiodic states and corresponding quantum states was treated in a practical way for the first time for systems with hard-walled potentials by Keller[7] and for systems with smooth potentials by Eastes[8] and by Noid.[9] Methods were presented there for calculating action integrals in nonseparable systems.[7-9] Since that time a variety of semiclassical quantization procedures have been introduced and are complementary to the original trajectory method (cf. ref. 3 for review). The quantum analog of classical quasiperiodic behavior seems fairly clear (e. g., [1-3]): (1) One expects regular eigenvalue sequences, each sequence corresponding to some mode being excited and each sequence with a smoothly varying spacing. In the terminology of Percival the spectrum is 'regular'.[10] Of course, in a system with many coordinates and hence with many eigenvalue sequences it may be very difficult or impractical to determine if there are regular sequences of eigenvalues, even if the behavior is actually quasiperiodic. (2) Secondly, one expects that in this quantum analog the wavefunction ψ (or $|\psi|^2$) will have a regular pattern, e. g.[3] Sometimes, as in the case of the 99-quasibound state Henon-Heiles system the wave pattern is rather complicated[11] but still regular.

In the quantum analog of classical chaotic states one expects irregular spectra, as Percival has noted[10] (no recognizable regular sequences of eigenvalues), and irregular patterns for $|\psi|^2$. One mechanism for generating these irregularities, we have noted, lies in "Fermi resonances", illustrated by avoided crossings in plots of eigenvalues versus perturbation parameters.[2, 4, 12, 13] We amplify on this point next:

We note that near-degeneracy of two eigenvalues in itself has no implications regarding energy transfer between two modes. However, if this near-degeneracy disappears when a suitable perturbation parameter is varied, i. e., if the two eigenvalues participate in an 'avoided crossing', a wavepacket constructed from a sum of two such states is associated with a quasiperiodic energy transfer between two modes. The two modes are reasonably well described by the two states away from the avoided corssing (e. g., cf Figs 4 and 5 of ref. 12). As discussed elsewhere,[2-4] isolated avoided crossings, like isolated classical resonances, do not yield chaos, but overlapping "avoided crossings" ("overlapping Fermi resonances") yield the two attributes of the chaotic behavior mentioned above--irregular spectra and irregular patterns for $|\psi|^2$. These avoided crossings are discussed in refs. 2-4 and 12-13.

We have also noted that classical chaos may be a necessary, but not a sufficient for "quantum chaos".[2] Sufficient conditions have been proposed (ref. 2 and references cited therein) for the case where a quantum Fermi resonance is generated by a classical resonance, (i. e., for a quantum resonance which occurs in a Hamiltonian having a classical resonance in the same phase space region). These conditions have been phrased in quantitatively and are[2] (1) that the width of the classocal resonances be larger than ħ; otherwise the splitting at the avoided crossing becomes very small, (2) that the overlap of neighboring resonances in action space be greater than ħ; otherwise there will not be a quantitatively significant overlap of the corresponding quantum Fermi res onances (avoided crossings), and (3) that the center of the resonance be sufficiently close[4b] to a relevant quantum state.

Insight into quantum mechanical avoided crossings and thereby into the associated quantum Fermi resonances, when generated by classical resonances,, has been obtained in a recent semiclassical analysis.[12, 14] The splitting of the eigenvalues in the vicinity of the avoided crossing was calculated in this way from data on classical trajectories and was found to be in reasonable agreement with that calculated numerically by diagonalizing a large basis set.[14] An interesting feature was that the classical trajectories themselves generated, in the 'primitive semiclassical' treatment, eigenvalue plots which intersect rather than avoid each other.[12, 14] The quantum mechanical splitting was associated, thereby, with a classically-forbidden process.[14, 15] Use of low order perturbation theory led to a functional form of a semiclassical Hamiltonian suitable for the latter analysis.[14] The various parameters in this Hamiltonian were then evaluated from the phase integrals associated with the two sets of trajectories and with the separatrix separating the two types of trajectories. The Hamiltonian led to a Mathieu's differential equation of fractional order for the semiclassical wavefunction, whose eigenvalues yielded the eigenvalues for the system.[14] A phase integral analysis of the problem has also been made.[15] Such studies delineate further the relation between these aspects of quantum and classical behavior.

RELATION OF AVOIDED CROSSINGS (FERMI RESONANCES) TO THEORIES OF INTRAMOLECULAR ENERGY TRANSFER

The concept of overlapping avoided crossings (overlapping quantum mechanical Fermi resonances) is related to a number of concepts that are used to interpret various types of experimental data. For example, in interpreting data on the dependence of the infrared multiphoton absorption cross-section for molecules as a function of absorbed energy, a series of one-quantum exchanges to other vibrational modes has been proposed to model the absorption coefficient at higher energies.[16] Such a model, involving the coupling of many modes by resonances, corresponds to "overlapping Fermi resonances".

A second example is in an interpretation of the linewidths in C_6H_6 in which a CH overtone vibration is excited and assumed to lose its energy by coupling to an HCC mode in a 2:1 Fermi resonance, and that this HCC mode is in turn, coupled to other nearly resonant modes.[17] Again, such a mechanism would correspond to the idea mentioned above of "overlapping Fermi resonances". If the linewidth of the high CH overtones is correctly modelled by this energy loss (T_1-type) mechanism, and if indeed the line is not inhomogeneously broadened, then the linewidth indicates an energy loss rate from the excited CH vibration in the subpicosecond region.

A third example of overlapping Fermi resonances may occur in Jensen's discussion, in this symposium, of the multiphoton ionization of atom.

In summary, a picture is beginning to emerge in which the quantum analog of classical quasiperiodic motion is, we believe, reasonably clear. The many successful examples of semiclassical quantization, beginning with the cited work of Keller,[7] Eastes[8] and Noid,[9] are consistent with this view. The quantum analog of classical chaotic behavior is also, we believe, becoming clearer, at least qualitatively. One route to this quantum chaos is that of 'overlapping' quantum mechanical Fermi resonances.

ACKNOWLEDGMENT

It is a pleasure to acknowledge the support of this research by a grant from the National Science Foundation. Contribution No. 6894 from the California Institute of Technology.

REFERENCES

1. R. A. Marcus, Laser Chem. (1983) (in press).
2. R. A. Marcus, Faraday Disc. Chem. Soc. 75:000 (1983).
3. D. W. Noid, M. L. Koszykowski, and R. A. Marcus, Annu. Rev. Phys. Chem., 32:267 (1981).
4. (a) R. A. Marcus, in: "Picosecond Phenomena III," K. B. Eisenthal, R. M. Hochstrasser, W. Kaiser, and A. Laubereau, eds., Reidel, Dordrecht (1982), p. 254;
   (b) R. A. Marcus, Ann. N. Y. Acad. Sci., 357:169 (1980);
   (c) R. A. Marcus, in: "Horizons of Quantum Chemistry," K. Fukui and B. Pullman, eds., Reidel, Dordrecht (1980), p. 107.
5. P. Rogers, D. C. Montague, J. P. Frank, S. C. Tyler, and F. S. Rowland, Chem. Phys. Lett. 89:9 (1982); P. J. Rogers, J. I. Selco, and F. S. Rowland, ibid. 97:313 (1983); but see S. P. Wrigley and B. S. Rabinovitch, ibid. 98:386 (1983).

6. V. Lopez and R. A. Marcus, Chem. Phys. Lett. 93:232 (1982).
7. J. B. Keller, Ann. Phys. 4:180 (1958).
8. W. Eastes and R. A. Marcus, J. Chem. Phys. 61:4301 (1974).
9. D. W. Noid and R. A. Marcus, J. Chem. Phys. 62:2119 (1975).
10. Cf. I. C. Percival, Adv. Chem. Phys. 36:1 (1977).
11. Cf. M. D. Feit, J. A. Fleck, Jr., and A. Steiger, J. Compt. Phys. 47:412 (1982).
12. D. W. Noid, M. L. Koszykowski, and R. A. Marcus, J. Chem. Phys. 78:4018 (1983).
13. M. V. Berry, Ann. Phys. N. Y. 131:163 (1981); P. J. Richens and M. V. Berry, Physica 2D:495 (1981); D. W. Noid, M. L. Koszykowski, M. Tabor, and R. A. Marcus, J. Chem. Phys. 72:6169 (1980); D. W. Noid, M. L. Koszykowski, and R. A. Marcus, Chem. Phys. Lett. 73:269 (1980); R. Ramaswamy and R. A. Marcus, J. Chem. Phys. 74:1379 (1981).
14. T. Uzer, D. W. Noid, and R. A. Marcus, J. Chem. Phys. 79:000 (1983).
15. J. N. L. Connor, T. Uzer, R. A. Marcus, and A. D. Smith (to be submitted).
16. J. Stone, E. Thiele, and M. F. Goodman, Chem. Phys. Lett. 71:171 (1980).
17. E. L. Sibert III, W. P. Reinhardt, and J. T. Hynes, Chem. Phys. Lett. 92:455 (1982).

SEMICLASSICAL ANALYSIS OF REGULAR AND IRREGULAR MODES OF QUANTUM COUPLED OSCILLATORS

Vincenzo Aquilanti, Simonetta Cavalli and Gaia Grossi

Dipartimento di Chimica dell'Università

06100 Perugia, Italy

ABSTRACT

A semiclassical analysis, involving asymptotic techniques and the explicit treatment of adiabatic and nonadiabatic behavior, has been developed for potential scattering, resonances and bound state problems in nonrelativistic time independent quantum mechanics. It is here applied to the Hénon-Heiles model, in order to assess the role of regular (normal or local) versus irregular modes in polyatomic molecules.

I. INTRODUCTION

Recent experimental progress has been made in the characterization of the behavior of highly excited, or even dissociating, states of polyatomic molecules. The traditional approach of chemical kinetics is based on statistical methods to deal with these complicated quantum mechanical problems, and concepts such as that of a structureless quasicontinuum of states is currently used. Largely stimulated by the lively discussions of transition from order to chaos, which are best formulated within classical mechanics, a search has recently started for some ordered features to show up in the quasicontinuum. This search is carried out by both experimentalists and theorists, the latter mainly looking for explicit quantum formulations of these problems.[1]

Because of the masses and the interactions involved, molecular behavior is typically a problem in semiclassical mechanics: quantum effects are too important to be neglected altogheter, but Planck's constant is definitely a parameter so small that appropriate asymptotic techniques can be effectively exploited. The paradigmatic example is the WKB approach to one dimensional problems, and it is useful both for bound states and for scattering.

The extension of the asymptotic approach to multidimensional nonseparable systems is not so straightforward: EBK quantization acts only on classical quasiperiodic trajectories, thus yielding only part of the spectrum, and cannot be generalized to scattering states.[2] Among the techniques developed for dealing explicitly with inelastic scattering and reactions, a generalization of the Born-Oppenheimer separation of nuclear and electronic motions has recently been proved very successful. It involves the search for a nearly separable variable, in terms of which the time independent Schrödinger equation reduces to an infinite set of coupled second order ordinary differential equations. Besides offering an effective computational scheme, the procedure allows to be implemented semiclassically, and each step is amenable to qualitative interpretation, such as is needed for deepening our insight of complicated quantum systems.

The asymptotic separation technique is sketched in Section II, and its implication for a semiclassical treatment of quantum problems are illustrated in Section III. Section IV summarizes the application of this method to a Hénon-Heiles potential, which is a typical model for the investigation of regular (normal or local) versus nonregular modes both in classical and in quantum mechanics.

II. ASYMPTOTIC SEPARABILITY AND ITS LOCAL BREAKDOWN

Regular, quasiperiodic behavior of a quantum system is definitely associated with separability of the equations of motion. How the converse, i.e. 'chaos' whatever may be its definition, can be associated with nonseparability, is a matter of current research. Separability, on the other hand, is always a manifestation of some symmetry: this, in quantum mechanics, corresponds to the existence of operators commuting with the Hamiltonian, and leads to the possibility of defining good quantum numbers. Although for intrinsically nonseparable problems separation cannot be carried out exactly (globally), it is nonetheless possible to find important examples where quasiseparability (approximate commuting operators, nearly good quantum numbers) can be obtained. In the description of diatomic molecules, for example, electronic, vibrational and rotational modes are progressively considered separately following a well established hierarchy, which allows to arrange modes according to characteristic frequencies. The underlying idea is that, if a mode is much slower than others, it can be considered as freezed, while studying the fast ones. Thus, in the familiar Born-Oppenheimer separation, internuclear distances are slow coordinates with respect to electron-nucleus and electron-electron ones. The whole of quantum chemistry capitalizes on this idea.

A key observation for fruitful generalizations is that, to achieve approximate separation, one employs, more or less rigorously, asymptotic expansions with respect to some parameters (mass ratios, frequency ratios). In the present investigations, we start from the consideration that it is often possible, for problems of definite chemical and physical interest, to find some representation which allows to obtain an approximate separation (at least locally), by expansions which are asymptotic in Planck's constant, treated as a small parameter: it is a natural choice, since this corresponds to what is commonly understood as the semiclassical regime. Although these approximately separated representations will fail somewhere, we found that the localization of failure may lead to a source of quantum irregular behavior and that the search for special asymptotic techniques for dealing with local nonseparability is particularly promising.

In general, within the framework of time independent nonrelativistic quantum mechanics, a Born-Oppenheimer type of separation can be attempted by introducing a suitable coordinate system[3]. Suppose the problem to be N-dimensional: in typical applications, one variable ρ is defined, in most cases as a radius of the N-hypersphere, properly parametrized by N-1 angles, collectively indicated by Ω . The Schrödinger equation can then be transformed into a coupled set of ordinary differential equations, the diagonal elements being $\varepsilon_n(\rho)$, the spectrum of an (N-1)-dimensional problem parametrically depending on ρ , with corresponding wavefunctions $\phi_n(\rho,\Omega)$.

Explicitly, an expansion of the exact wavefunction as

$$\Psi(\rho,\Omega) = \sum_n F_n(\rho)\phi_n(\rho,\Omega) \tag{1}$$

leads to the exact set, in matrix notation[4]

$$-\frac{\hbar^2}{2\mu}\left[\underline{1}\frac{d}{d\rho} + \underline{P}(\rho)\right]^2 \vec{F}(\rho) = \left[\underline{\varepsilon}(\rho) - \underline{1}\,E\right]\vec{F}(\rho) \tag{2}$$

where μ is an appropriate reduced mass, depending on the definition of ρ . The elements of the diagonal matrix $\underline{\varepsilon}(\rho)$ are referred to as the adiabatic potential energy curves. The matrix $\underline{P}(\rho)$, which has elements

$$P_{nm}(\rho) = \int d\Omega\, \phi_m^*(\rho,\Omega)\,\frac{\partial}{\partial\rho}\,\phi_n(\rho,\Omega) = -P_{mn}(\rho) \tag{3}$$

is responsible for the nonadiabatic, or diabatic, coupling. At a given total energy E, one has to introduce the proper (scattering or bound states) boundary conditions and solve for the unknown channel functions $F_n(\rho)$. The $\underline{\varepsilon}$ and $\underline{P}$ matrices, which are obtained by solving an N-1 dimensional problem, are in principle infinitely dimensional: the success of the procedure from the point of view of its use in practical calculations is determined by its rate of convergence upon truncation.

The above manipulation of Schrödinger equation is particularly useful for a discussion of properties of systems from the point of view of aymptotic methods. It is immediate to see from (2) that whenever elements of $\underline{P}$ are small, since $\hbar^2/2\mu$ is a small parameter, equations adiabatically decouple into one-dimensional problems for the effective potentials $\varepsilon_n(\rho)$. In turn, these problems can be analyzed by the Liouville-Green WKB technique, which requires special care whenever $\varepsilon_n(\rho) = E$: but this problem is to be considered as effectively solved by the method of comparison equations. It is important to realize that proper coordinate choices may lead to wide regions of ρ space where this decoupling is very effective: in such a case, it is possible to compute semiclassically bound or resonance states and scattering properties.[5]

Wavefunctions are then adiabatically given by a single term of expansion (1), and approximate quantum numbers can be assigned: however strictly the class of regular, quasiperiodic modes are defined in quantum mechanics, these states definitely appear to belong to it! As will be illustrated in the following, the success of the procedure critically depends on how apprpriate the definition of the ρ coordinate is,to exhibit as more localized as possible any breakdown of approximate separability.[6] This breakdown is measured by the $\underline{P}$ matrix, and therefore a study of its analytical structure is an important step of the present program. In fact, eq.(2) shows that around any poles of $\underline{P}$ matrix elements sufficiently close to the real ρ axis their neglect is not warranted, however small $\hbar^2/2\mu$ is considered. In the following, therefore, we will sketch examples where these features of $\underline{P}$ matrix have been characterized. Around these features, adiabatic conditions fail, and several terms of comparable magnitude may contribute to (1), leading to departure from regular behavior.

III. THE TRANSITION BETWEEN NORMAL AND LOCAL MODES: THE SEMICLASSICAL PENDULUM

Strongly nonadiabatic behavior is often localized where the actual character of systems changes drastically. So, when the interaction between two atoms is considered as a function of internuclear distance R, it is found[7] that the transition between the typical behavior of separated atoms and that of a diatomic molecule is often localized around sharp maxima in elements of a $\underline{P}(R)$ matrix. These maxima correspond to poles near the real R axis in a proper analytic continuation of $\underline{P}(R)$.

Several examples can be put forward in order to show that transitions between modes, i.e. a local breakdown of adiabaticity, typically take place at well defined characteristic features of the potential. For problems involving more than two bodies,several investigations[8] have identified as a good candidate for near separability a hyperradial variable, ρ . Low values of ρ correspond to closeness of all particles, and the various possible rearrangement channels correspond to large ρ. It has recently been shown that,[9] at least for the simplified situation that the three particles are constrained to be on a line, a rearrangement process, such as a chemical reaction, can be described in a time independent picture as the transition between two types of modes, one corresponding to an intermediate complex (transition state) which may dissociate into channels corresponding to reactants and products: the transition can be described adiabatically, nonadiabaticity being important only along a line in the potential energy surface (the ridge) which separates the valleys of reactants and products. Implementing semiclassically these ideas, it has been possible to obtain not only qualitative descriptions, but also quantitative results for resonance positions and widths, and for interference effects in the probability for reactive collisions. A recent classical study[10] of these problems points at a connection between chaotic behavior and temporary trapping in the transition state: it is tempting to remark that our semiclassical analysis leads to a correlation between classical temporary trapping and the strongly nonadiabatic quantum behavior which substantially increases the lifetime of the transition state. Again, a connection between local nonseparability and irregular modes is emerging.

As the simplest model where these aspects of transition between modes are exhibited, let's consider the familiar one-dimensional pendulum: classically,[11] the separatrix trajectory marks the transition between low energy oscillatory motion (a model for what in different contexts, would be referred to as local, vibrating, or librating modes) and high energy rotating (normal, precessing) modes. These modes are completely described in terms of ellyptic functions within classical mechanics.

Quantum mechanically, the pendulum problem is solved within the theory of periodic second order differential equations, specifically in terms of Mathieu funtions:[12] it is interesting to note that, since in quantum mechanics things are never so neat as in classical mechanics, a sharp separation between modes is not possible, and this poses a 'connection problem' for Mathieu funtions, a problem which is being actively investigated. We find that a promising approach is via a Liouville-Green technique: in the physical language, this of course corresponds to a WKB, and therefore semiclassical, technique, as witnessed by the appearance of ellyptic integrals in the asymptotic approximations to Mathieu functions.

The details of this approach are here omitted: its implications for the analysis of modes of coupled oscillators will be illustrated by the following discussion of quantum modes in the Hénon-Heiles potential.

IV. ADIABATIC-DIABATIC SEMICLASSICAL ANALYSIS FOR THE HENON-HEILES POTENTIAL

Extensive studies have well characterized classical, semiclassical and quantum mechanical behavior for a particle in a two dimensional potential originally introduced by Hénon and Heiles [13-15]. For this potential, as a model of coupled oscillators, various quasi-periodic modes have been classified, and states non EBK quantizable have also been indicated as possible examples of quantum chaotic behavior. The analysis sketched in previous Sections is particularly straightforward for this potential, yet sufficiently informative both for qualitative purposes and for quantitative semiclassical quantization prescriptions.

For the Hénon-Heiles model of two non linear coupled oscillators, it is convenient the use of polar coordinates

$$V(\rho,\theta) = \frac{1}{2}\rho^2 + \frac{\lambda\rho^3}{3}\cos 3\theta \tag{4}$$

As a matter of fact, an adiabatic analysis carried out by Child and Shapiro [16] in Cartesian coordinates fails to explcit the proper symmetries of the system, and to lead to a localization of ranges of nonadiabatic behavior. On the contrary, the polar representation (4) immediately suggests the individuation of ρ as the quasiseparable variable: at low ρ , the behavior of a bidimensional harmonic oscillator is expected, deviations occurring as ρ increases, being more marked the larger the parameter λ. Fig.1 shows ridge and valley bottom profiles for $\lambda = 80^{-1/2}$: this potential has a C_{3v} symmetry which leads to three different types of states, labelled as A_1, A_2 and E. The construction of adiabatic potential energy curves as a function of ρ for this system can be carried out analytically: in fact, inspection of eq.(4) reveals that the one dimensional quantum mechanical problem to be solved in θ at fixed ρ is that of the quantum mechanical pendulum described in the previous Section. The relationship between ρ and the q parameter which appears in Mathieu differential equation [12] is

$$q = \frac{4}{27}\lambda\rho^5$$

The C_{3v} symmetry is easily enforced by requiring a proper periodicity of Mathieu function: it leads in a simple manner to A_1, A_2 and E separation, in contrast with some current quantization procedures which fail to remove A_1 and A_2 degeneracies. For example, adiabatic potential curves corresponding to A_2 symmetries can be written in terms of odd Mathieu eigenvalues $b_n(q)$ corresponding to $se_n(q,\theta)$ functions:

$$\varepsilon_n(\rho) = \frac{9}{8\rho^2} b_n(q) + \frac{1}{2}\rho^2 \tag{5}$$

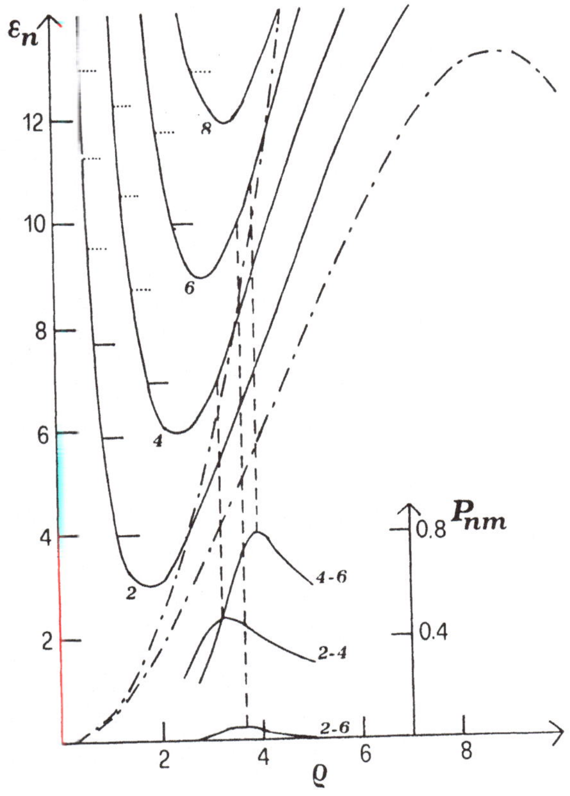

Fig. 1. For the Hénon-Heiles potential with $\lambda = 80^{-1/2}$ [eq.(4): upper broken curve, ridge profile (θ = 0, $2\pi/3$, $4\pi/3$); lower broken curve, valley bottom profile ($\theta = \pi/3$, π , $5\pi/3$)], adiabatic potential curves $\varepsilon_n(\rho)$[eq.(5)] and corresponding nonadiabatic coupling matrix elements $P_{mn}(\rho)$ [eq.(6)] as a function of radial coordinate ρ for A_2 symmetry. Positions of levels indicated by continuous segments for those identified as quasiperiodic (Ref.14) and by dotted segments for those not identified as quasiperiodic.

These curves are reported in Fig.1. It is also straightforward to compute $\underline{P}$ matrix elements by some manipulation of Mathieu functions: those reported in Fig.1 were obtained using the formula

$$P_{nm}(\rho) = \frac{dq}{d\rho} P_{nm}(q) = \frac{20}{27} \lambda \rho^4 \int_0^{2\pi} d\theta \; se_m(q,\theta) \frac{d}{dq} se_n(q,\theta) \qquad (6)$$

Results similar to those in Fig.1 for A_2 are easily obtained for A_1 and E symmetries and for any λ values. They show that all computed quantum levels can be arranged within each supporting $\varepsilon_n(\rho)$ curves in proper sequences, and approximate quantum numbers can be assigned. In contrast with usual assignments in terms of harmonic oscillator states, which are 'good' only at low ρ and do not distinguish states whose degenerations are removed by a breaking of spherical symmetry, the adiabatic curves ε_n account for the smooth change in character of the angular motion when the ridge is crossed.

Actually, the role of ridge effects, as stressed in different contexts, appears to be decisive also for this type of problems. From the discussion of the previous Section, we find that the classical analog of ridge effects is to be individuated as the separatrix trajectory: therefore, a key for the understanding of most of the qualitative features of this and similar problems is the realization that the border line between different qualitative aspects of quasi-periodic motions (such as for example normal or local modes), is often marked by an energy dependent ridge effect.

Accordingly, it is not surprising to find, as Fig.1 shows, that nonadiabatic coupling matrix elements between states peak at ridge. The analysis for this behavior is straightforward: here, it suffices to remark, using the language of previous Sections, that $P_{nm}(\rho)$ will be largest at ridge because the requirement for asymptotic separability (adiabatic behavior with respect to the 'slow' variable ρ) fails there since angular motion is also 'slow'. From a classical point of view, this is easily seen: it is in fact known that for the separatrix trajectory it takes an infinite time to complete an oscillation.[17]

An adiabatic approximation (i.e. a single term in expansion (1)) is not only qualitatively justified far from ridge for characterizing the nature of quantum states, but has been found to be also excellent for quantitative semiclassical quantization. For the lowest two states in Fig.1, for example, by one dimensional WKB (here essentially a Bohr-Sommerfeld rule) we obtain 3.9837 and 5.8840, to be compared with exact 3.9858 and 5.8815 respectively.[15] This is significantly better than any previous semiclassical recipe, but of course the procedure will fail for higher states when nonadiabatic effects associated with the ridge come into play and invalidate the assumption of separability.

Close inspection of graphs such as those in Fig.1 reveals an avoided crossing structure of adiabatic states near the ridge:[5] this structure strongly suggests that diabatic behavior be quantitatively described by extensions of the Landau-Zener approach, i.e. essentially by a WKB treatment of two states problems. Work along this line is in progress.

This avoided crossing structure, which marks the separation between modes for a given system (fixed λ), should not be confused with accidental isolated avoided crossings in the dependence of energy levels as a function of the perturbation parameter λ .[18] These isolated features are sometimes indicated as possible sources of quantum chaotic behavior, and they can be easily discussed within the present approach.

That transition from quasiperiodic to nonquasiperiodic (chaotic?) quantum states [14] be associated to a progressive extension of wavefunctions beyond the ridge is qualitatively suggested by Fig.1: the method allows however quantitative assessments, by the explicit consideration of approximate (i.e. adiabatic or diabatic semiclassical) and exact wavefunctions.

Along these lines, mode transitions between local and normal behavior are discussed most naturally.[19] The method has also been extended to a description of dissociation phenomena: for the present Hénon-Heiles example, this involves a construction of adiabatic potential energy curves at larger ρ values than those in Fig.1. Regularity and possible mode selectivity in decay modes are currently being investigated, with reference to recent interesting work:[20] as stated at the beginning of this contribution, this aspect is central in chemistry and molecular physics, because of experimental implications, and represents an important motivation for further research.

AKNOWLEDGEMENTS

Thanks are due to Professor Howard S. Taylor for stimulating discussions and preprints on his work.

REFERENCES

1. R. S. Berry, J.Chem.Phys. 78, 3976 (1983); R. E. Wyatt, G.Hose and H. S. Taylor, Phys.Rev.A 28, in press (1983); and references therein.
2. References to papers and remarks on semiclassical quantization can be found in P. Pechukas, J.Chem.Phys. 78, 3999 (1983).
3. Applications to atomic physics reviewed in U. Fano, Rep.Progr. Phys. 46, 97 (1983), and to nuclear physics by Yu. F. Smirnov

and K. V. Shitikova, Sov.J.Part.Nucl. 8, 344 (1977).
4. H. Klar, Phys.Rev.A 15, 1452 (1977).
5. V.Aquilanti, G.Grossi and A.Laganà, Chem.Phys.Lett. 93, 174 (1982); V.Aquilanti, S.Cavalli and A.Laganà, Chem.Phys.Lett. 93, 179 (1982); V.Aquilanti, S.Cavalli, G.Grossi and A.Laganà, J.Mol.Structure 93, 319 (1983) and in press.
6. The asymptotic nature of this procedure can be proven following, for example, S. F. Feshchenko, N. I. Shkil', and L. D. Nikolenko, "Asymptotic methods in the theory of linear differential equations", Elsevier, New York (1967).
7. V.Aquilanti, G.Grossi and F.Pirani, unpublished results, based on the general theory developed in V.Aquilanti and G.Grossi, J.Chem.Phys. 73, 1165 (1980).
8. See Ref.3 and also V.Aquilanti, G.Grossi and A.Laganà, J.Chem. Phys. 76, 1587 (1982).
9. The relevant literature is referred to in Ref.5.
10. Ch. Schlier, in: "Energy storage and redistribution in molecules", J. Hinze, ed., Plenum, New York (1983).
11. B. V. Chirikov, Phys.Reports 52, 263 (1979).
12. S. Flügge, "Practical quantum mechanics", Springer, Berlin (1974); W. Barrett, Phil.Trans.R.Soc.Lond. A 301, 75, 81, 99, 115, 137 (1981). Note formal analogies with parametric resonance in classical mechanics, see e.g. L. D. Landau and E. M. Lifschitz, "Mechanics", Pergamon, New York (1976).
13. D. W. Noid, M. L. Koszykowski and R. A. Marcus, Ann.Rev.Phys. Chem. 32, 267 (1981).
14. G. Hose and H. S. Taylor, J.Chem.Phys. 76, 5356 (1982).
15. C. Jaffé and W. P. Reinhardt, J.Chem.Phys. 77, 5191 (1982).
16. M. Shapiro and M. S. Child, J.Chem.Phys. 76, 6176 (1982).
17. W. I. Arnol'd, "Mathematical methods of classical mechanics", Springer, Berlin (1978).
18. R. Ramaswamy, Chem.Phys. 76, 15 (1983), and refernces therein.
19. The example discussed here has been limited to states labelled as Q_I in Ref.14, but can be extended to Q_{II} modes.
20. B. A. Waite and W. H. Miller, J.Chem.Phys. 74, 3910 (1981); Y. Y. Bay, G. Hose, C. W. McCurdy and H. S. Taylor, to be published.

QUANTUM CHAOS AS BASIS FOR STATISTICAL MECHANICS

N.G. van Kampen

Institute for Theoretical Physics of the University at Utrecht
P.O.Box 80.006, 3508 TA Utrecht, The Netherlands

ABSTRACT

Quantum chaos cannot be defined through correspondence with classical chaos, but only through its function as a basis for statistical behavior. The foundations of quantum statistics are outlined and the steps that rely on chaos are emphasized. Finally a simplified model of dissociation is given as an example.

1. INTRODUCTION

Classical chaos is a property of equations of the type[1])

$$\dot{x}_i = f_i(x_1,x_2,\ldots,x_n) \quad (i=1,2,\ldots,n) \ . \tag{1}$$

They may be dissipative and have attractors - which is the case of interest for turbulence. Or they may be of Hamilton form, having a conserved energy and an invariant measure[2]) - which is the case we are interested in. The study of chaos in Hamilton systems is inspired by the hope to understand how irreversibility and stochastic behavior arise. Although there is no unique definition of even classical chaos it is clear that, just as attractors and ergodicity, it is an asymptotic property of the long time behavior of the solutions of (1). Strictly speaking, if one knows the solutions merely in a finite time interval $o \leqslant t \leqslant T$ one cannot tell whether there is chaos, although of course finite time computer calculations produce valuable suggestions. When the equations (1) can be integrated they are deemed to be non-chaotic[3]).

Quantum mechanics is formulated in terms of operators in a Hilbert space and the time evolution is given by the Schrödinger equation for the state vector $\psi(t)$,

$$\dot{\psi} = -(i/\hbar)\, H\psi \, , \tag{2}$$

which is a very different mathematical structure than (1). The question is whether nonetheless one can identify for equations of type (2) a property that corresponds to classical chaos.

For a finite, closed, isolated system, the general solution of (2) can be written in terms of eigenvalues and eigenfunctions of H:

$$\psi(t) = \sum_n c_n \phi_n e^{-itE_n/\hbar} \, , \tag{3}$$

with arbitrary constants c_n. The expectation value of any observable A is

$$\langle A \rangle_t = \sum_{n,m} c_n^* c_m A_{nm} e^{it(E_n - E_m)/\hbar} \, . \tag{4}$$

From the fact that this is a multiperiodic function such as occurs in classical integrable systems it has sometimes been concluded that no quantum chaos exists. On the other hand, the general idea of correspondence suggests that some analog of classical chaos should survive in quantum theory.

In section 2, however, I shall argue that the correspondence principle cannot serve to define quantum chaos. The reason is that the classical limit does not commute with the long-time limit needed for defining chaos. I therefore propose the definition: Quantum chaos is that property that causes a quantum system to behave statistically, i.e., in such a way that its evolution can be described by equations for probabilities (rather than probability amplitudes). Hence the study of quantum chaos amounts to investigating the foundations of statistical mechanics of quantum systems - which will be done in section 3.

Thus, the two questions in classical mechanics, viz., how to define chaos, and how to prove that it leads to a statistical description, are now short-circuited into one, because I do not see a way of defining quantum chaos by itself. Anyway this is what one wants to know, for instance for describing the dissociation of complex molecules[4]). In section 4 a simple model for dissociation is worked out as an illustration.

If (2) involves particles with spin there is no corresponding classical system. If there are no spins the operator H can be written as an expression in the coordinate and momentum operators q_k, p_k:

$$H = \Phi_\hbar(q,p) \ . \tag{5}$$

In the 'expression' $\Phi_\hbar$ the order of factors q,p matters; they can be interchanged only at the expense of adding compensating terms involving $\hbar$. Every observable A is also an expression in q,p and the evolution can be described by time-dependent operators q(t), p(t) obeying

$$\dot{q}_k = (i/\hbar)\left[\Phi_\hbar(q,p),q_k\right] , \quad p_k = (i/\hbar)\left[\Phi_\hbar(q,p),p_k\right] \ . \tag{6}$$

If H has the form (5) one may define a classical system by choosing as Hamilton function $\Phi_0(q,p)$, which is obtained from (5) by setting $\hbar = 0$. Then there is first the formal correspondence[5] based on the fact that (6), in the limit $\hbar \to 0$ reduces to the equations of motion of this classical system, which in terms of Poisson brackets are

$$\dot{q}_k = -\left\{\Phi_0(q,p),q_k\right\} , \quad \dot{p}_k = -\left\{\Phi_0(q,p),p_k\right\} \ . \tag{7}$$

It should be emphasized that, although this leads from a given quantum system (5) to a unique classical system, the converse is not true. If only the function $\Phi_0(q,p)$ is given there is an infinity of expressions $\Phi_\hbar(q,p)$ that reduce to it in the limit $\hbar \to 0$. Moreover one could always add a spin. The process called 'quantizing a classical Hamilton system' is not unique and does not correspond to anything in Nature, which is fundamentally quantum mechanical.

This correspondence was called 'formal', because the relation of q,p to physical observations is also drastically changed. In (7) they are directly observable parameters specifying the state of the system, while in (6) they are operators, of which only the expectation values in a given state are amenable to observations. The more physical formulation of correspondence[6]) states that at high quantum numbers the quantum system behaves in the same way as the classical one. The reason is that at high quantum numbers there are solutions of (2) having the form of localized wave packets whose positions move approximately as given by (7). Alternatively it may be regarded as a statement about the solutions in the limit that $\hbar$ goes to zero while the energy is kept fixed. Physical correspondence is therefore a statement about the asymptotic domain of high quantum numbers. Moreover, since it refers to wave packets, it is not just concerned with eigenvalues of H, but involves the corresponding eigenfunctions as well[7]).

Various attempts have been made to base a definition of quantum chaos on correspondence[8]), the idea being that a quantum system is chaotic if its classical limit is chaotic. Thus one takes a classical chaotic system, 'quantitizes' it and investigates for instance the properties of its spectrum. The first or formal correspondence leads to studying the sequence of levels starting from the ground state; the second, physical correspondence leads to an asymptotic investigation of E_n and ϕ_n for large n. The following consideration, however, induces me to conclude that these attempts are misguided.

Roughly, the reason is that any wave packet (3) will ultimately spread out and thereby loose its resemblance to the solution of (7). More precisely, chaos is an asymptotic property for $t \to \infty$ while correspondence refers to $\hbar \to 0$, and these two limits do not commute. I cannot demonstrate this for chaos but take in its stead another long time property: ergodicity. The time average of (4) is

$$\lim_{T\to\infty} \frac{1}{T} \int_0^T \langle A \rangle_t \, dt = \sum_n |c_n|^2 A_{nn} \, . \tag{8}$$

The phase relations among the c_n have disappeared and only the probability distribution over the eigenstates ϕ_n survives. This fact is the quantummechanical ergodic theorem.

But how long a time T is needed for the average of (4) to be virtually equal to its limiting value (8)? Consider the dependence of the level distance δE on $\hbar$. For a system of N particles in a finite volume near a fixed E one has $\delta E \sim \hbar^{3N}$. Hence to kill the oscillating terms in (4) one must have at least

$$T \sim \hbar/\delta E \sim \hbar^{-3N+1} \, . \tag{9}$$

Thus for $\hbar \to 0$ the limit (8) becomes illusory; it has therefore no bearing on any property of the classical system. The long time limits in the classical and in the quantum mechanical system are not related.

Numerically one finds that T is much longer than the age of the universe[9]) so that (8) cannot have any physical meaning. Any time interval that may interest an observer is far too short to sense the discreteness of the spectrum. The long-time behavior observed in experiments, as well as that in the corresponding classical system, must be based on an entirely different description, in which the spectrum is treated as quasi-continuous. This cannot, however, be done by simply replacing sums with integrals, because the eigenfunctions belonging to successive levels are wildly different[10]). It is therefore necessary to disentangle the sequence of eigenvalues and eigenstates into sub-sequences according to their physical aspect. This is the underlying idea of the program outlined in the next section.

3. STATISTICAL MECHANICS OF QUANTUM SYSTEMS[11-14])

Consider a closed, finite many-body system having a tremendously dense energy spectrum[9]). Any measurement or preparation of the system has an energy uncertainty $\Delta E \gg \delta E$. The ensuing wave function $\psi(t)$ is some superposition of all eigenstates within ΔE, so that $|\psi(t)|^2$ varies with t. Thus, for a many-body system it is not true that an energy measurement forces the system into a stationary state.

Let A be another macroscopic observable. One may take for ψ a superposition of eigenstates ϕ_n inside ΔE, which is at $t = o$ almost an eigenfunction of A corresponding to a precise value of A. Then $\langle\psi(t)|A|\psi(t)\rangle$ is the evolution of A in a system whose energy is determined within the macroscopic uncertainty ΔE, and for which the initial value of A is also determined within some margin ΔA.

To make this more explicit first define a coarse-grained energy operator $\{H\}$ in the following way. Subdivide the energy scale in intervals (α) of length ΔE; the eigenstates in each interval span a subspace of the total Hilbert space, corresponding to a classical energy shell. Now $\{H\}$ will be defined as that operator that is diagonal in the original energy representation, but has in each shell (α) one and the same eigenvalue E_α, for which one may take the mid-point of the interval (α). Thus $\{H\}$ is highly degenerate. It cannot be used to describe the time evolution of the system, but it can serve as the operator that corresponds to an energy measurement.

Next take an operator A, of which we suppose that it varies slowly, so that the commutator $[H,A]$ is small. Then the matrix elements A_{nm} in the energy representation are concentrated in a narrow ribbon along the diagonal. As a result most of them connect two eigenstates ϕ_n, ϕ_m belonging to the same subspace (α). Only a few of the A_{nm} connect two different subspaces (α), (α'); these will be neglected. The remaining operator commutes with $\{H\}$ and can be diagonalized simultaneously with it by applying a suitable unitary transformation in each subspace (α). In this new representation one can again collect the eigenvalues of A into coarse-grained intervals (β) of length ΔA, thus subdividing our linear subspaces (α) into smaller ones (α,β). The coarse-grained operator $\{A\}$ is again defined by replacing the eigenvalues in each sub-sequence (α,β) by some intermediate value A_β.

In this way one can continue if there are more observables, B,C,... . The end result is a subdivision of Hilbert space in linear subspaces, which correspond to classical phase cells and will be labelled by J. Observations are represented by the coarse-grained operators {H},{A},{B},...; these commute and have well-defined values E_J, A_J, B_J,... in each phase cell J. Each phase cell has a large number g_J of dimensions and may be spanned by an arbitrarily chosen orthonormal set $\xi_{J\iota}$ ($\iota=1,2,\dots,g$). The conditions are that the original spectrum is very dense, and that all commutators of the original H,A,B,... are small.

The wave function $\psi(t)$ can be expanded in the set ξ,

$$\psi(t) = \sum_{J\iota} b_{J\iota}(t)\xi_{J\iota} \; . \tag{10}$$

The evolution of the coefficients is given by

$$b_{J\iota}(t) = \sum_{J'\iota'} \langle J\iota|U_t|J'\iota'\rangle\, b_{J'\iota'}(o) \; , \tag{11}$$

where $U_t = \exp[-itH/\hbar]$. The probability to be in phase cell J at time t is

$$P_J(t) = \sum_{\iota} |b_{J\iota}(t)|^2$$

$$= \sum_{\iota J'\iota' J''\iota''} \langle J\iota|U_t|J'\iota'\rangle^* \langle J\iota|U_t|J''\iota''\rangle\, b^*_{J'\iota'}(o)\, b_{J''\iota''}(o). \tag{12}$$

This is exact but expresses $P_J(t)$ in terms of <u>all</u> coefficients at $t = o$. For a stochastic description it is necessary that (12) reduces to an equation for the probabilities alone:

$$P_J(t) = \sum_{J'} T_t(J|J')P_{J'}(o) \; . \tag{13}$$

In order that that is true one has to make the following two assumptions of the nature of molecular chaos.

(i) The phase factors of the matrix elements $\langle J\iota|U_t|J'\iota'\rangle$ vary so chaotically with ι' that all terms in the double sum (12) cancel except the diagonal ones,

$$P_J(t) = \sum_{\iota J'\iota'} |\langle J\iota|U_t|J'\iota'\rangle|^2\, |b_{J'\iota'}(o)|^2 \; . \tag{14}$$

(ii) The sizes of these diagonal elements vary so chaotically within each phase cell that they may be replaced with their average size:

$$P_J(t) = \sum_{J'} \left[\frac{1}{g_{J'}} \sum_{\iota\iota'} |\langle J\iota|U_t|J'\iota'\rangle|^2\right] \sum_{\iota'} |b_{J'\iota'}(o)|^2 . \tag{15}$$

This is the desired equation (13).

Actually both assumptions are linked together, because each one separately is not invariant for a change of the orthogonal set $\xi_{J\iota}$ inside each phase cell J. Together they assert that the transition probability from J' to J is insensitive to the details of the state ψ within J'. This corresponds to the molecular chaos assumption of classical statistical mechanics[14,15]).

This outline of the foundations of quantum statistics is merely a program and leaves us with a number of questions. First and foremost is the characterization of the set of observables A,B,... for which our chaos assumptions are true. If one takes too many the phase cells are too small to rely on the cancellation of terms in (12); if one takes too few there will be correlations which also obviate the transition form (12) to (15). The precise set is determined by the structure of the system - not by what is 'relevant' for the observer. Secondly one must characterize the systems for which such a set exists. Thirdly one should understand the nature of the approximation and how higher correction can be obtained. These three questions have not yet be answered even for classical kinetic theory.

4. EXAMPLE

Consider the following simple model for dissociation of a diatomic molecule. A Schrödinger particle moves in one dimension $r > o$ in the potential

$$V(r) = - W\theta(a-r) + B\delta(r-a) . \tag{16}$$

The Schrödinger equation is

$$\frac{d^2\phi}{dr^2} + \left\{k^2 - \frac{2\mu}{\hbar^2} V(r)\right\}\phi = o , \quad k^2 = \frac{2\mu}{\hbar^2} E . \tag{17}$$

Moreover $\phi(o) = o$ and we also set $\phi(L) = o$ for some large L to make the spectrum discrete. Although this is not a many-body system it can serve to demonstrate the main features described in the preceding section.

The wave vector in the interior is

$$K = \sqrt{k^2 + 2\mu W/\hbar^2} . \tag{18}$$

The eigenfunctions are[16])

$$\phi_n(r) = \sqrt{2/L}\,\sin(k_n r+\eta_n) \qquad (a < r < L) \tag{19}$$

$$= \sqrt{2/L}\,\frac{\sin(k_n a+\eta_n)}{\sin K_n a}\,\sin K_n r \qquad (o < r < a). \tag{20}$$

Here $\eta_n = \eta(k_n)$ with $\eta(k)$ defined by

$$k\cot(ka+\eta) = K\cot Ka + \frac{2\mu B}{\hbar^2} \tag{21}$$

and k_n by $k_n L + \eta(k_n) = n\pi$ $(n=1,2,\dots)$. As the last term is large, the cotangent on the left must be large unless

$$K \approx K_M = M\pi/a \qquad (M=1,2,\dots)\ . \tag{22}$$

The imaginary roots of (21) give the bound states. The amplitude of the interior wave (20) is small except in these bound states and near the resonances (22)[17]). The widths of the resonances are[18])

$$\gamma_M = \frac{\hbar^4}{4\mu^2 B^2 a}\,K_M k_M\ . \tag{23}$$

For A we take the operator that observes whether the particle is in the interior: $A = \theta(a-r)$. Its matrix elements are

$$A_{nm} = \frac{2}{L}\,\frac{k_m\cos(k_m a+\eta_m)\sin(k_n a+\eta_n)-k_n\cos(k_n a+\eta_n)\sin(k_m a+\eta_m)}{K_n^2-K_m^2}\ , \tag{24}$$

Owing to the denominator they do indeed decrease away from the diagonal[19]). Hence it is possible to construct eigenfunctions of A by combining the ϕ_n in a small interval ΔE. Of course, A is degenerate with two eigenvalues o and 1; we choose for the eigenfunctions with eigenvalue 1

$$\chi_M(r) = \sqrt{2/a}\,\theta(a-r)\,\sin(M\pi r/a)\ . \tag{25}$$

Each is a wave packet of functions ϕ_n in a small interval around $K_n \approx M\pi/a$ with width (23). In fact, the expansion coefficients are

$$\langle\phi_n|\chi_M\rangle = \frac{2(-1)^M K_M}{\sqrt{La}}\,\frac{\sin(k_n a+\eta_n)}{K_M^2-K_n^2}\ , \tag{26}$$

which is small unless K_n is within a range of K_M indicated by (23).

Thus we have diagonalized A simultaneously with {H}.

Now suppose our molecule is originally in its ground state ϕ_0. (To simplify the discussion we assume that it has no other bound states.) Let it undergo a collision with a duration τ. After the collision it is left in a state ψ that is a superposition of the ϕ_n in an energy range $\hbar/\tau$ - which we suppose large compared with the binding energy. Then ψ contains numerous ϕ_n, but only these near the resonances enter, because the others have virtually no overlap with ϕ_0. On the other hand, all ϕ_n near a resonance M are in the interior (o,a) practically the same function as χ_M in (25), so that

$$\psi = c_o\phi_o + \sum c_M\chi_M \, . \tag{27}$$

Each bunch of ϕ_n around one resonance constitutes one of our phase cells; the index M has the role of J, and the ϕ_n themselves may serve as our orthonormal set $\xi_{J\iota}$. The non-resonant ϕ_n constitute additional phase cells, in which A has the value zero - but we shall not need them.

From this initial state the wave function $\psi(t)$ evolves according to (12), in which

$$b_{J\iota}(o) \leftrightarrow c_M \langle\phi_n|\chi_M\rangle \, . \tag{28}$$

The terms with $J' \neq J''$ involve the rapidly varying time factors $\exp[i(E_{M'}-E_{M''})t/\hbar]$ and therefore disappear. Hence the phase relations among the c_M are irrelevant and one makes no error on replacing (27) with an ensemble of excited states with probabilities $P_M(o) = |c_M|^2$. The evolution of this ensemble is given by an equation for the probabilities alone

$$P_M(t) = P_M(o) \exp[-\Gamma_M t] \, , \tag{29}$$

where the Γ_M are the familiar transition probabilities, easily found from (27).

This is the equation (13) (without the reverse transitions, which we have not included here). Hence we have shown that our diatomic molecule, after being excited by the collision, can be approximately described by a statistical distribution over the internal states M, each of which has a certain probability per unit time to dissociate.

NOTES

1) Non-autonomous equations can of course be cast in this form by introducing one additional variable $x_{n+1} = t$. We shall not consider non-autonomous quantum systems, see e.g. T. Hogg and

and B.A. Huberman, Phys. Rev. Letters 48, 711 (1982); G. Casati and I. Guarneri, Phys. Rev. Letters 50, 640 (1983).

2) It should be absolutely continuous with respect to the Lebesgue measure in x space, and finite on each energy surface. The damped harmonic oscillator can be cast in Hamilton form, with a conserved quantity (not the usual energy of course) and a conserved measure, which however is not finite.

3) Some of the numerous surveys of classical chaos are: J. Ford, in: Fundamental Problems in Statistical Mechanics 3 (E.G.D. Cohen ed., North-Holland, Amsterdam 1975) and R.H.G. Helleman, in: idem 5 (1980).

4) See the reviews by S.A. Rice in: Photoselective Chemistry I (= Adv. Chem. Phys. 47; J. Jortner et al. eds., Wiley-Interscience, New York 1981), and by D.W. Noid, M.L. Koszykowski, and R.A. Marcus, Annual Review of Physical Chemistry 32, 267 (1981).

5) W. Heisenberg, Die physikalischen Prinzipien der Quantentheorie (Hirzel, Leipzig 1930) p. 78; R.T. Prosser, J. Mathem. Phys. 24, 548 (1983).

6) N. Bohr, Collected Works III (North-Holland, Amsterdam 1976) p. 576; A. Sommerfeld, Atombau und Spektrallinien (4. Aufl., Vieweg, Braunschweig 1924) ch. 5.

7) The principal tool for investigating this limit is the WKB method, see e.g. M.V. Berry and K.E. Mount, Reports on Progress in Physics 35, 315 (1972); M.V. Berry and M. Tabor, Proc. Roy. Soc. Lond. A346, 101 (1976); M. Tabor, Physica D6, 195 (1983); M. Gutzwiller, Physica D5, 183 (1982).

8) Summarized in K.K. Lehmann, G.J. Scherer, and W. Klempener, J. Chem. Phys. 77, 2853 (1982).

9) For a cubic centimeter of hydrogen at 1 atmosphere the number of levels in an energy range corresponding to one microdegree is roughly $10^{10^{20}}$; hence also T.

10) This is already clear for a single particle in a rectangular box: if one lists the eigenstates according to increasing values of

$$E_{\ell mn} = \left(\hbar^2\pi^2/2\mu\right)\left(\ell^2/a^2+m^2/b^2+n^2/c^2\right) \tag{30}$$

one sees that the direction of the momentum varies wildly. Of course in this soluble case they can be disentangled and the spectrum would be called 'regular' by I.C. Percival, in: Adv. Chem. Phys. 36, 1 (1977).

11) J. v. Neumann, Z. Phys. 57, 30 (1929); Mathematische Grundlagen der Quantenmechanik (Springer, Berlin 1932) p. 212 ff.

12) N.G. van Kampen, Physica 20, 603 (1954); Fortschritte der Physik 4, 405 (1956); Fundamental Problems in Statistical Mechanics (E.G.D. Cohen, ed., North-Holland, Amsterdam 1962).

13) P. Bocchieri and A. Loinger, Phys. Rev. 114, 948 (1959); G.M. Prosperi and A. Scotti, J. Mathem. Phys. 1, 218 (1960).

14) R. Jancel, Les fondements de la mécanique statistique classique et quantique (Gauthier-Villars, Paris 1963); I.E. Farquhar, Ergodic Theory in Statistical Mechanics (Interscience, London 1964).

15) E.P. Wigner, J. Chem. Phys. 22, 1912 (1954).

16) The normalization is for large L; the next approximation replaces L by $L+\eta'(k_n)$. See H.A. Kramers, Quantum Mechanics (North-Holland, Amsterdam 1957) p. 306.

17) J.M. Blatt and V.F. Weisskopf, Theoretical Nuclear Physics (Wiley, New York 1952) p. 382.

18) This is the width in the k-scale; those in the scales of K and E differ by trivial factors, which we ignore in the discussion.

19) Perhaps not so fast as to be convincing, but that could be remedied by taking for A a smoothed version of the step function θ.

CONTINUAL OBSERVATIONS IN QUANTUM MECHANICS

A. Barchielli, L. Lanz and G.M. Prosperi*

Dipartimento di Fisica dell'Università, Milano
Istituto Nazionale di Fisica Nucleare, Sezione di Milano
Via Celoria, 16, 20133 Milano, Italy

1. INTRODUCTION

In Quantum Mechanics, as well known, due to the occurrence of the so called interference terms, any statement has to refer to a definite experiment performed on an object, i.e. to the response of a specific apparatus. No unambiguous meaning can be attached to statements on the value of any quantity independently of an explicit measurement of it.

On the contrary in classical physics the state of the system is specified in terms of a set of variables

$$\underline{z} \equiv (z^1, z^2, \dots) \qquad (1.1)$$

to which definite values are ascribed at any time, independently of an actual observation. The variables $\underline{z}$ are said to specify the state of the system and are often assumed to evolve in time deterministically, according to a system of differential equations

$$\frac{dz^i(t)}{dt} = f^i(z^1, z^2, \dots) \qquad (1.2)$$

Usually it is intended that the quantum description applies to small systems as particles, atoms or molecules and that the classical description applies to large bodies. Since, however, as emphasized by Bohr,

* Referred by G.M. Prosperi

for the interpretation of Quantum Mechanics it is essential that the experimental equipments and the results of the experiments are described in classical terms and, since on the other side a large body is made by particles, atoms and molecules, the problem of a correct understanding of the relations between the two kind of description is crucial.

A first solution of the problem essentially relies on the difference of scale to which the two mentioned descriptions refer. The methods of the Quantum Statistical Mechanics applied to systems of a very large number of elementary constituents should provide pactically well determined values for the so called macroscopic quantities, at least in principle, and enable us to derive evolution equations for such quantities of the form (1.2) by using appropriate approximations. A reasonably satisfactory treatment of the measurement process in such terms can be actually given at various level of mathematical rigour[1,2].

However, such a kind of solution does not seem to be sufficient. In fact it amounts to consider Quantum Mechanics as the really fundamental theory and Classical Physics as a kind of approximation. But, if the use of the classical description for a large body is essential for the interpretation of Quantum Mechanics itself, the possibility of a classical description should appear at a much more fundamental level in any theory which has the pretence of applying to small and to large bodies as well.

A less striking contrast between the quantum and the classical description is encountered if, following Ludwig[3], one adopts a more general characterization of what we mean by classical description, renouncing to a deterministic equation like (1.2) as an exact equation and replacing it by a statistical law for the allowed trajectories $\underline{\Xi}(t)$.

In this order of ideas it is interesting to consider a generalization of the formalism of Quantum Mechanics which allows to handle observations which have as outcome the full story of a set of variables during a certain time interval rather than a single set of values at a definite time.

Obviously such a generalization should be interesting also under other respects. Also at the microscopic level it is not difficult to find examples of experiments the result of which is most naturally expressed as the progressive development of a process. Think e.g. of an elementary particle experiment using a track chamber or any large assemblage of detectors.

All attempts at handling the mentioned situation in the context of ordinary text book formalism have met insuperable difficulties. Such difficulties have been discussed in the literature and are often referred as the Zeno paradox[4].

It should be remembered however that the ordinary text book formalism (based on the correspondence between observables and selfadjoint operators) can be made internally consistent only at the price of introducing very unnatural and unrealistic assumptions on the form of the interaction between the object and the apparatus. It exists a much more satisfactory formalism [5-8] which gives a first generalization of the ordinary one and overcomes its difficulties. Such first degree generalized formalism is based on the two fundamental ideas of effect valued measure (e.v.m.) and operation valued measure (o.v.m.). It turns out that in its framework also the problem of constructing observables whose "values" are elements of an appropriate space of trajectories has a satisfactory solution [9-13] and bring to the consideration of what can be called an operation valued stochastic processes (O.V.S.P.).

Perhaps the class of O.V.S.P.'s we are presently able to construct is not sufficiently large to enable us to handle a realistic macroscopic description of the physical world. However, some simplified model can already be treated.

The paper is mainly devoted to an introduction of the formalism of the O.V.S.P.'s following essentially the lines at ref. 10. A brief summary of the formalism of the e.v.m.'s and o.v.m.'s and a discussion of its consistency* is premitted.

2. GENERALIZED FORMULATION OF QUANTUM MECHANICS

In this section we shall revue the generalized formalism for Quantum Mechanics mentioned in the introduction.

We begin to recall that by the term effect a bounded, selfadjoint operator is meant with the following property:

$$0 \leq \hat{F} \leq 1 \tag{2.1}$$

and by effect valued measure (e.v.m.) on $\mathbb{R}^p$ an application $\hat{F}(T)$ from $\mathcal{B}(\mathbb{R}^p)$ into the family of the effects in H such that

$$\hat{F}\left(\bigcup_{j=1}^{\infty} T_j\right) = \sum_{j=1}^{\infty} \hat{F}(T_j) \quad \text{for} \quad T_i \cap T_j = \emptyset, \; i \neq j \tag{2.2}$$

Here $\mathcal{B}(\mathbb{R}^p)$ denotes the class of the Borel subsets of $\mathbb{R}^p$ and H the Hilbert space relative to the system. Similarly by operation we mean a linear mapping $\mathcal{F}$ of the space of the trace class operators $T(H)$ into itself which is positive and trace decreasing, i.e. which is such that

$$\mathcal{F}\hat{X} \geq 0 \quad \text{and} \quad \mathrm{Tr}(\mathcal{F}\hat{X}) \leq \mathrm{Tr}\,\hat{X} \quad \text{for} \quad \hat{X} \geq 0 \tag{2.3}$$

---

* In this discussion we have followed ref. 14 in particular.

and by operation valued measure (o.v.m.) an application $\mathcal{J}(T)$ from $\mathcal{B}(\mathbb{R}^P)$ into the family of operations, such that

$$\mathcal{J}\Big(\bigcup_{j=1}^{\infty} T_j\Big)=\sum_{j=1}^{\infty}\mathcal{J}(T_j) \quad \text{for} \quad T_i\cap T_j=\emptyset\ ,\quad i\neq j \tag{2.4}$$

Actually we shall assume in the following that the operation $\mathcal{J}(T)$ is not only positive but also completely positive.*
Note that a projection operator $\hat{E}$ is a particular kind of e.v.m..However, being $\hat{E}(T)$ an idempotent operator, from (2.2) it follows:

$$\hat{E}(T)\hat{E}(S)=\hat{E}(T\cap S)=\hat{E}(S)\hat{E}(T)\ , \tag{2.5}$$

whilst in general

$$\hat{F}(T)\hat{F}(S)\neq\hat{F}(S)\hat{F}(T)\ . \tag{2.6}$$

Furthermore from the well known fact that the dual space of $T(H)$ is the space $B(H)$ of the bounded operators in H , it follows that to any operation $\mathcal{J}$ (o.v.m.$\mathcal{J}(T)$) an effect (an e.v.m.) is related. Such effect is defined by the equation

$$\mathrm{Tr}(\hat{F}\hat{X})=\mathrm{Tr}(\mathcal{J}\hat{X}) \quad \text{or} \quad \mathrm{Tr}[\hat{F}(T)\hat{X}]=\mathrm{Tr}[\mathcal{J}(T)\hat{X}]\ . \tag{2.7}$$

One can also write

$$\hat{F}=\mathcal{J}^T\hat{I} \quad \text{or} \quad \hat{F}(T)=\mathcal{J}^T(T)\hat{I}\ . \tag{2.8}$$

$\mathcal{J}^T$ being the traspose mapping defined by

$$\mathrm{Tr}[\hat{Y}(\mathcal{J}\hat{X})]=\mathrm{Tr}[(\mathcal{J}^T\hat{Y})\hat{X}],\ \forall\hat{X}\in T(H),\ \forall\hat{Y}\in B(H) \tag{2.9}$$

and $\hat{I}$ the identity in H .
It can be shown that a completely positive map $\mathcal{J}$ must have the form

$$\mathcal{J}\hat{X}=\sum_s\hat{\Gamma}_s\hat{X}\hat{\Gamma}_s^+ \quad \text{with} \quad \hat{\Gamma}_s\in B(H) \tag{2.10}$$

In order this mapping be an operation we need furthermore (cf.eq.(2.3))

$$\sum_s\hat{\Gamma}_s^+\hat{\Gamma}_s\leq\hat{I}\ ; \tag{2.11}$$

---

*As well known by this terminology it is meant that $\mathcal{J}(T)$ remains positive when interpreted as a mapping over $T(H\otimes C^n)$ for every n .

then, the effect related to $\mathcal{J}$ is given by

$$\hat{F} = \sum_s \hat{\Gamma}_s^+ \hat{\Gamma}_s \ . \tag{2.12}$$

Note that the correspondence between operations and effects (o.v. and e.v. measures) is not one to one, but there are infinitely many operations (o.v.measures) that correspond to the same effect (e.v.m.). Given an effect $\hat{F}$ a particular simple example of operation which corresponds to $\hat{F}$ is given by

$$\mathcal{J}\hat{X} = \hat{F}^{1/2}\hat{X}\hat{F}^{1/2}. \tag{2.13}$$

In the particular case in which $\hat{F}$ is a projection operator this becomes

$$\mathcal{J}\hat{X} = \hat{E}\hat{X}\hat{E} \ . \tag{2.13'}$$

Similarly given any e.v.m. $\hat{F}(T)$ a particularly simple example of o.v.m. related to $\hat{F}(T)$ is given by

$$\mathcal{J}(T)\hat{X} = \int_T \hat{F}^{1/2}(d\lambda)\hat{X}\hat{F}^{1/2}(d\lambda). \tag{2.14}$$

In the following it shall be particularly interesting for us the case in which the e.v.m. $\hat{F}(T)$ derives from a density of effect, i.e. the case in which we can write

$$\hat{F}(d\lambda) = \hat{f}(\lambda)d\mu(\lambda), \tag{2.15}$$

$\mu(T)$ being a positive numerical measure. In such a case eq. (2.14) can be written as

$$\mathcal{J}(T)\hat{X} = \int_T d\mu(\lambda)\,\hat{f}^{1/2}(\lambda)\hat{X}\hat{f}^{1/2}(\lambda). \tag{2.16}$$

On the contrary in the case of purely discrete projection valued measure $\hat{E}(T) = \sum_{\lambda_j \in T} \hat{E}_{\lambda_j}$ we have

$$\mathcal{J}(T)\hat{X} = \sum_{\lambda_j \in T} \hat{E}_{\lambda_j}\hat{X}\hat{E}_{\lambda_j} \ . \tag{2.16'}$$

After the above definitions the following postulates, which replace the corresponding ones of ordinary formulation of Quantum Mechanics, can be introduced

a) To any observable A (any set of compatible observables $A_1, A_2, \ldots, A_p$) a normalized e.v.m. $\hat{F}^A(T)$ in $\mathcal{B}(\mathbb{R})$ $(\mathcal{B}(\mathbb{R}^p))$ is associated.

b) If the system is prepared in the state W at the time t = 0, the probability of observing a value belonging to $T\in\mathcal{B}(\mathbb{R})$ for A [to $T\in\mathcal{B}(\mathbb{R}^p)$ for $A\equiv(A_1,A_2,\dots,A_p)$] at the time t is given by

$$P(A\in T,t|W)=\mathrm{Tr}[\hat{F}^A(T,t)\hat{W}]\,. \tag{2.17}$$

c) To a given apparatus or procedure S_A for the measurement of the quantity A a normalized o.v.m. $\mathcal{F}_{S_A}(T)$ is associated which is related to $\hat{F}^A(T)$ by the equation

$$\hat{F}^A(T)=\mathcal{F}^T_{S_A}(T)\hat{I}\,. \tag{2.18}$$

If at the time t_0 the procedure S_A has been performed and a value $A\in T$ has been found, the state of the system is modified in the following way

$$\hat{W}\longrightarrow\mathcal{F}_{S_A}(T,t_0)\hat{W}/\mathrm{Tr}[\mathcal{F}_{S_A}(T,t_0)\hat{W}]\,. \tag{2.19}$$

Note that by the terms normalized e.v.m. and o.v.m. we mean that $\hat{F}^A(T)$ and $\mathcal{F}_{S_A}(T)$ must satisfy the obvious conditions

$$\hat{F}^A(\mathbb{R})=\hat{I} \qquad \text{and} \qquad \mathrm{Tr}\{\mathcal{F}_{S_A}(\mathbb{R})\hat{X}\}=\mathrm{Tr}\,\hat{X}\,. \tag{2.20}$$

Note also that we have adopted the Heisenberg picture and set

$$\hat{F}^A(T,t)=e^{i\hat{H}t}\hat{F}^A(T)\,e^{-i\hat{H}t}\,, \tag{2.21a}$$

$$\mathcal{F}_{S_A}(T,t)\hat{X}=e^{i\hat{H}t}\{\mathcal{F}_{S_A}(T)(e^{-i\hat{H}t}\hat{X}\,e^{i\hat{H}t})\}\,e^{-i\hat{H}t}\,. \tag{2.21b}$$

In the case of an ordinary observable with a purely discrete spectrum $\mathcal{F}(T)$ may be assumed of the form (2.16') and, if T reduces to a single point λ_j, eq.(2.19) becomes

$$\hat{W}\longrightarrow\hat{E}_{\lambda_j}(t_0)\hat{W}\hat{E}_{\lambda_j}(t_0)/\mathrm{Tr}[\hat{E}_{\lambda_j}(t_0)\hat{W}]\,. \tag{2.19'}$$

This equation is often considered as the expression of the reduction postulate in the usual formulation. Note however that eq.(2.19) applies as well to the case of a discrete and of a continuous spectrum.

The need for generalizing ordinary Quantum Mechanics in the way we have summarized here is discussed in the next section, which is devoted to an explicit consideration of the apparatus. The discussion provides also the method by which, in principle, we could actually build the e.v.m.

$\hat{F}^A(T)$ and the o.v.m. $\mathcal{F}_{S_A}(T)$ corresponding to a specific measuring procedure, once the e.v.m. corresponding to the significant properties of the apparatus is given.

We close this section by giving two very simple examples of e.v.m.. Both concern a one-dimensional particle. The first one is defined by

$$\hat{F}_x(T) = \int_T d\mu(x)\hat{f}(x) \qquad (2.22a)$$

with

$$\hat{f}(x) = \exp\left[-\alpha(x-\hat{q})^2\right] \quad \text{and} \quad d\mu(x) = \left(\frac{\alpha}{\pi}\right)^{1/2} dx \qquad (2.22b)$$

and corresponds to a kind of "coarse grained position". Similarly, the second one is defined by

$$\hat{F}_{xp}(T) = \int_T \frac{dx\,dp}{2\pi\hbar}\hat{f}(x,p) \qquad (2.23a)$$

with

$$\hat{f}(x,p) = C\exp\left[\frac{i}{\hbar}(p\hat{q}-x\hat{p})\right]\exp\left[-\alpha(\hat{q}^2+\varkappa^2\hat{p}^2)\right]\exp\left[-\frac{i}{\hbar}(p\hat{q}-x\hat{p})\right] =$$

$$= C\exp\left\{-\alpha\left[(x-\hat{q})^2+\varkappa^2(p-\hat{p})^2\right]\right\}, \qquad (2.23b)$$

$$C^{-1} = \mathrm{Tr}\left\{\exp\left[-\alpha(\hat{q}^2+\varkappa^2\hat{p}^2)\right]\right\}, \qquad (2.23c)$$

and corresponds to a coarse grained simultaneous measurement of position and momentum. The symbols $\hat{q}$ and $\hat{p}$ in eq.s (2.22) and (2.23) stay for the ordinary position and momentum operators.

3. THE ROLE OF THE APPARATUS

Now we want to introduce explicitly in the treatment the apparatus by which a certain observation on the system is performed.

In its most general acceptation by apparatus we mean a second system which interacts for a certain time with the object and which is affected by this interaction in an appreciable way.

Denoting the object by I and the apparatus by II we write the Hamiltonian of the compound system as

$$\hat{H} = \hat{H}_I + \hat{H}_{II} + \hat{H}_{int} = \hat{H}_0 + \hat{H}_{int} \; . \tag{3.1}$$

Then we can assimilate the interaction between I and II to a scattering process and assume that the limit

$$\lim_{\substack{t'' \to +\infty \\ t' \to -\infty}} e^{i\hat{H}_0 t''} e^{-i\hat{H}(t''-t')} e^{-i\hat{H}_0 t'} = \hat{U} \tag{3.2}$$

exists in the strong or in the weak sense.

We denote by A_{II} the position of an index or any other quantity by which we measure the modifications that occurre in II and by $\hat{F}^A_{II}(T)$ the e.v.m. corresponding to A_{II} according to postulate a). We also assume that II is in the state W_{II} and I in the state W_I before the interaction, that at the time t_0 the interaction is finished and that then the observation of A_{II} is performed. We obtain

$$P(A_{II} \in T, t_0 | W_I W_{II}) = Tr\{\hat{F}^A_{II}(T,t_0)\hat{U}\hat{W}_I\hat{W}_{II}\hat{U}^{+}\} , \tag{3.3}$$

where we have used the interaction picture, i.e. we have written

$$\hat{F}^A_{II}(T,t) = exp(i\hat{H}_{II}t)\hat{F}^A_{II}(T)\, exp(-i\hat{H}_{II}t) .$$

Taking into account the relation

$$exp(-i\hat{H}_0 t)\hat{U} = \hat{U}\, exp(-i\hat{H}_0 t) , \tag{3.4}$$

which follows from eq. (3.2), we can rewrite the right-hand side of eq. (3.3) as

$$P(A_{II} \in T | W_I W_{II}) = Tr^I\{\hat{F}^A_I(T,t_0)\hat{W}_I\} = P(A_I \in T, t_0 | W_I) . \tag{3.5}$$

Here

$$\hat{F}^A_I(T,t_0) = e^{i\hat{H}_I t_0}\, Tr^{II}\{\hat{W}^{1/2}_{II} e^{i\hat{H}_{II}t_0}\hat{U}^{+}\hat{F}^A_{II}(T)\hat{U} e^{-i\hat{H}_{II}t_0}\hat{W}^{1/2}_{II}\} e^{-i\hat{H}_I t_0} \tag{3.6}$$

is obviously an e.v.m. and Tr, Tr^I and Tr^{II} denote the trace operations performed on $H = H_1 \otimes H_2$, H_1, H_2 which are the Hilbert spaces respectively of the compound system, of system I alone and of system II alone.

Eq. (3.5) shows that the observation of the quantity A_{II} on II at time t_0 after the interaction can be equivalently described as the observation of the quantity A_I on I corresponding to the e.v.m. $\hat{F}_I^A(T)$ and essentially it proves the consistency of the postulates a) and b) with a reasonably general characterization of the apparatus.
Note that, even if $\hat{F}_{II}^A(T)$ is supposed to be a projection valued measure, in general $\hat{F}_I^A(T)$ turns out to be an effect valued measure unless very special and unrealistic assumptions on $\hat{U}$ are adopted. So the original postulates of the ordinary formulation are actually inconsistent.

Assume now to perform a second observation on I at a subsequent time t. If we do not introduce the new apparatus explicitly in the treatment we can write

$$P(B_I \in S, t; A_{II} \in T, t_0 | \hat{W}_I \hat{W}_{II}) = Tr\{\hat{F}_I^B(S,t)\hat{F}_{II}^A(T,t_0)\hat{U}\hat{W}_I\hat{W}_{II}\hat{U}^+\} \qquad (3.7)$$

for the joint probability of observing $A_{II} \in T$ at t_0 and $B_I \in S$ at t. Here B_I denotes the new observable and $\hat{F}_I^B(S)$ the corresponding e.v.m.; we have taken into account the fact that $\hat{F}_I^B(S,t) = \exp(i\hat{H}_I t)\hat{F}_I^B(T)\exp(-i\hat{H}_I t)$ commutes with $\hat{F}_{II}^A(T,t_0)$ even for $t \neq t_0$.
Eq. (3.7) can be rewritten in the form

$$P(B_I \in S, t; A_{II} \in T, t_0 | \hat{W}_I \hat{W}_{II}) = Tr^I\{\hat{F}_I^B(S,t)\mathcal{J}_I^A(T,t_0)\hat{W}_I\}, \qquad (3.8)$$

where

$$\begin{aligned}\mathcal{J}_I^A(T,t_0)\hat{W}_I &= Tr^{II}\{\hat{F}_{II}^A(T,t_0)\hat{U}\hat{W}_I\hat{W}_{II}\hat{U}^+\} = \\ &= Tr^{II}\{[\hat{F}_{II}^A(T,t_0)]^{\frac{1}{2}}\hat{U}\hat{W}_{II}^{\frac{1}{2}}\hat{W}_I\hat{W}_{II}^{\frac{1}{2}}\hat{U}^+[\hat{F}_{II}^A(T,t_0)]^{\frac{1}{2}}\}\end{aligned} \qquad (3.9)$$

defines an o.v.m.. Note that

$$\hat{F}_I^A(T,t_0) = [\mathcal{J}_I^A(T,t_0)]^T\hat{I}. \qquad (3.10)$$

Finally from eq.s (3.5) and (3.8) we obtain

$$P(B_I \in S, t | A_I \in T, t_0; W_I) = \frac{Tr^I\{\hat{F}_I^B(S,t)\mathcal{J}_I^A(T,t_0)\hat{W}_I\}}{Tr^I\{\mathcal{J}_I^A(T,t_0)\hat{W}_I\}} \qquad (3.11)$$

for the probability of observing $B_I \in S$ at t conditioned by having observed $A_I \in T$ at t_0.

Eq. (3.11) shows that postulate c) can be actually understood as a consequence of a) and b) when applied to the compound system I + II. Again the assumptions we need in order that $\mathcal{J}_I^A(T)$ be of the form corresponding to eq. (2.19') are very special and even more unrealistic than those required in order that $\hat{F}_I^A(T)$ be a projection valued measure.

Finally note that eq.s (3.6) and (3.9) provide explicit expressions for e.v.m. and o.v.m. corresponding to a given measuring procedure in terms of the characteristics of the apparatus, once the physical meaning of the formal observable A_{II} corresponding to $\hat{F}_{II}(T)$ is known. On the other side the choice of $\hat{F}_{II}^A(T)$ should be practically obvious on the basis of the correspondence principle if II is a macroscopic body.

4. CONTINUAL OBSERVATIONS AND OPERATION VALUED STOCHASTIC PROCESSES

In the context of the generalized formulation of Quantum Mechanics recalled in §§ 2 and 3 let us consider a system prepared in the state W, an observable A and an apparatus S_A for observing A.

According to eq.s (2.17) and (2.22) the probability for observing $A \in T_0$ at the time t_0 is given by

$$P(A\in T_0, t_0 | W) = Tr\{\hat{F}_A(T_0,t_0)\hat{W}\} = Tr\{\mathcal{J}_{S_A}(T_0,t_0)\hat{W}\} \qquad (4.1)$$

and according to (2.19) the conditional probability for observing $A \in T_1$ at t_1, if $A \in T_0$ has been observed at t_0, by

$$P(A\in T_1, t_1 | A \in T_0, t_0; W) = Tr\{\hat{F}_A(T_1,t_1)\mathcal{J}_{S_A}(T_0,t_0)\hat{W}\} / Tr\{\mathcal{J}_{S_A}(T_0,t_0)\hat{W}\} =$$

$$= Tr\{\mathcal{J}_{S_A}(T_1,t_1)\mathcal{J}_{S_A}(T_0,t_0)\hat{W}\} / Tr\{\mathcal{J}_{S_A}(T_0,t_0)\hat{W}\} . \qquad (4.2)$$

Then the joint probability for observing $A \in T_0$ at t_0 and $A \in T_1$ at t_1 can be written as

$$P(A\in T_1,t_1;A\in T_0,t_0|W)=P(A\in T_0,t_0|W)P(A\in T_1,t_1|A\in T_0,t_0;W)=$$

$$=\mathrm{Tr}\{\mathcal{J}_{S_A}(T_1,t_1)\mathcal{J}_{S_A}(T_0,t_0)\hat{W}\} \qquad (4.3)$$

and, more in general, the joint probability for observing a sequence of results at $t_0 < t_1 < \ldots < t_N$ as

$$P(A\in T_N,t_N;\ldots;A\in T_1,t_1;A\in T_0,t_0|W)=$$

$$=\mathrm{Tr}\{\mathcal{J}_{S_A}(T_N,t_N)\ldots\mathcal{J}_{S_A}(T_1,t_1)\mathcal{J}_{S_A}(T_0,t_0)\hat{W}\}\;. \qquad (4.4)$$

Clearly eq. (4.4) is the generalization of the well known Wigner equation

$$P(A=\lambda_{j_N},t_N;\ldots;A=\lambda_{j_0},t_0|W)=$$

$$=\mathrm{Tr}\{\hat{E}_{\lambda_{j_N}}(t_N)\ldots\hat{E}_{\lambda_{j_0}}(t_0)\hat{W}\hat{E}_{\lambda_{j_0}}(t_0)\ldots\hat{E}_{\lambda_{j_N}}(t_N)\} \qquad (4.5)$$

which refers to an observable with purely discrete spectrum and to the related o.v.m. defined by eq. (2.16').

Note that

$$\mathcal{J}(T_N,t_N;\ldots;T_0,t_0)=\mathcal{J}_{S_A}(T_N,t_N)\ldots\mathcal{J}_{S_A}(T_0,t_0) \qquad (4.6)$$

and

$$\hat{F}(T_0,t_0;\ldots;T_N,t_N)=\mathcal{J}^{T}_{S_A}(T_0,t_0)\ldots\mathcal{J}^{T}_{S_A}(T_N,t_N)\hat{I} \qquad (4.7)$$

are operations and effects respectively and that they generate an o.v.m. and an e.v.m. on $\mathcal{B}(\mathbb{R}^{N+1})$. Using such notations eq. (4.3) can be written as

$$P(A\in T_N,t_N;\ldots;A\in T_0,t_0|W)=$$

$$=\mathrm{Tr}\{\mathcal{J}(T_N,t_N;\ldots;T_0,t_0)\hat{W}\}=\mathrm{Tr}\{\hat{F}(T_N,t_N;\ldots;T_0,t_0)\hat{W}\} \qquad (4.8)$$

and the conditional probability of observing $A\in T_{M+1}$ at t_{M+1}, ..., $A\in T_N$ at t_N, if $A\in T_0$ at t_0, ..., $A\in T_M$ at t_M has been observed, as

$$P(A\in T_N,t_N;\ldots;A\in T_{M+1},t_{M+1}\,|\,A\in T_M,t_M;\ldots;A\in T_0,t_0;W)= \tag{4.9}$$

$$=\mathrm{Tr}\{\hat{F}(T_N,t_N;\ldots;T_{M+1},t_{M+1})\mathcal{F}(T_M,t_M;\ldots;T_0,t_0)\hat{W}\}/\mathrm{Tr}\{\mathcal{F}(T_M,t_M;\ldots;T_0,t_0)\hat{W}\}.$$

So in the generalized formulation of Q.M. a sequence of observations can be put on an equal footing to a single observation at a definite time.

Note that, in the case of eq. (2.19'), eq. (4.7) becomes

$$\hat{F}(\lambda_{j_N},t_N;\ldots;\lambda_{j_0},t_0)=\hat{E}_{j_N}(t_N)\ldots\hat{E}_{j_1}(t_1)\hat{E}_{j_0}(t_0)\hat{E}_{j_1}(t_1)\ldots\hat{E}_{j_N}(t_N), \tag{4.10}$$

but only in the trivial case in which A is a constant of the motion the various factors in this equation commute and $\hat{F}$ becomes a projection operator. Note also that from the definitions (4.6) and (4.7) it follows

$$\mathcal{F}(T_N,t_N;\ldots;T_0,t_0)=\mathcal{F}(T_N,t_N;\ldots;T_{M+1},t_{M+1})\mathcal{F}(T_M,t_M;\ldots;T_0,t_0). \tag{4.11}$$

The results obtained above for a sequence of instantaneous observations suggest the extension to observations which last for a finite interval of time (t_0, t_1) and have as outcome the determination of the complete trajectory $\underline{x}(t)\equiv(x^1(t),x^2(t),\ldots,x^m(t))$ for a set of variables during that interval.

We shall denote by Y the space of all apriori admittable trajectories $\underline{x}(t)$ (which may be assumed extended to the entire t - axis for mathematical convenience), by Σ an appropriate σ-algebra of subsets of Y and by $\Sigma_{t_0}^{t_1}$ the portion of Σ whose elements correspond to certain requirements on the trajectories only in the time interval (t_0, t_1).

We then assume that for any given interval (t_0, t_1) an o.v.m. $\mathcal{F}(t_1,t_0;M)$ on $\Sigma_{t_0}^{t_1}$ and a related e.v.m. $\hat{F}(t_1,t_0;M)=\mathcal{F}^T(t_1,t_0;M)\hat{I}$ exist with the following properties:

1) The probability that the experiment gives a result $\underline{x}(t)\in M\in\Sigma_{t_0}^{t_1}$ is expressed by

$$P(M|W,t_0)=\mathrm{Tr}\{\hat{F}(t_1,t_0;M)\hat{W}\}=\mathrm{Tr}\{\mathcal{F}(t_1,t_0;M)\hat{W}\}, \tag{4.12}$$

if no observation is performed after the preparation in the state W before the time t_0.

2) The composition law

$$\mathcal{F}(t_2,t_0;N\cap M)=\mathcal{F}(t_2,t_1;N)\mathcal{F}(t_1,t_0;M) \tag{4.13}$$

holds for $M\in\Sigma_{t_0}^{t_1}$ and $N\in\Sigma_{t_1}^{t_2}$ (note that $N\cap M\in\Sigma_{t_0}^{t_2}$)
3) The conditional probability of finding $\underline{x}(t)\in N\in\Sigma_{t_1}^{t_2}$, if $\underline{x}(t)\in$ $\in M\in\Sigma_{t_0}^{t_1}$ has been observed, is given by

$$P(N|M;W,t_0)=$$

$$=\mathrm{Tr}\{\hat{F}(t_2,t_1;N)\mathcal{F}(t_1,t_0;M)\hat{W}\}/\mathrm{Tr}\{\mathcal{F}(t_1,t_0;M)\hat{W}\}. \tag{4.14}$$

4) The time translation equation

$$\mathcal{F}(t_1+\tau,t_0+\tau;M_\tau)=\mathcal{T}(\tau)\mathcal{F}(t_1,t_0;M)\mathcal{T}(-\tau) \tag{4.15}$$

holds with

$$M_\tau=\{\underline{x}(t):\underline{x}(t)=\underline{x}'(t-\tau),\ \underline{x}'(t)\in M\}, \tag{4.16}$$

$$\mathcal{T}(\tau)\hat{X}=\exp(i\hat{H}\tau)\hat{X}\exp(-i\hat{H}\tau). \tag{4.17}$$

5) $\mathcal{F}(t_1,t_0;M)$ is normalized; i.e., if we set $\mathcal{G}(t_1,t_0)=\mathcal{F}(t_1,t_0;Y)$ the equation

$$\mathrm{Tr}[\mathcal{G}(t_1,t_0)\hat{X}]=\mathrm{Tr}\hat{X} \tag{4.18}$$

holds.

Note that eq. (4.13) corresponds to eq. (4.11) and establishes the order between subsequent observations, whilst eq. (4.15) corresponds to eq. (2.21b) and to the use of the Heisenberg picture.

Note also that for M = Y eq. (4.14) becomes

$$P(N|Y;W,t_0)=\mathrm{Tr}\{\hat{F}(t_2,t_1;N)\mathcal{G}(t_1,t_0)\hat{W}\},\ N\in\Sigma_{t_1}^{t_2}, \tag{4.19}$$

so the mapping $\mathcal{G}(t_1,t_0)$ describes the modification produced on the state of the system by the action of the apparatus when no notice is taken of the result, briefly it describes the <u>disturbance</u> by the apparatus.

The assumptions we have introduced above have a physical motivation. In order to give them a precise mathematical meaning it is however necessary to specify some more the nature of the space Y and of the σ-algebra Σ. The most natural framework to do so seems to be that of the so called stochastic processes, better of the generalized stochastic processes [15,16]. In this order of ideas we denote by E a space of test functions of the form $\underline{h}(t) \equiv (h_1(t), h_2(t), \ldots, h_m(t))$ and identify Y by E^*, the algebraic dual space of E. Then for any given $\underline{h}(t) \in E$ we define the time average

$$x_h = \int dt\, h_\alpha(t)\, x^\alpha(t) \tag{4.20}$$

(summation over repeated greek indices has to be understood) and consider the subsets of E^* of the form

$$C(\underline{h}^{(1)}, \ldots, \underline{h}^{(\ell)}; B) = \{\underline{x}(t) \in E^* : (x_{h^{(1)}}, \ldots, x_{h^{(\ell)}}) \in B\}, \tag{4.21}$$

where ℓ is an arbitrary integer, $\underline{h}^{(1)}, \ldots, \underline{h}^{(\ell)}$ are elements of E and B is a Borel set in $\mathbb{R}^\ell$. The subsets $C(\underline{h}^{(1)}, \ldots \underline{h}^{(\ell)}; B)$ are called cylinder sets. We identify Σ with the σ-algebra generated by the cylinder sets and consequently $\Sigma_{t_0}^{t_1}$ with the σ-subalgebra generated by the cylinder sets corresponding to test functions with support in (t_0, t_1).

With the above definitions the triad

$$\{E^*, \Sigma_{t_0}^{t_1}, P(\cdot\, | W, t_0)\} \tag{4.22}$$

realizes what is called a generalized stochastic process in the mathematical literature*. Similarly we may call

$$\{E^*, \Sigma_{t_0}^{t_1}, \mathcal{J}(t_1, t_0; \cdot)\} \tag{4.23}$$

an operation valued stochastic process (O.V.S.P.).

---

* The term generalized refers to making use of the time average (4.20) rather than of the values of $\underline{x}(t)$ at definite times in eq. (4.21). The stochastic process becomes an ordinary one when the $h^{(i)}(t)$'s may be replaced by δ-functions.

5. CHARACTERISTIC FUNCTIONAL AND CONSTRUCTION OF A CLASS OF O.V.S.P.'s

In order to be sure that the formalism we have introduced in the preceding section is not empty we have to show that a class of O.V.S.P.'s which satisfies eq.s (4.13), (4.15) and (4.18) does actually exist and it is physically significant. For actually constructing such a class of O.V.S.P.'s we find convenient to start from the concept of characteristic functional and characteristic operator.

The characteristic functional for the g.s.p. (4.22) is defined by

$$L(t_1 t_0;[\underline{\xi}(t)]|W)=\int dP([\underline{x}(t)]|W,t_0)\exp\{i\int_{t_0}^{t_1}dt\,\xi_\alpha(t)\,x^\alpha(t)\}, \qquad (5.1)$$

for every $\underline{\xi}(t)\in E$, in obvious notations.

In terms of such an object the density of probability for the quantities $x_{\underline{h}^{(1)}},\dots,x_{\underline{h}^{(l)}}$ can be immediately evaluated as

$$p(x_1,\underline{h}^{(1)};\dots;x_l,\underline{h}^{(l)}|W,t_0)\equiv\int dP([\underline{x}(t)]|W,t_0)\,\delta(x_1-x_{\underline{h}^{(1)}})\cdots\delta(x_l-x_{\underline{h}^{(l)}})=$$

$$=\frac{1}{(2\pi)^l}\int dk_1\dots dk_l\exp\{-i\sum_{j=1}^{l}k_jx_j\}L\Big(t_1,t_0;\Big[\sum_{j=1}^{l}k_j\,\underline{h}^{(j)}(t)\Big]\Big|W\Big). \qquad (5.2)$$

Similarly we have for the momenta

$$\langle x_{\underline{h}^{(1)}}\cdots x_{\underline{h}^{(l)}}\rangle=(-i)^l\frac{\partial^l}{\partial k_1\cdots\partial k_l}L\Big(t_1,t_0;\Big[\sum_j k_j\,\underline{h}^{(j)}(t)\Big]\Big|W\Big)\Big|_{k_1,\dots,k_l=0}=$$

$$=(-i)^l\int dt^{(1)}\dots dt^{(l)}\,h^{(1)}_{\alpha_1}(t^{(1)})\dots h^{(l)}_{\alpha_l}(t^{(l)})\frac{\delta^l L(t_1,t_0;[\underline{\xi}(t)]|W)}{\delta\xi_{\alpha_1}(t^{(1)})\cdots\delta\xi_{\alpha_l}(t^{(l)})}\Big|_{\underline{\xi}=\underline{0}} \qquad (5.3a)$$

or even, symbolically,

$$\langle x^{\alpha_1}(t^{(1)})\dots x^{\alpha_l}(t^{(l)})\rangle=(-i)^l\frac{\delta^l L(t_1,t_0;[\underline{\xi}(t)]|W)}{\delta\xi_{\alpha_1}(t^{(1)})\cdots\delta\xi_{\alpha_l}(t^{(l)})}\Big|_{\underline{\xi}=\underline{0}} \qquad (5.3b)$$

Note that when L is known the original distribution of probability $P(\cdot\,|W,t_0)$ can be reobtained by the equations (5.2) and

$$P(C(\underline{h}^{(1)},\ldots,\underline{h}^{(\ell)};B)|W,t_0)=\int_B dx_1\cdots dx_\ell\, p(x_1,\underline{h}^{(1)};\ldots;x_\ell,\underline{h}^{(\ell)}|W,t_0) \quad (5.4)$$

for the cylinder sets and, then, extended to the whole $\Sigma_{t_0}^{t_1}$ [16].

A role analogous to the characteristic functional is played for O.V.S.P. (4.23) by the characteristic operator which can be defined as

$$\mathcal{G}(t_1,t_0;[\underline{\xi}(t)])=\int d\mathcal{F}(t_1,t_0;[\underline{x}(t)])\exp\{i\int_{t_0}^{t_1}dt\,\xi_\alpha(t)x^\alpha(t)\}. \quad (5.5)$$

In terms of $\mathcal{G}$ we can introduce the density of operation

$$f(t_1,t_0;x_1,\underline{h}^{(1)};\ldots;x_\ell,\underline{h}^{(\ell)})=$$
$$=\frac{1}{(2\pi)^\ell}\int dk_1\cdots dk_\ell\exp(-i\sum_{j=1}^{\ell}k_jx_j)\mathcal{G}(t_1,t_0;[\sum_j k_j\underline{h}^{(j)}(t)]) \quad (5.6a)$$

and write

$$\mathcal{F}(t_1,t_0;C(\underline{h}^{(1)},\ldots,\underline{h}^{(\ell)};B))=\int_B dx_1\cdots dx_\ell\, f(t_1,t_0;x_1,\underline{h}^{(1)};\ldots;x_\ell,\underline{h}^{(\ell)}) \quad (5.6b)$$

Obviously we have

$$L(t_1,t_0;[\underline{\xi}(t)]|W)=Tr\{\mathcal{G}(t_1,t_0;[\underline{\xi}(t)])\hat{W}\} \quad (5.7)$$

and

$$p(x_1,\underline{h}^{(1)};\ldots;x_\ell,\underline{h}^{(\ell)}|W)=Tr\{f(t_1,t_0;x_1,\underline{h}^{(1)};\ldots;x_\ell,\underline{h}^{(\ell)})\hat{W}\}. \quad (5.8)$$

Note that

$$\mathcal{G}(t_1,t_0;[\underline{0}])=\mathcal{F}(t_1,t_0;E^*)=\mathcal{G}(t_1,t_0) \quad (5.9)$$

and that in terms of $\mathcal{G}$ eq. (4.13) becomes

$$\mathcal{G}(t_2,t_0;[\underline{\xi}(t)])=\mathcal{G}(t_2,t_1;[\underline{\xi}(t)])\mathcal{G}(t_1,t_0;[\underline{\xi}(t)]). \quad (5.10)$$

We now assume that eq. (5.10) can be put in differential form and write

$$\frac{\partial}{\partial t}\mathcal{G}(t;t_0;[\underline{\xi}]) = \mathcal{K}(t;[\underline{\xi}])\mathcal{G}(t,t_0;[\underline{\xi}]), \tag{5.11}$$

where we have set

$$\mathcal{K}(t;[\underline{\xi}]) = \frac{\partial}{\partial t'}\mathcal{G}(t',t;[\underline{\xi}])\Big|_{t'=t} . \tag{5.12}$$

Together with the obvious identity

$$\mathcal{G}(t_0,t_0,[\underline{\xi}]) = 1 \tag{5.13}$$

eq. (5.11) determines $\mathcal{G}$ in the form

$$\mathcal{G}(t_1,t_0;[\underline{\xi}]) = T\exp\left\{\int_{t_0}^{t_1} dt\, \mathcal{K}(t;[\underline{\xi}])\right\}, \tag{5.14}$$

where T denotes the time ordering prescription. The O.V.S.P. can be then completely reconstructed in terms of $\mathcal{K}$.

Reversing the point of view let us now prescribe an operator $\mathcal{K}$ a priori and construct $\mathcal{F}(t_1,t_0;\cdot)$ using eq.s (5.14) and (5.6) assuming only the right hand side of (5.6a) to be convergent. We find that the resulting $\mathcal{F}(t_1,t_0;\cdot)$ has authomatically all the required properties with the exception of positivity, normalization and time translation transformation.

In order to obtain that also the last three properties be satisfied, we make the ansatz

$$\mathcal{K}(t;[\underline{\xi}]) = \mathcal{L}(t) + i\,\xi_\alpha(t)\mathcal{R}^\alpha(t) - \frac{1}{4\gamma}\xi_\alpha(t)\Delta^{\alpha\beta}\xi_\beta(t), \tag{5.15}$$

where $\Delta^{\alpha\beta}$ is a numerical positive matrix with determinant equal to 1 and γ is a positive constant. With this ansatz the right-hand side of (5.6a) becomes a gaussian integral, it is certainly convergent and for $t_1 - t_0$ infinitesimum it can be even explicitly evaluated. Then let us begin by setting $\underline{\xi}(t) = \underline{c}$ in eq. (5.14). We obtain

$$\mathcal{G}(t_1,t_0) = T\exp\left\{\int_{t_0}^{t_1} dt\, \mathcal{L}(t)\right\} . \tag{5.16}$$

Complete positivity and normalization of $\mathcal{F}(t_1,t_0;\cdot)$ require that even $\mathcal{G}(t_1,t_0)$ is completly positive and trace preserving (cf. eq. (4.18)). On the other side the form that the operator $\mathcal{L}(t)$ must have, in order the right hand side of (5.16) to have these properties, has been studied in the literature in a different context [17, 18, 6]. In the case of a bounded $\mathcal{L}(t)$ it is found that

$$\mathcal{L}(t)\hat{X} = -i[\hat{K}(t),\hat{X}] - \frac{1}{2}\sum_s\Big(\hat{V}_s^+(t)\hat{V}_s(t)\hat{X} + \hat{X}\hat{V}_s^+(t)\hat{V}_s(t) - 2\hat{V}_s(t)\hat{X}\hat{V}_s^+(t)\Big), \tag{5.17}$$

with $\hat{K}(t)=\hat{K}^+(t)$ and $\hat{V}_1(t), \hat{V}_2(t), \ldots$ bounded operators in H. Eq. (5.17) is the infinitesimal counterpart of eq. (2.10) and it is a sufficient condition for the above properties to be verified even if $\hat{K}, \hat{V}_1, \ldots$ are not bounded. We shall assume it valid in any case.

Now once that $\mathcal{L}(t)$ has the form (5.17) it is possible to show that, in order the right-hand side of (5.6a) to be positive, the equation has to hold

$$\mathcal{R}^\alpha(t)\hat{X} = \frac{1}{2}\left(\hat{R}^\alpha(t)\hat{X} + \hat{X}\hat{R}^\alpha(t)^+\right), \tag{5.18}$$

with

$$\hat{R}^\alpha(t) = \sum_s c_s^\alpha \hat{V}_s(t) + c_0^\alpha \hat{I}\ ; \tag{5.19}$$

moreover

$$\alpha_{rs} = \delta_{rs} - \frac{\gamma}{2} c_r^{\alpha} \Delta^{-1}_{\alpha\beta} c_s^{\beta} \qquad \left(\Delta^{-1}_{\alpha\beta}\Delta^{\beta\gamma} = \delta_\alpha{}^\gamma\right) \tag{5.20}$$

must be a non negative matrix. We refer to ref. 10 for the proof.

Finally note that, for a $\mathcal{K}$ of the type (5.15), eq. (4.15) becomes

$$\mathcal{T}(\tau)\mathcal{L}(t)\mathcal{T}(-\tau) = \mathcal{L}(t+\tau) \qquad \mathcal{T}(\tau)\mathcal{R}^\alpha(t)\mathcal{T}(-\tau) = \mathcal{R}^\alpha(t+\tau) \tag{5.21}$$

and such equations are authomatically verified if $\hat{K}(t), \hat{V}_s(t), \hat{R}^\alpha(t)$ are understood as ordinary Heisenberg picture operators

$$\hat{K}(t) = \exp(i\hat{H}t)\hat{K}\exp(-i\hat{H}t), \ldots \tag{5.22}$$

In conclusion for any $\mathcal{K}$ of the form specified by eq.s (5.15) and (5.17 - 5.21) an O.V.S.P. exists and can be explicitly constructed.

To clarify the physical meaning of the class of O.V.S.P.'s we have constructed it is convenient to start from two particular cases.

First let us assume $\underline{x}(t)$ one-dimensional and set

$$\hat{V}_1 = \sqrt{\frac{\gamma}{2}}\,\hat{A} \quad , \quad \hat{R} = \hat{A} \quad ,$$

$$\hat{K} = \hat{V}_2 = \hat{V}_3 = \cdots = 0 \quad , \quad \Delta^{11} = 1 \quad , \tag{5.23}$$

with $\hat{A}^{+} = \hat{A}$. We have

$$\mathcal{K}(t, \xi(t)) = -\frac{\gamma}{4}[\hat{A}(t),[\hat{A}(t),\cdot]] + \frac{i}{2}\xi(t)\{\hat{A}(t),\cdot\} - \frac{1}{4\gamma}\xi^2(t). \tag{5.24}$$

Then we can introduce the test function

$$h_\varepsilon(t') = \begin{cases} \frac{1}{\varepsilon} & \text{for} \quad t' \in (t, t+\varepsilon) \\ 0 & \text{otherwise,} \end{cases} \tag{5.25}$$

(for which x_{h_ε} coincides with the ordinary time average $x_{h_\varepsilon} = \frac{1}{\varepsilon} \cdot \int_t^{t+\varepsilon} dt' x(t')$) and consider the <u>elementary operation</u>

$$f(t, t+\varepsilon; x, h_\varepsilon) \simeq \frac{1}{2\pi}\int dk \, \exp(-ikx)\exp[\varepsilon\,\mathcal{K}(t; \tfrac{1}{\varepsilon}k)] \tag{5.26}$$

for $\varepsilon \to 0$. Evaluating the integral we obtain (cf. ref. 10)

$$f(t, t+\varepsilon; x, h_\varepsilon)\hat{W} \simeq \hat{F}_\varepsilon^{\frac{1}{2}}(x,t)\,\hat{W}\,\hat{F}_\varepsilon^{\frac{1}{2}}(x,t) \quad , \tag{5.27}$$

where $\hat{F}_\varepsilon(x,t)$ denotes the <u>elementary effect</u>

$$\hat{F}_\varepsilon(x,t) = f^T(t, t+\varepsilon; x, h_\varepsilon)\hat{I} \simeq \sqrt{\frac{\gamma\varepsilon}{\pi}}\exp\left[-\varepsilon\gamma(x - \hat{A}(t))^2\right]. \tag{5.28}$$

Eq.s (5.27) and (5.28) can be compared with (2.16) and (2.22). They show that the O.V.S.F. provided by eq. (5.24) can be understood as corresponding to the limit of a sequence of coarse grained observations of the quantity related to the operator $\hat{A}$ in ordinary Q.M.. Note that as $\varepsilon \to 0$ the observations become progressively poorer and only at this price a finite limit can exist (cf.ref. 9, 12 in this connection). In particular, if $\hat{A}$ is identified with the ordinary position operator $\hat{q}$ for a particle on the real axis, the O.V.S.P. corresponds to a coarse grained observation of the " continuous motion " of such a particle.

Secondly let us assume n > 1 and set

$$\hat{V}_\alpha=\sqrt{\frac{\gamma}{2}}\,\Delta^{-\frac{1}{2}}_{\alpha\beta}\hat{A}^\beta , \quad \hat{R}^\alpha=\hat{A}^\alpha , \quad \alpha=1,2,\dots,n ,$$
$$\hat{K}=\hat{V}_{n+1}=\hat{V}_{n+2}=\dots=0 , \tag{5.29}$$

with $(\hat{A}^\mu)^+=\hat{A}^\mu$. This time we have

$$\mathcal{K}(t;\underline{\xi}(t))=-\frac{\gamma}{4}\Delta^{-1}_{\mu\nu}[\hat{A}^\mu(t),[\hat{A}^\nu(t),\cdot]]+$$
$$+\frac{i}{2}\xi_\mu(t)\{\hat{A}^\mu(t),\cdot\}-\frac{1}{4\gamma}\xi_\mu(t)\Delta^{\mu\nu}\xi_\nu(t) \tag{5.30}$$

and

$$\rho(t,t+\varepsilon;x^1,h_\varepsilon\underline{e}^{(1)};\dots;x^n,h_\varepsilon\underline{e}^{(n)})\hat{W}\simeq\frac{1}{(2\pi)^n}\int dk_1\dots dk_n \exp(-ik_\mu x^\mu)\cdot$$
$$\cdot\exp[\varepsilon\,\mathcal{K}(t;\frac{1}{\varepsilon}k_\mu\underline{e}^{(\mu)})]\hat{W}\simeq\hat{f}^{\frac{1}{2}}_\varepsilon(\underline{x},t)\hat{W}\hat{f}^{\frac{1}{2}}_\varepsilon(x,t) \tag{5.31}$$

with $e^{(\mu)}_\nu=\delta_\nu^{\ \mu}$ and

$$\hat{f}_\varepsilon(\underline{x},t)\simeq\left(\frac{\gamma\varepsilon}{\pi}\right)^{\frac{n}{2}}\exp[-\gamma\varepsilon(x^\mu-\hat{A}^\mu(t))\Delta^{-1}_{\mu\nu}(x^\nu-\hat{A}^\nu(t))] . \tag{5.32}$$

The O.V.S.P. corresponding to eq. (5.30) can be so interpreted as related to the limit of a sequence of a coarse grained simultaneous observations of the quantities associated to $\hat{A}^1,\dots,\hat{A}^n$. This is true whether the $\hat{A}^\mu$ commute each other or not, being the order of the operators in (5.32) immaterial for infinitesimal ε . In particular for n = 2 $\hat{A}^1$ and $\hat{A}^2$ can be identified with the operators $\hat{q}$ and $\hat{p}$ in eq. (2.23) and we may talk of observation of trajectories in the classic phase space of a particle.

Coming back to the general case, if we set

$$\hat{R}^{\mu} = \hat{A}^{\mu} + i\,\hat{B}^{\mu} \tag{5.33}$$

with $(\hat{A}^{\mu})^{\dagger} = \hat{A}^{\mu}$, $(\hat{B}^{\mu})^{\dagger} = \hat{B}^{\mu}$, we obtain

$$f(t+\varepsilon, t; x^{1}, h_{\varepsilon}\,\underline{\ell}^{(1)}; \ldots; x^{m}, h_{\varepsilon}\,\underline{\ell}^{(m)})\hat{W} \simeq \exp[\varepsilon\,\mathcal{L}'(t)] \cdot$$

$$\cdot \left[\hat{U}_{\varepsilon}(\underline{x},t)\hat{F}_{\varepsilon}^{\frac{1}{2}}(\underline{x},t)\hat{W}\hat{F}_{\varepsilon}^{\frac{1}{2}}(\underline{x},t)\hat{U}_{\varepsilon}^{\dagger}(\underline{x},t)\right] \tag{5.34}$$

where

$$\hat{F}_{\varepsilon}(\underline{x},t) = \left(\frac{\gamma\varepsilon}{\pi}\right)^{m}\hat{\Omega}_{\varepsilon}(\underline{x},t) \cdot$$

$$\cdot \exp\left[-\gamma\varepsilon(x^{\mu} - \hat{A}^{\mu}(t))\Delta^{-1}_{\mu\nu}(x^{\nu} - \hat{A}^{\nu}(t))\right]\hat{\Omega}_{\varepsilon}^{\dagger}(\underline{x},t). \tag{5.35}$$

and $\mathcal{L}'(t)$, $\hat{U}_{\varepsilon}(x,t)$ and $\hat{\Omega}_{\varepsilon}(\underline{x},t)$ are complicate expressions which we do not give explicitly here. We only mention that $\mathcal{L}'(t)$ is an operator in $T(H)$ of the same general form as $\mathcal{L}$ (and so $\exp[\varepsilon\mathcal{L}'(t)]$ is completely positive and trace preserving) that $\hat{U}_{\varepsilon}$ is a unitary operator in H and finally that both $\hat{U}_{\varepsilon}$ and $\hat{\Omega}_{\varepsilon}$ reduce to the identity for $\hat{B}^{1} = \cdots = \hat{B}^{m} = 0$. The qualitative meaning of eq.s (5.34) and (5.35) is then clearly similar to that of eq.s (5.31) and (5.32). We note however that $\hat{F}_{\varepsilon}$ has now the structure of a kind of distorted gaussian and that f is not longer of the simple form corresponding to eq. (2.16) and in general maps pure states into mixtures.

As an example before closing the section, we want to evaluate the first momenta, for the O.V.S.P. defined by eq. (5.24).

We note first that from eq.s (5.14) and (5.15) it follows

$$\frac{\delta\mathcal{G}(t_{1},t_{0};[\xi])}{\delta\xi(t)} = \mathcal{G}(t_{1},t;[\xi])\frac{\partial\mathcal{K}(t;\xi(t))}{\partial\xi(t)}\mathcal{G}(t,t_{0};[\xi]). \tag{5.36}$$

Then from (5.3b), (5.7), (5.9) and (4.18) one obtains

$$\langle x(t)\rangle = \mathrm{Tr}\left(\hat{A}(t)\mathcal{G}(t,t_{0})\hat{W}\right), \tag{5.37a}$$

$$\langle x(t)x(t')\rangle = \frac{1}{2\gamma}\delta(t-t') + \frac{1}{2}\mathrm{Tr}\Big(\vartheta(t-t')\hat{A}(t)\mathcal{G}(t,t')\cdot$$
$$\cdot\{\hat{A}(t'),\mathcal{G}(t',t_0)\hat{W}\} +$$
$$+\vartheta(t'-t)\hat{A}(t')\mathcal{G}(t',t)\{\hat{A}(t),\mathcal{G}(t,t_0)\hat{W}\}\Big). \qquad (5.37b)$$

The occurrence of δ-terms in eq. (5.37b) and in the higher momenta shows that the stochastic process (4.22) is a generalized one. The test functions cannot be replaced by δ-functions.

6. CONCLUSIONS

As simple practical applications of the formalism presented in the preceding sections, we may mention a treatment of the observation of the trajectory of a particle in the ordinary or in the phase space [9], a treatment of the quantum Brownian motion and, more interesting, a simplified analysis of a gravitational wave detector [11].

In general, in the spirit of Quantum Mechanics any particular O.V.S.P. should correspond to the use of a specific equipment on the system and the use of one or another equipment during a certain time interval should be a matter of choice of the experimenter.

However as we mentioned in the introduction the formalism opens even the possibility of reconciling the classical description of the physical world with Quantum Mechanics at a fundamental level. In order to do so it should be necessary to construct an O.V.S.P. for the trajectories $\underline{z}(t)$ of the variables $\underline{z} \equiv (z^1, z^2, \dots)$, ideally specifying the macroscopic state of the whole universe. We should treat then such variables as continually observed during the whole interval $(-\infty, +\infty)$ excluding in this way any uncompatible observation and introducing a permanent modification in the fundamental dynamics. Naturally, since today we believe that the most fundamental description of the physical world must be given in the formalism of Quantum Field Theory, it is in this framework that the mentioned O.V.S.P. must be constructed. Furthermore, since in this order of ideas we should think of $z^1, z^2, \dots$ as having definite values at any time independently of the way we practically observe them all reference frames have to be equivalent and the family of the operations $\{\mathcal{F}(t_1,t_0;M)\}$ must transform into itself under the Galilei or the Poincaré group. Finally, the perturbation of the dynamics due to the continual observation of $\underline{z}$ should not spoil the conservation rules, which play an important role in macroscopic physics. In fact, the fundamental differential equations (1.2) of macrophysics are simply continuity equations supplemented by fenomenological laws reexpressing the "forces" in terms of the fundamental variables.

Presently we have not yet succeeded in constructing realistic examples of O.V.S.P.'s in field theory with the last mentioned properties and it is not clear at all if such O.V.S.P.'s can exist in the class we have been able to characterize. However, many aspects of the involved problematic can be illustrated on the simplified model of a universe described in terms of a single selfinteracting quantum field. The details shall be given elsewhere (see also ref. 10).

REFERENCES

1. G.M.Prosperi, Models of the Measuring Process and of the Macro Theories, in Lecture Notes in Physics 29, Springer, Berlin (1973), p.163.
2. B.d'Espagnat ed., Proc. Enrico Fermi School, course IL, Academic Press, New York (1971). M.M.Yanase, M.Namiki, S.Machida ed., Selected Papers on the Theory of Measurement in Quantum Mechanics, Phys.Soc. Japan, Tokyo (1978). Wheeler, Zureck, Quantum Theory of Measurement, Princeton University Press (1983). For recent contributions see e.g. the Proceedings of the " International Symposium on Foundations of Quantum Mechanics" (Tokyo, August 29-31 1983) (to appear) and: S.Machida and M.Namiki, Theory of Measurement in Quantum Mechanics, Prog. Theor. Phys. 63: 1457 (1980); 63: 1833 (1980); B.Misra and I.Prigogine, A Letter on Measurement Processes and Irreversibility, to appear in Proceedings of the "Workshop on Fundamental Logical Concepts of Measurement" (Torino, September 1983).
3. G.Ludwig, Measuring and Preparing Processes, in Lecture Notes in Physics 29, Springer, Berlin (1973),p.122; Makroskopische Systeme und Quantenmechanik, in Notes in Math.Phys., Marburg (1972).
4. B.Misra and E.C.G.Sudarshan, The Zeno's Paradox in Quantum Theory, J.Math.Phys. 18:756 (1977).
5. G.Ludwig, Deutung des Begriffs "Physikalische Theorie" und axiomatische Grundlegung der Hibertraumstruktur der Quantenmechanik durch Hauptsätze des Messens, Lecture Notes in Physics 4, Springer, Berlin (1970); Foundation of Quantum Mechanics, Springer, Berlin (1982).
6. E.B.Davies, Quantum Theory of Open Systems, Academic Press, London (1976).
7. A.S.Holevo, Probabilistic and Statistical Aspects of Quantum Theory, North Holland, Amsterdam (1982).
8. E.Prugovecki, Stochastic Quantum Mechanics and Quantum Spacetime, Reidel, Dordrecht and Boston (1983).

9. A. Barchielli, L. Lanz and G. M. Prosperi, A Model for the Macroscopic Description and Continual Observations in Quantum Mechanics, Nuovo Cimento 72B : 79 (1982).
10. A. Barchielli, L. Lanz and G. M. Prosperi, Statistics of Continuous Trajectories in Quantum Mechanics : Operation - Valued Stochastic Processes, Found. of Physics 13 : 779 (1983).
11. A. Barchielli, Continual Measurements for Quantum Cpen Systems, Nuovo Cimento 74B : 113 (1983); Continual Observations in Quantum Mechanics and the Problem of Macroscopic Description, to appear in: Proceedings of the "Workshop on Fundamental Logical Concepts of Measurement" (Torino, Sept. 1983).
12. G. M. Prosperi, The Quantum Measurement Process and the Observation of Continuous Trajectories, to appear in Lecture Notes in Physics, Springer, Berlin.
13. G. Lupieri, Generalized Stochastic Processes and Continual Observations in Quantum Mechanics, J. Math. Phys. 24 : 2329 (1983); A. Rimini, Quantum Mechanics for Macroscopic Bodies: a Reformulation of the Milan Approach, preprint, Salerno (1983).
14. K. Kraus, Operations and Effects in the Hilbert Space Formulation of Quantum Theory, in Lecture Notes in Physics 29, Springer, Berlin (1973), p. 206.
15. I. M. Gel'fand and N. Ya. Vilenkin, Gneralized Functions, Applications of Harmonic Analysis, vol. 4, Academic Press, New York and London (1964).
16. M. C. Reed, Functional Analysis and Probability Theory, in Lecture Notes in Physics 25, Springer, Berlin (1973), p. 2.
17. V. Gorini, A. Kossakowski and E. C. G. Sudarshan, Completely Positive Dynamical Semigroups of N-level Systems, J. Math. Phys. 17: 821 (1976).
18. G. Lindblad, On the Generators of Quantum Dynamical Semigroups, Commun. Math. Phys. 48 : 119 (1976).

CLASSICAL QUANTIZATION: TWO POSSIBLE APPROACHES

Otto E. Rössler

Institut für Physikalische und Chemie
Universität Tübingen
7400 Tübingen 1, F.R.G.

Abstract

Two classical results (Archimedean quiver; time-reversal invariance) are reconsidered under the aspect of possibly yielding a limit to observation from within. A historical question concerning a related idea of Maxwell is posed.

1. INTRODUCTION

The advent of dynamical chaos theory (cf. [1,2]) has the side effect of rekindling interest in some well-known classical results, obtained more than one hundred years ago in a not fully dynamical context. Specifically, it is suddenly admissible to be fascinated once more by the interaction of three frictionless billiard balls on a two-dimensional bounded surface, for example. One knows that the solution is chaotic ("strongly mixing" or "Bernouilli" are other, almost equivalent terms [3,2]). One expects (in analogy to what holds true for strange attractors [4]) that the trajectories will not fill the energy surface (or an invariant subset of it, respectively) with equal density everywhere. That is, probabilistic concepts acquire a fundamental importance even in this non-statistical problem already. Thirdly, one cannot say immediately how many positive Lyapunov characteristic exponents (that is, what degree of chaos [5]) to expect in this three-particle problem. In other words, there are a number of simple questions waiting to be worked out even in this apparently trivial situation.

There are, of course, also some tougher dynamical questions waiting to be answered in the field of multiple-particle classical mechanics. For example, there is the well-known result from statistical mechanics that viscosity in an ideal gas is independent of pressure, that is, particle number (cf. [6]). It would be nice to study this problem dynamically and find the simplest qualitative analogue. Another, equally counter-intuitive result in many-particle statistical mechanics concerns Brownian motion, specifically the fact that mean displacement of a large particle is independent from its mass (and in fact only depends, apart from the relative radius of the particle, on the mean velocity of the surrounding billiard balls; cf. [7], for example).

In the following, two even older results from classical mechanics will be given a second look. The aim will be to convince the reader that classical, multiple-particle chaos may imply results that approach the quantum reality.

2. THE "CUTTING-OPEN" EFFECT IN HAMILTONIAN SYSTEMS

Hamiltonian systems ideally consist of balls and springs. The springs may be either invisible and short-range (built into the walls of the colliding balls) or long-range like cushions that extend toward the fringe of the universe. Any Hamiltonian system has a center of gravity which is invariant. Only if an infinite-mass particle is involved, on the other hand, can one be sure that there exists a particle that is always at rest.

By the equipartition theorem, all n - 1 internal kinetic energies of an n-particle system are equal. So are the mean external kinetic energies of all groups of particles that are somehow bound together (Brownian particles). In fact, any arbitrarily defined subset of particles has this unit mean kinetic energy. For example, also n - 1 remaining particles, after one particle has been singled out, jointly possess the unit mean kinetic energy. Only the whole set (n-particles) has zero kinetic energy, that is, possesses a motionless center of gravity.

All of this corresponds to a 'cutting-open' effect: whenever a subset is singled out, it suddenly acquires the unit mean kinetic energy. In other words, the effect occurs whenever a decision to single out subsystem (for the purpose of making a measurement, say) is made. Suppose, for example, the residual set of n - 1 particles contains a subsystem called 'physicists' who wants to make a measurement on the n-th particle. Then he will by definition be unable to faithfully record the objective motion of the n-th particle because his own motion (that of the giant Brownian particle of which he is part) cannot be separated from that of the object to be measured. There will be an 'energy uncertainty' of 1/2 kT per collision dimension [8].

There are some open problems concerning the interpretation of this result. Firstly, from the viewpoint of eternity (equilibrium theory), situations in which there is an open subsystem that could make measurements are negligible. Secondly, there is the strong temperature dependence of the effect. Thirdly, there is the giant size of the effect. Finally, the effect as formulated concerns mean values only and not concrete individual events. Still, a quantitative description of the average indivdiual case should be possible.

Possibly, the two counter-intuitive (buffering-type) mechanisms mentioned in the Introduction will again appear in the ultimate dynamical formulation, generating a new invariance in the process. While all of this is only a hope, it appears that it is too early to exclude the possibility that Archimedes' law momentum conservation entails a universal limit to internal observation.

3. ZERO AVERAGE-MOMENTUM INITIAL CONDITIONS AND TIME'S ARROW

Zero-momentum initial conditions in Hamiltonian systems are special. When optically recording all motions that occur after a system has been released from such an ('only potential energies are nonzero') initial condition, one obtains a movie that is identical for the two time directions. (Curiously, the trajectories in state space do <u>not</u> coincide in the two time directions. The reason is that momentum is defined differently with respect to the two ends of the time axis.) Thus, if a Hamiltonian system is released from a zero-momentum initial condition that is far-from-equilibrium, the approach toward equilibrium is identical in both time directions. A very simple special example possessing just two degrees of freedom (but showing nevertheless typical 'gradient-controlled' macroscopic behavior as a transient in both directions of time) was considered recently [9].

The implications of this simple assumption (zero initial kinetic energies) are not yet exhausted. Suppose there is a third particle which is to be 'observed' by the two-particle system for which such an initial condition was assumed. Then the two otherwise identical temporal versions of the latter system will make differing measurements. In the absence of an overriding directionality in time (which certainly is lacking in the simple, small-n Hamiltonian systems considered here), this implies the existence of an ambiguity of measurement.

To repeat: the outcome of measurement is not unique if the object to be measured is released from an ordinary (non-zero momentum) initial condition the very moment the measuring subsystem is itself released from a zero-momentum initial condition. The reason is that one does not know which of the two 'temporal twins' is more legitimate than the other.

One possible way to generalize the result is to drop the requirement of zero-momentum initial conditions for the whole observing subsystem and replace it by a relaxed assumption. Since the observing subsystem usually is macroscopic, it makes sense to assume that it is released, not from an exactly zero-momentum initial condition, but from one in which only microscopically non-zero momenta are permitted. In this case, virtually all subsystems containing a moderately large number of particles would have a near-zero combined momentum initially (corresponding to a condition of 'local equilibrium'). More precisely, the initial conditions should be such that they do not imply a spontaneous decrease of entropy in either direction of time. (Such freshly prepared, non selected initial conditions correspond to what Boltzmann [10] called the assumption of molecular chaos.) Even more precisely, the system is assumed to be released in a far-from-equilibrium initial condition from which entropy increases in the same way in both directions of time.

Such an initial condition is just as 'neutral' in its specifying one direction of time over the other as was the exactly zero-momentum initial condition assumed previously. Therefore, a microscopic third process released at the same moment will again appear different to the two microscopic observers that are both specified by the initial condition. Even if either of these (macroscopically indistinguishable) observers makes a perfectly accurate measurement, the two outcomes will be different.

The main implication seems to be that without an a priori knowledge of the 'true' direction of time, and of whether oneself is complying with it or running against it, it is impossible to make an accurate microscopic measurement in classical mechanics.

What remains to be worked out is again a quantitative formulation. Hereby a special problem arises. If it was possible to force the microscopic system to be measured into a zero-momentum initial condition before making the measurement, any uncertainty would disappear. Therefore, it is interesting to see that such a zero-momentum initial condition is impossible to obtain for the object to be observed in the present case. The reason is that the zero-average-momentum initial condition of the measuring subsystem automatically implies a certain finite tolerance in which to put precisely the zero of time. This is because a zero-average-momentum initial condition is equivalent to an all-momenta-not-exactly-simultaneously-zero initial condition. Hence, any possible initial condition of the subsystem to be measured is equivalent to a non-zero-momentum initial condition.

4. DISCUSSION

Two different classical results were presented from a new angle. They were selected because they are somewhat counter-intuitive and might (just might) help find a dynamical formulation for the qualitative behavior of the finite-n classical mechanical systems which implies the existence of a universal limit to observation from the inside. Chances are, of course, that neither approach leads anywhere.

Why should one entertain such classical hopes at all? There are two powerful reasons. One stems from chaos theory. Chaotic systems require an exponential increase in the accuracy of computation if an error catastrophe is to be avoided (cf. [21]). Whether chaos plays any role in microscopic physics - for example, on the level of the Schrödinger equation - is so far an open question. But if it does, the empirical finding of energy conservation even in large systems over fairly long time spans suddenly implies that nature does her computations with a virtually unlimited number of digits. The fact that the quantum world appears probabilistic to us then no longer exempts nature from having to perform exact, deterministic calculations on a more basic level. The main asset of the Copenhagen interpretation (that there is 'no need to worry') thereby gets lost.

The second powerful reason stems from discrete mathematics. Gödel [11] showed that limits to the accessibility from the inside of certain facts are unavoidable in sufficiently complex number-theoretic systems. Certain universal automata (reversible cellular automata) are idential in their dynamics to a certain class of stroboscopically observed Hamiltonian systems [12]. Hence if Gödel's results apply to finite automata, they also apply to certain Hamiltonian systems at least.

Recently, Finkelstein [13], quoting earlier work of Moore [14] and Zwick [15], successfully specified a condition under which an internal limit to observability applies in certain finite automata. The connection to continuous Hamiltonian systems has yet to be drawn.

One could therefore say that the time is ripe for another attempt to explain quantum mechanics classically - as a 'from within' phenomenon. The least that can come out from a future theory of 'physics from within' [8] (or 'endophysics' [16]) is that apart from quantum mechanics, a second theory of similar structure exists classically which then has to be painstakingly integrated into quantum mechanics once more. It was with the hope to avoid such disaster that the above simplistic scenarios were proposed for re-consideration.

Let me close with a historical riddle. Maxwell [17, p. 154] uttered the conviction that microscopic motions are necessarily 'impalpable' to us (as macroscopic subsystems). We will never be able to lay hold of one to stop it, he said. This prediction of a limit to microscopic observability (and controllability) was then apparently forgotten by the turn of the century. Unfortunately, Maxwell did not give any account of the reasoning that lead him to his claim. Therefore, a third alternative to the two approaches proposed above can be indicated: to search through Maxwell's notes in Cambridge and find the quantitative answer there.

Acknowledgments

I thank Joe Ford and Giulio Casati for their encouragement. Discussions with John Kosak, Ilya Prigogine, Mitch Feigenbaum and Oliver Penrose were also very helpful.

REFERENCES

1. O. E. Rössler, Chaos, in: Structural Stability in Physics, eds. W. Güttinger and H. Eikemaier, pp. 290-309, Springer-Verlag, New York - Berlin (1979).
2. J. Ford, How random is a coin toss? Physics Today, 4:40-47 (1983).
3. Ya. G. Sinai, On the foundations of the ergodic hypothesis for a dynamical system of statistical mechanics, Sov. Math. Dokl., 4:1818-1822 (1963).
4. J. D. Farmer, Information dimension and the probabilistic structure of chaos, Z. Naturforsch., 37a:1304-1325 (1982).
5. O. E. Rössler, The chaotic hierarchy, Z. Naturforsch., 38a: 788-801 (1983).
6. J. C. Maxwell, Comments on the dynamical theory of the gases, I. Philos. Mag., 19:19-32 (1860).
7. F. Reif, Statistical Physics, Berkeley Physics Course Vol. 5, McGraw-Hill, N.Y. (1965).
8. O. E. Rössler, Chaos and chemistry, in: Nonlinear Phenomena in Chemical Dynamics, eds. C. Vidal and A. Pacault, pp. 79-87, Springer-Verlag, New York - Berlin (1981).
9. O. E. Rössler, Macroscopic behavior in a simple chaotic Hamiltonian system, in: 1982 Sitges Conference, ed. L. Garrido, Lecture Notes in Physics, 179:67-77 (1983).
10. L. Boltzmann, Lectures on Gas Theory, Part I, Section 1.3, University of California Press, Berkeley (1964); Barth, Leipzig (1896).
11. K. Gödel, On Formally Undecidable Properties, Basic Books, N.Y. (1962); trans. of: On formally undecidable theorems of Principia Mathematica and related systems, Monatshefte f. Math. u. Physik., 38:173-188 (1931).

12. E. Fredkin, Digital information mechanics, M.I.T. Preprint, "draft" (1983).
13. D. Finkelstein and S. R. Finkelstein, Computer interactivity simulates quantum complexity, Int. J. Theor. Phys., 22, (1983), in press.
14. E. F. Moore, Gedanken experiments on sequential machines, in: Automata Studies, eds. C. E. Shannon and J. McCarthy, Princeton University Press, Princeton (1956).
15. M. Zwick, Quantum measurement and Gödel's proof, Speculations Science Technol., 1:135 (1978).
16. D. Finkelstein, Personal communication, July(1983).
17. J. C. Maxwell, Theory of Heat, Allen and Unwin, London (1872).

QUANTUM CHAOS AND THE THEORY OF MEASUREMENT

Willis E. Lamb, Jr.

University of Arizona
Tucson, Arizona, USA

This paper gives an indication of what I think should be the meaning of the phrase "quantum chaos" (QC for short). It is in no way intended to express a consensus of the views of other participants at the Conference Como 83, or its organizers. QC suggests to me a concern with the transcription into Quantum Mechanics of the kinds of dynamical problems discussed at Como 77 (where the order term "stochastic" was used rather than the catchy and more modern word "chaos").

I am not going to discuss the eigenvalues of random matrices, or the nodal lines of the stationary state wave functions of various billiard problems. We heard a lot about such work at Como 83. These are, of course, highly important and interesting problems in their own right. Stationary state solutions of the Schrödinger equation have had great importance for spectroscopic and structural phenomena during the historical development of quantum physics. However, they only form a set of measure zero of the set of all (time dependent) solutions, and my present interest lies more with the time evolution of systems than with their stationary states.

In Como 77, many classical hyperspace trajectories x(t) were represented pictorially by delicate and intricate patterns on surfaces of section. Surprisingly few quantum transcriptions of such problems were considered at Como 83. In a few cases, treatments involving a diffusion through a space, characterizing semi-classical stationary states, were reported. Jensen, for instance, talked on the application of such methods to the dissociation of high Rydberg states of hydrogen-like atoms by microwave fields. He is planning to make calculations for this problem using the time dependent Schrödinger equation. One reason for the small number of such calculations, particularly for problems of more than one degree of freedom, is that they are very time consuming. In Tucson, Sami Shakir, Milivoj Belic, Jerome Moloney and I have made such calculations for one-dimensional atoms and molecules, as well as for a two-dimensional Morse roto-vibrator. We are using a middle-aged Data General Eclipse minicomputer which has especially good interactive graphics facilities. The small amount of addressable memory, (32K), for such machines has been a serious limit, but virtual memory computers now coming on the market should be much better adapted for this kind of calculation.

Classical and Quantum Mechanics are devised to give models of natural phenomena. The two disciplines have irreconcilable differences, and there is no way in which one theory reduces to the other in any limiting process. It is certainly too naive to think only about taking the limit of quantum mechanics as Planck's constant approaches the essential singularity at value zero. Quantum Mechanics <u>has</u> to be formulated in terms of probabilities. It is true that Classical Mechanics <u>can</u> be reformulated in terms of distributions in phase space, but the underlying theory is non-probabilistic. Of course, in some limiting cases the classical and quantum treatments of a problem may give essentially equivalent results. I will return to this matter at the end of the paper.

The two theories also differ in respect to the conceptually attainable degree of isolation of a dynamical system. A classical system can be regarded as being completely isolated from the rest of the universe. Any observation of the system would involve some intervention from outside the system, but the structure of the theory is such that the effect of this is easily ignored. In Quantum Mechanics, on the other hand, the action of any observer involves a radical change in the specification of the system of interest. One can calculate the time evolution of the wave function of any system if its Hamiltonian is known. A measurement can only be made by coupling this system to another dynamical system. Then one has a larger and completely new and different problem to solve. At best, one could hope to follow the subsequent time development of the wave function for the combined systems. Even if the two systems are subsequently decoupled from each other, the proper description of the state is given by the wave function for the combination of the two systems. It is no longer possible to assign a wave function to the original system of interest. After its return to isolation, the description of the system of interest can only be in terms of a density matrix, i.e., a <u>mixture</u> of wave functions each having a probability which can be calculated from the two system wave function. Hence, one cannot assign a single wave function to a system after it has been observed. It might be possible to carry out a process of preparation of a new state of the system of interest, but that would represent a fresh start which would wipe out all memory of the past. A measurement process is very different from the preparation of a new state. Still less can one define a wave function for a system which is being continually observed. The classical physicists at Como 77 did not consider that observation of their systems presented any problem.

An understanding of the problem of measurement in Quantum Mechanics is essential for a satisfactory discussion of QC. Unfortunately, this subject is poorly understood and very controversial. Quantum Mechanics has had many successes. It seems to be very difficult to "understand" it without thinking about making measurements. Unfortunately, Quantum Mechanics gives no clue about how measurements are to be made. One or more additional assumptions have to be made. In Dirac's "The Principles of Quantum Mechanics", there is no indication of how measurements are to be made but, in Chapter 1, there is a hypothesis (which, for a one degree of freedom problem, I transcribe as follows:) that if a measurement of an observable is made, the state of the system will be found to be the eigenstate of observable corresponding to the result obtained in the measurement.

Von Neumann's "Mathematical Foundations of Quantum Mechanics", Chapter III, introduced an assumption, often called the von Neumann (wave packet) reduction hypothesis, which is roughly equivalent to the one proposed by Dirac. At the end of his final Chapter VI, von Neumann also gave a model for position measurement, but some highly unrealistic assumptions were made about the acceptable forms for Hamiltonians. Furthermore, he neglected to point out that the result of his calculation does not conform to the reduction hypothesis except in an unrealistic limiting case.

For various reasons, I am not happy with these assumptions about the measurement process in Quantum Mechanics. I will base the following discussion on the ideas of measurement theory as partially described in References [1,2,3] and further developed in my Leigh Page Prize Lectures at Yale in 1982. The central theme is that any discussion of measurement should be accompanied by a physically realistic description of the interaction of the measuring apparatus and the system of interest in dynamical terms. The reduction hypothesis of von Neumann is replaced by assumptions involving the availability of suitable Hamiltonians, and the final break away from a quantum to a classical description.

Suppose that we have a system of interest whose wave function is known at the time of measurement t_m. To make a measurement of the value of some dynamical variable for this system requires more than merely thinking or talking about making one. Some hardware and an experimental procedure is required. Corresponding to the hardware and its method of use will be a model Hamiltonian which appears in the Schrödinger wave equation. The following arrangement might seem reasonable: we will use a quantum mechanical measuring instrument, called the "meter system", which is initially in a known pure case state. We will couple the meter to the system for a certain time. After separation, the state of the meter is determined by a method to be indicated below. We hope that the resulting information will tell us something about the state of the system of interest at time t_m. As always in Quantum Mechanics, it is necessary to repeat the whole process of the state preparation and measurement in order to get the maximal amount of statistical information.

Before they are joined together, the two systems are described by Hamiltonians H_s and H_m, respectively. Then the system of interest and the meter system are coupled together. The starting wave function of the combined system is a simple (outer) product of the wave functions for the two systems. At t_m, the two systems are brought into contract, as described by a Hamiltonian $H_s + H_m + H_i$, for a certain time interval. After the interaction, the combined system will still be in a pure case state, but its wave function will, in general, not be a simple product, but a sum of such terms. Each term will have a coefficient which is a rapidly varying complex function of time. Afterwards, the meter is taken away and some measurement is made on its state. We hope thereby to learn something about the state of the system at the time of the measurement. Obviously, the kind of information we can get about the system will depend on the Hamiltonians for the system, meter, and the interaction between them. We will also have to specify what kind of measurement procedure has been involved in reading the meter. Any time delays will scramble the additive terms making up the wave function for the two system problem, and in effect make a mixture out of what would otherwise be pure case state. It should be noted that there is a danger that we will run into a circularity problem, especially if we have to use other meter systems to measure the state of the first meter. However, we will be able to avoid that difficulty because we are willing to destroy any knowledge of the state of the meter during the state recongition procedures which are to be used. Some kind of Stern-Gerlach apparatus will be used to determine as much as can be learned about the state of the meter. This information will be used to partially unmix the density matrix of the system of interest. The whole procedure will have to be repeated many times on suitable replica systems in order to build up good statistical data.

At the time of separation of the meter from the system, neither the system of interest or the meter can be separately described by pure case wave functions. The proper description for each system involves the use of a density matrix. There is no longer a wave function for either of the two systems, only a probability distribution of wave functions. One speaks of a "mixed" case. It is interesting that density matrices were introduced into Quantum Mechanics in three independent papers by Landau, von Neumann and Dirac. None of these authors made much use of density matrices in discussions of measurement theory.

One sees that the complications of the measurement process will have serious consequences for the attempt to transfer classical chaotic dynamics into the quantum domain. After one measurement, not only is the wave function of a system changed, but there is no longer a definite wave function. Any attempt to explore experimentally the microscopically intricate patterns on a surface of section will ultimately be frustrated by the chaos introduced in a measurement process. One might say that the chaos of Classical Mechanics tries to anticipate the radically different kinds of chaos which will be imposed on it in the quantum domain.

I will now examine a series of simple illustrative problems whose chaotic aspects, or lack of them, will be discussed from both classical quantum viewpoints.

(1) Free particle in one dimension. The classical solution, $X(t) = x_0 + v_0t$, can hardly be said to exhibit any chaos. In Quantum Mechanics, a simple solution for the same problem has a wave function in the form of a Gaussian error function whose centroid follows the classical trajectory. The width of the packet changes in time in a way which can be understood in terms of a simple uncertainty principle argument. At this point, we already have chaos! For two reasons: (a) use of probabilistic interpretation of Quantum Mechanics, and (b) measurements are required to follow the motion. Each measurement adds incoherence to the statistical density matrix description of the particle's state. A watched system becomes increasingly chaotic. Chaos of the types (a) and (b) will be found in the subsequent examples which begin to show more nearly the familiar kinds of stochastic behavior.

(2) Step potential in one dimension (with $V(x) = 0$ for $x < 0$ and $V(x) = V_0 > 0$ for $x > 0$. There can be extreme sensitivity to initial conditions for a particle with a kinetic energy nearly equal to V_0 moving from $x < 0$ toward $x = 0$. This may represent one of the simplest systems for which the preconditions for chaos are met. In the Quantum Mechanics of the same problem, a wave packet travelling to the right and crossing the discontinuity will be split into waves travelling to the right and left. The probability interpretation of Quantum Mechanics will provide a simple, albeit primitive, example of the bifurcation phenomena which underlie chaotic behavior. We will refer to the chaos of this paragraph, whether classical or quantum as being of type (c). It is characterized by extreme sensitivity to initial conditions in the classical problem, and to splitting of wave packets in the quantum problem.

(3) A hill potential $V(x)$ where $V(x) > 0$ is an even function of x. This case is similar to case (2). In addition, there would be extreme sensitivity to initial conditions for a particle placed at rest near to the top of the hill. In the corresponding quantum

problem, a single peaked wave packet would split into wave packets travelling to the right and left. Again, we have a basis for chaotic behavior.

(4) Forced simple harmonic oscillator. This is probably the most complicated one degree of freedom problem for which there is a very close connection between the classical and quantum treatments. This follows from the linearity of the problem. The quantum Heisenberg picture equations of motion for x and p have exactly the same form as the classical ones (which, incidentally, can be solved in terms of quadratures involving the time dependent force F(t)). Frictional terms are neglected. (If one wanted to consider only conservative dynamical systems, a damping term would not be permitted. I am primarily interested in atoms and molecules where such terms are not appropriate. It would be less convenient if the external force also had to be ruled out. However, one can easily convert the problem to a conservative one of two or more degrees of freedom. If the added degrees of freedom are associated with very large masses, the system can be designed so that the original coordinate x obeys with high precision the desired forced simple harmonic oscillator equation of motion.) The classical problem shows no great sensitivity to initial conditions, and the quantum problem has the chaos associated with measurement and the probabilistic interpretation of the wave function. This problem gives a good model for a gravitational wave detector. It should already be apparent that I do not look with favor on the current treatment of "quantum non-demoltion observations". Work on application of my ideas on the quantum theory of measurement to the forced simple harmonic oscillator system will be published elsewhere.

(5) Monochromatically forced anharmonic oscillator (Duffing problem). See Reference [4] for a partial treatment of the classical problem. The Newtonian equation of motion is

$$\ddot{x} + \omega^2 x + \mu x^3 = F \cos \nu t$$

where a nonlinear restoring force in x^3 with coefficient μ has been added to the equation describing the corresponding simple harmonic oscillator. For simplicity, we will consider that the particle starts from rest at $x = 0$ at $t = 0$. In the linear case the motion can be expressed as a product of two sine functions of time: one with a high frequency $(\nu + \omega)/2$ and the other with a low frequency $(\nu - \omega)/2$ providing amplitude modulation of the high frequency carrier wave. A resonance occurs when $\nu = \omega$, and the displacement eventually increases linearly with time to an indefinitely large value. For ν very close to ω, the largest amplitude attained would have a very large value, inversely proportional to $1/|\nu - \omega|$.

For small values of the nonlinearity parameter μ, one finds that the motion is generally rather similar to that found for $\mu = 0$, with a slightly changed value of ω. However, important qualitative changes take place in the neighborhood of resonance. The unforced anharmonic oscillator has periodic solutions. If the nonlinearity parameter μ is negative, the "soft spring case", the period of such a motion increases with the amplitude of the motion. Consider a case where the forcing term has a frequency slightly lower than the small amplitude frequency ω. The motion will be approximately what it would be in example (4), except that as the amplitude of the motion builds up, the effective resonance frequency of the corresponding unforced motion decreases because of the finite value of the nonlinearity parameter μ. It may happen that a nonlinear resonance phenomenon occurs, and the amplitude of the forced motion will become considerably larger than would otherwise be expected. We have been considering that the motion started from rest at $t = 0$. For fixed values of ν, ω, F and μ, when the conditions are right for the so-called "jump" phenomenon to occur, there will be a high degree of sensitivity to the initial conditions (x_0 and $\dot{x}_0$) for the classical motion.

The above jump phenomenon does not exhaust the interesting nonlinear phenomena shown by the Duffing problem. However, the jump phenomenon is easy to treat analytically, by the method of slowly varying amplitude and phase, and can be regarded as the simplest starting point for the chaos in more complicated problems.

The possibilities for disorder multiply rapidly as the dimensionality of the system increases. There are now resonance frequencies for each small amplitude normal mode, and many associated jump phenomena. Furthermore, there are nonlinear coupling terms in the Hamiltonian involving three or more modes which lead effectively to many driving forces on each degree of freedom at combination tone frequencies (parametric resonance). Many of these terms will be able to produce jump phenomena of the sort met in the Duffing problems. The opportunities for stochastic or chaotic behavior are very rich.

The quantum version of the Duffing problem will have a wave packet solution which, at a certain time, may split rather unexpectedly into two parts. These packets move in the ways corresponding to the two kinds of classical trajectories, one with jump and one without. This behavior is seen clearly in the computer integration of the wave equation for a one-dimensional forced Morse oscillator made by Walker and Preston (Reference [5]). I have made numerical studies of the Heisenberg equations of motion for a Duffing oscillator. The expectation value $\langle x(t) \rangle$ for a forced system initially in the ground state of the unforced anharmonic osillator behaves for a certain amount of time in a very simple way, closely following the motion of the classical Duffing problem. Once conditions are met for the jump phenomenon, the value $\langle x(t) \rangle$ can be regarded as getting contributions from the two wave packet constituents of the total wave function, and it can differ dramatically from the classical function x(t). In a sense, the system can be regarded as a mixture of two systems following two very different classical trajectories rather than one.

Of course, in some limiting cases, the classical and quantum treatments of a problem may give essentially equivalent results. Quantum Mechanics is not well suited for giving a discussion of the motion of the Moon, but it can be made to do so. The description could not be in terms of a wave function, but only in terms of a density matrix whose time dependence would have to be calculated by rather laboriously taking into account all of the known and unknown perturbations considered by astronomers, as well as those associated with the observations made by millions of people who have occasion to look at the Moon. The calculation would be very complicated to make, but it would certainly give a very strongly peaked probability distribution, and it is extremely unlikely that the Moon will be found very far from its expected position. On the other hand, the dynamical effects of observation on a system on a more atomic scale will be far from negligible. Such observations are hard to make, and hence they are hardly ever attempted. There is much current interest in a case which falls on the borderline between small and large, such as a gravity wave detector of such exquisite mechanical and electrical design that the calculations of the sort outlined above will have to be made if the utmost information is to be extracted from the measurements. A similarly difficult problem is faced by those who want to reach beyond the "quantum limit" in communication theory.

The trajectories which are met in problems of nonlinear classical dynamics such as the Henon-Heiles problem have a great fascination for may people. The classical theory in principle permits an unending subdivision of the regions on surfaces of section. More realistically, in any given problem, one has to expect that the possibility of experimentally following the finest details of the motion would be seriously affected if there were disturbances from a noisy environment. Furthermore, at some level of precision, the inevitable quantum disturbances will smear out the fine structure. This would happen in two ways:

(1) from the disturbances caused by whatever observations are being made on the system;
(2) because of the breakdown of the concept of trajectories brought about by the probability interpretation of Quantum Mechanics and the wave packet splitting which we met in the quantum theory of the Duffing problem and its generalizations.

I have been trying to give a meaning to the title of this Conference. I think the term "Quantum Chaos" is a rather unfortunate choice, but I suppose that it will survive my objections.

When Quantum Mechanics needs to be applied to any classical problem, it has an overwhelming effect on its causality. This is even more true of classical problems having stochastic features, and the smearing effects of the quantization will eventually overwhelm the classical chaos. If it is quantum, there is more than enough chaos!

REFERENCES

1. W. E. Lamb, Jr., "An Operational Interpretation of Nonrelativistic Quantum Mechanics", Physics Today, 22:23-28 (1969).
2. W. E. Lamb, Jr., "Von Neumann's Reduction of the Wave Function", pp. 297-303, in: "The Centrality of Science and Absolute Values", Proceedings of Fourth International Conference on the Unity of Science, ICF, Inc., New York (1976).
3. W. E. Lamb, Jr., "Remarks on the Interpretation of Quantum Mechanics", pp. 1-8, in: "The Ta-Yu Wu Festschrift: Science of Matter", ed. S. Fujita, Gordon and Breach, N.Y. (1979).
4. M. Borenstein and W. E. Lamb, Jr., "Classical Laser", Phys. Rev., A5:1298-1311 (1972).
5. B. R. Walker and R. K. Preston, "Quantum Versus Classical Dynamics in the Treatment of Multiple Photon Excitation of the Anharmonic Oscillator", J. Chem. Phys., 67:2017-2028 (1977).

BOHR'S "PHENOMENON" AND "LAW WITHOUT LAW"

John Archibald Wheeler

Center for Theoretical Physics
University of Texas
Austin, Texas 78712

COMO AND COMPLEMENTARITY

Como means much to science. Como means 1977 and 1983. Then Giulio Casati and Joseph Ford brought together distinguished colleagues to lay out the streets of a rapidly growing district of physics and help prepare it to be a new city of its own. Como means 1800. Then Alessandro Volta gave us the pile. He shocked us into a new world of electricity. But Como also means 1927.

This hall is the place; the 1927 centennial of Volta's death, the occasion. Q. Majorana is president of the conference. H. A. Lorentz, white haired father-figure of physics, presides as vice-president. Wolfgang Pauli, Werner Heisenberg, H. A. Kramers, C. G. Darwin, Max von Laue, Enrico Fermi, Emilio Segre, and many another hero of the quantum era is in the audience.

Niels Bohr rises to give his address, the central feature of the meeting. For two years neither public statement nor paper has issued from him. Ever since the revolutionary discoveries in modern quantum theory by Heisenberg and Schrödinger, he has been discussing with Heisenberg, Kramers, Oskar Klein, Leon Rosenfeld, and other colleagues at Copenhagen the physical interpretation of the new formalism.

Earlier in 1927, Heisenberg has sent in from Copenhagen his famous paper on indeterminism, the term which he then used and we still use in preference to "uncertainty." He added in proof that "...I have overlooked essential points in the course of several discussions in this paper" ... "brought to my attention" by

Professor Bohr, who has achieved "a point of view which permits an essential deepening and sharpening of the analysis of quantum-mechanical correlations attempted in this work" [1,2].

As Bohr speaks, "He had," Rosenfeld tells us [2], "the greatest misgivings about presenting his conception of complementarity to the community of physicists in a state which he judged immature, but he yielded to the advice of his more practically-minded brother Harald ... and even consented to write up a brief account...", just before departing for Como.

The central lesson of quantum theory, as Bohr expounds it, is complementarity [3,2]. As he later put the idea [4,2], "any given application of classical concepts precludes the simultaneous use of other classical concepts which in a different connection are equally necessary for the elucidation of the phenomenon." He subsequently added [5,2], "...atomic phenomena under different experimental conditions must be termed complementary in the sense that each is well defined and that together they exhaust all definable knowledge about the object concerned." The way of thinking symbolized by complementarity is so important in Bohr's view that every school child should be taught it.

It took time to learn the language of complementarity. In the beginning it seemed clear enough. The famous double-slit experiment, for example, allows us to register interference fringes. Modified, it allows us in principle to measure the lateral kick delivered to the photographic plate and thus--it would at first sight seem--tell "through which slit" the photon came. It is clearly impossible to register at the same time with full precision [6,2] both the "which slit" information and the information necessary to build up the standard two-slit interference pattern. Thus the principle of indeterminism forbids the simultaneous sharp determination of both the lateral coordinate of the plate and its lateral momentum. In brief, indeterminism enforces complementarity.

Still deeper insight into what complementarity says and means comes from Bohr's analysis [2] of the idealized experiment proposed by Einstein, Podolsky and Rosen [7,2] in 1935 (simplified by Bohm [8] in 1951) and from a variety of other experiments (1978) which, like the EPR experiment itself, we have learned to recognize as "delayed-choice" experiments [9]. They illustrate the same point as the EPR experiment [10] but in clearer form. None is simpler than the double-slit one in its delayed-choice version.

Shall we determine the lateral position of the photographic plate--as we must if we want to develop a <u>two</u>-slit interference pattern--or shall we determine its lateral momentum--as we must if we want to know "from which slit" the photon comes? We delay our choice to the very last instant, after the photon has already passed through the doubly-slit screen, and just before it hits the plate. Shall we say that "we decide whether the photon shall pass through one hole, or both holes, after it has already traversed the screen"? No, Bohr taught us. We have no right to speak about what the photon "does" between point of entry and point of registration.

BOHR'S "PHENOMENON"

We have to recognize a new animal in the zoology of nature: beyond particle, beyond field of force, beyond spacetime geometry, the elementary quantum phenomenon. This great smoky dragon, Bohr's phenomenon, has its tail sharply localized at the point of entry to the apparatus. Its teeth are sharply localized where it bites the grain of photographic emulsion. In between it is utterly cloud-like, localized neither in space nor in time. The search for the message of the quantum is a continuing pilgrimage, marked so far by three great way-stations: indeterminism, complementarity, phenomenon.

In today's words, Bohr's point--and the central point of quantum theory--can be put into a single, simple sentence. "No elementary phenomenon is a phenomenon until it is a registered ("observed"; "indelibly recorded") phenomenon." It is wrong to speak of the "route" of the photon in the experiment of the beam splitter. It is wrong to attribute a tangibility to the photon in all its travel from the point of entry to its last instant of flight. We do better to conceive of the photon, not as a messenger passing through space, but as a message registered at a point in space. A phenomenon is not yet a phenomenon until it has been brought to a close by an irreversible act of amplification such as the blackening of a grain of silver bromide emulsion or the triggering of a photodetector.

Nature at the quantum level is not a machine that goes its inexorable way. It is wrong to think of the universe as "sitting out there." Instead, what answer we get depends on what question we put, what experiment we arrange, what registering device we choose. We are inescapably involved in bringing about that which appears to be happening.

Nowhere better than in simple examples does one see the twin attributes of the elementary quantum phenomenon. One is the choice of question asked, of apparatus employed, of type of measurement made. The other is the chance character of the "reply."

If the world is so crazy and full of chance, why doesn't it look crazy and full of chance? Because many a chance "yes, no" adds up to the definiteness, the apparent definiteness, of "how much."

From the single dot of color on an impressionist painting, the pupil of our eye receives in the glance of a second 50,000 photons. Each photon is accidental in its direction and time of arrival. The quanta in that hail of information are so numerous that they give the impression of perfect steadiness of illumination.

Chance--one attribute of the elementary quantum phenomenon--is nature's way to build definiteness. What about the other attribute, choice?

The elementary quantum phenomenon has much of the character of an elementary act of creation. The phenomenon that we call "a photon from a distant quasar" reaches into the present from billions of years in the past. It is wrong to think of that past as "already existing" in all detail. The "past" is theory. The past has no existence except as it is recorded in the present. By deciding what questions our quantum registering equipment shall put in the present, we have an undeniable choice in what we have the right to say about the past.

What we call reality consists of a few iron posts of observation between which we fill in by an elaborate papier-maché construction of imagination and theory.

Spacetime in the prequantum dispensation was a great record parchment. This sheet, this continuum, this carrier of all that is, was and shall be, had its definite structure with its curves, waves and ripples; and on this great page every event, like a glued down grain of sand, had its determinate place. In this frozen picture, a far-fetching modification is forced by the quantum. What we have the right to say of past spacetime, and past events, is decided by choices--of what measurement to carry out--made in the near past and now. The phenomena called into being by these decisions reach backward in time in their consequences, back even to the earliest days of the universe.

Registering equipment operating in the here and now has an undeniable part in bringing about that which appears to have happened. Useful as it is under everyday circumstances to say that the world exists "out there" independent of us, that view can no longer be upheld. There is a strange sense in which this is a "participatory universe."

We cannot speak in these terms without a caution and a question. The caution: "consciousness" has nothing whatsoever to do with the quantum process. We are dealing with an event that makes itself known by an irreversible act of amplification, by an indelible record, an act of registration. Does that record subsequently enter into the "consciousness" of some person, some animal or some computer? Is that the first step in translating the measurement into "meaning"--meaning regarded as the "joint product of all the evidence that is available to those who communicate"? Then that is a separate part of the story, important but not to be confused with "quantum phenomenon."

From this caution we turn to the question: If the elementary quantum process is an act of creation, is an act of creation of any other kind required to bring into being all that is?

At first sight, no question could seem more ridiculous. How fantastic the disproportion seems between the microscopic scale of the typical quantum phenomenon and the gigantic reach of the universe! Disproportion, however, we have learned, does not give us the right to dismiss. Else how would we have discovered that the heat of the carload of molten pig iron goes back for its explanation to the random motions of billions of microscopic atoms and the shape of the elephant to the message on a microscopic strand of DNA? Is the term "big bang" merely a shorthand way to describe the cumulative consequence of billions upon billions of elementary acts of observer-participancy reaching back into the past?

Are elementary quantum phenomena, those untouchable, indivisible acts of creation, indeed the building material of all that is? Beyond particles, beyond fields of force, beyond geometry, beyond space and time themselves, is the ultimate constituent the still more etheral act of observer-participancy?

TOWARDS LAW WITHOUT LAW [11]

So far as we can see today, the laws of physics cannot have existed from everlasting to everlasting. They must have come into being at the big bang (Fig. 1). There were no gears and pinions,

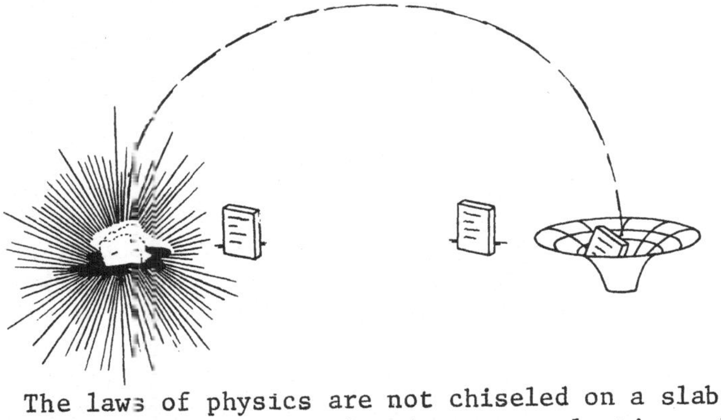

Fig. 1. The laws of physics are not chiseled on a slab of granite that stands from everlasting to everlasting. Those regularities had to come into being at the big bang and be obliterated in the gravitational collapse that takes place at the center of a black hole.

no Swiss watchmakers to put things together, not even a pre-existing plan. If this assessment is correct, every law of physics must be at bottom like the second law of thermodynamics, higgledy-piggledy in character, based on blind chance, insofar as it is not mere mathematical identity.

There is no simpler illustration of the second law than the way molecules distribute themselves between two regions in proportion to the volumes of those two regions. Every heat engineer knows he can design his heat engine reliably and accurately on the foundation of the second law. Run alongside one of the molecules, however, and ask it what it thinks of the second law. It will laugh at us. It never heard of the second law. It does what it wants. All the same, a collection of billions upon billions of such molecules obeys the second law with all the accuracy one could want. Is it possible that every law of physics, pushed to the extreme, will be found to have the character of the second law of thermodynamics, be statistical and approximate, not mathematically perfect and precise? Is physics in the end "law without law"?

Nothing seems at first sight more violently to conflict with this idea than all the beautiful structure of the three great field theories of our age, electrodynamics, geometrodynamics, and chromodynamics. They are the fruit of hundreds of experiments, scores of gifted investigators and a century of labor. Impressive treatises spell out the physics and mathematics of all three theories. How can anyone possibly imagine all this richness coming out of a higgledy-piggledy origin?

Only a principle of organization which is no organization at all would seem to offer itself. In all of mathematics, nothing of this kind more obviously offers itself than the principle that "the boundary of a boundary is zero." Moreover, all three great field theories of physics use this principle twice over, once in the form that "the one-dimensional boundary of the two-dimensional boundary of a three-dimensional region is zero," and again in the form that "the two-dimensional boundary of the three-dimensional boundary of a three-dimensional region is zero." This circumstance would seem to give us some reassurance that we are talking sense when we think of almost all of physics being founded on almost nothing.

To spell out what has just been said in the context of electrodynamics would be to say anew what is already widely known to every student of that subject. To illustrate the boundary of a boundary in the context of chromodynamics would lead to too much technical detail for our purpose. Therefore, I shall illustrate the meaning of those mysterious words, "the boundary of a boundary is zero," in a theory of intermediate difficulty, great interest and widely recognized beauty, Einstein's general relativity, his geometric theory of gravitation, his geometrodynamics.

Every word I utter, every drawing I make, every conclusion I draw, will presuppose the existence of a spacetime manifold. How absolutely inconsistent this is with the spirit of "law without law," one can well say. What was the point of that working hypothesis if not to reason that at bottom there can be no spacetime manifold? The reply is straightforward. Exactly by examining the implications of "the boundary of a boundary is zero," we seek a little guidance, a few clues, a bit of insight into an ongoing and still unraveled detective story: how to build up almost everything out of almost nothing.

Why should boundaries occupy a prominent place in formulating a law of physics? Quantum theory, in the shape of the Aharonov-Bohm experiment [12] (Fig. 2) gives a clue. The electron interference pattern is shifted by an amount proportional to the flux of magnetic field through the space between the two paths. In other words, the phase difference around the closed circuit formed by taking one route out, the other route back, measures the magnetic flux. The phase difference in a wave, as evaluated around a loop, provides as well in electromagnetism and in gauge-field theory as in every other form of field theory a way to define and measure the fields [13]. Therefore we are open to believe that no field has any existence or significance except as it is defined by such phase differences around a loop. To adopt this point of view is to make loops--the one-dimensional boundary of two-dimensional regions--a central element in the bookkeeping of nature.

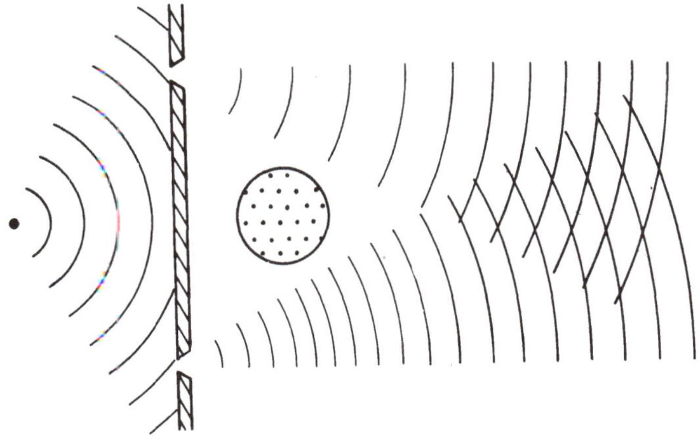

Fig. 2. Double-slit electron-interference experiment as affected by the presence of a magnetic field, confined to the indicated region, and directed normal to the plane of the paper.

"THE BOUNDARY OF A BOUNDARY IS ZERO" ILLUSTRATED IN GEOMETRY

Einstein's 1915 and still standard "general relativity," or geometric theory of gravity, or "geometrodynamics" has two central principles: (1) space tells mass how to move; and (2) mass tells space how to curve.

Energy can be considered to be the decisive quantity in governing the action of mass on geometry. But the energy of a system of particles and fields is only one component of the 4-vector, p, of energy and momentum. Its value depends on the frame of reference of the observer. In other words, it depends on the unit vector, u, tangent to the worldline of the observer,

$$E = -u^{\alpha} p_{\alpha} \quad .$$

The amount of energy in a given 3-dimensional region $d^3\Sigma$ or "slice" through spacetime also depends on the orientation of that region. If that region is turned tangent to the worldlines of all the particles then no matter how energetic they are none will cross it. Thus the amount of energy depends both on the observer and on the region observed:

$$E = u^{\alpha} T_{\alpha}{}^{\beta} d\Sigma_{\beta} \quad .$$

This equation defines the stress energy tensor, $T_{\alpha}{}^{\beta}$. It is the source term in Einstein's field equation.

There is a great difference between electromagnetism and gravity. In the case of electromagnetism we speak about the force on one particle,

$$D^2x^{\alpha}/D\tau^2 = (e/m)F^{\alpha}{}_{\beta}(dx^{\beta}/d\tau) \quad .$$

In the case of gravitation there is no force on one particle:

$$D^2x^{\alpha}/D\tau^2 = 0 \quad .$$

It was Einstein's first step in explaining gravity to get rid of gravity.

Einstein tells us that physics is only then simple when it is analyzed locally. Instead of talking about the acceleration of an object relative to the faraway center of the Earth he tells us to look at the acceleration of nearby test objects relative to each other. The physics name for Einstein's local measure of gravitation is "tide producing acceleration." The mathematics name is "geodesic deviation" or spacetime curvature.

The separation, η^α, between two nearby test masses is given in terms of the curvature of spacetime, $R^\alpha{}_{\beta\eta\nu}$, by the "equation of geodesic deviation,"

$$D^2\eta^\alpha/D\tau^2 = -R^\alpha{}_{\beta\eta\nu}(dx^\beta/d\tau)\eta^\nu(dx^\nu/d\tau) \quad .$$

This equation is as important for gravitation as is the Lorentz equation for electromagnetism. It opens the door to experimental methods for measuring spacetime curvature.

It takes considerable thinking to see how to get each component of the curvature individually by measurements of the relative separations of nearby test particles. However, in the end, the language of physics is equivalent to the language of geometry. In geometric language, the curvature is measured by the rotation at a vector transported parallel to itself around a closed curve. The alteration of a vector, V^α, on parallel transport around a little parallelepipedal circuit bounded by two vectors, $\Delta_1 x^\alpha$ and $\Delta_2 x^\alpha$, is

$$\delta V^\alpha = -R^{\alpha\beta}{}_{\mu\nu} V_\beta \Delta_1 x^\mu \Delta_2 x^\nu \quad .$$

This equation provided the geometrical definition of the curvature tensor. For example, to find the component $R^1{}_{023}$ of the curvature tensor we take a unit vector pointing in the timelike or zero direction and carry it parallel to itself around a circuit of unit extent in the y, z surface and at the end measure how much the tip of the vector has been swung in the x direction.

Einstein's field equation reads

$$G_\alpha{}^\beta = 8\pi T_\alpha{}^\beta \quad .$$

Here the quantity on the right is the stress energy tensor. On the left, the quantity that appears is not the full Riemann curvature tensor but the Einstein curvature tensor of which typical components are

$$G^0{}_0 = -(R^{12}{}_{12} + R^{23}{}_{23} + R^{31}{}_{31}) \quad ,$$

$$G^1{}_1 = -(R^{02}{}_{02} + R^{03}{}_{03} + R^{23}{}_{23}) \quad ,$$

$$G^0{}_1 = R^{02}{}_{12} + R^{03}{}_{13} \quad ,$$

$$G^1{}_2 = R^{10}{}_{20} + R^{13}{}_{23} \quad .$$

To understand the meaning of these strange equations is the central point of what follows.

In gravitation as well as in electromagnetism we used to look at the source as primary and the field as secondary. The source (electric charge or mass energy) "knew" that it wanted to be conserved, and the field ran along behind as its slave, obedient to its wish. Today we regard the field as primary and the source as secondary. Without the field to govern it, the source would not know what to do. It would not even exist.

When two gigantic spaceships smash into each other, much is destroyed. One quantity, we know, is conserved, the energy-momentum 4-vector. What master is so powerful that it can hold those two mighty spaceships in straightline motion before they hit and see to the conservation law in the crash itself? Space! Space grips them both. Space, right where they are, enforces the conservation of momentum and energy.

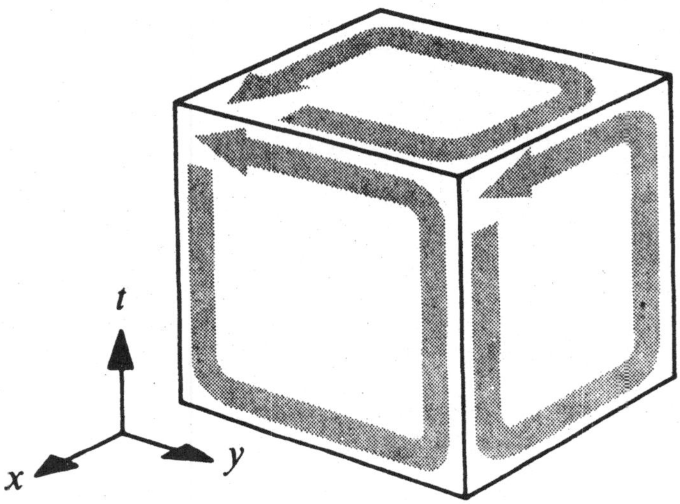

Fig. 3. The 1-dimensional boundary of one face of the cube is the collection of four directed segments or edges. The 2-dimensional boundary is the collection of all six faces. When we add up the four edges for each of the six faces, or twenty-four edges altogether, we get zero because the edges cancel in pairs. Thus the 1-dimensional boundary of the 2-dimensional boundary of the 3-dimensional cube is zero $\partial\partial = 0$.

We used to think of the universe, figuratively speaking, as made out of gears and pinions and put together with the sophistication of an elaborate watch. Today, what we know of the big bang and gravitational collapse encourages us to think just the opposite. We see no other way for the laws of physics to come into being except higgledy-piggledy chance. The hypothesis of "law without law" suggests to us that almost everything comes into being out of almost nothing. Of "almost nothing" principles, it is difficult to think of one more natural than the central principle of algebraic topology: the boundary of a boundary is zero, $\partial\partial = 0$. The principle $\partial\partial = 0$ lies at the very heart of electrodynamics, geometrodynamics and chromodynamics. In each field theory it occurs twice, once at the 1-2-3-dimensional level and again at the 2-3-4-dimensional level (Fig. 3).

The idea $\partial\partial = 0$ applies to curvature. We subject a vector in imagination to parallel transport around one face of a little cube in spacetime. It undergoes a rotation. However, the sum of the rotations associated with all six faces is zero. "The rotational diagram closes." The alteration in the vector V^α in parallel transport around the front face of the cube is

$$\delta V^\alpha = -R^{\alpha\beta}{}_{31}(x,y + \tfrac{1}{2}\,\Delta y,z)V_\beta \Delta x^3 \Delta x^1 \quad .$$

The combination of the effects on V^α produced by going around the front and the back face is

$$-(\partial R^{\alpha\beta}{}_{31}/\partial x^2)V_\beta \Delta x^1 \Delta x^2 \Delta x^3 \quad .$$

The vanishing of the totalized rotation for all six faces gives the so-called full Bianchi identity,

$$(\partial R^{\alpha\beta}{}_{23}/\partial x^1) + (\partial R^{\alpha\beta}{}_{31}/\partial x^2) + (\partial R^{\alpha\beta}{}_{12}/\partial x^3) = 0 \quad .$$

How are we to "wire up" stress energy, our source, to spacetime geometry, our field, so as to guarantee conservation of the source, and do this automatically, without any "gears and pinions and watchworks"? By applying the principle $\partial\partial = 0$ at the 2-3-4-dimensional level.

We consider a region of spacetime, a 4-cube $\Delta x\Delta y\Delta z\Delta t$. We demand that there shall be no creation of source in the given 3-dimensional region in the given time (Fig. 4). That means that the amount of energy-momentum in the 3-cube $\Delta x\Delta y\Delta z$ at the time $t + \tfrac{1}{2}\Delta t$ is to be equal to the amount in the same region at the time $t - \tfrac{1}{2}\Delta t$ with corrections for what flows across the six surfaces ($\Delta x\Delta y\Delta t$ at $z + \tfrac{1}{2}\Delta z$ and five others) of the 4-cube in the intervening time and with no other corrections.

We want zero creation of source in the 4-cube $\Delta x\Delta y\Delta z\Delta t$. That means a zero total for the energy-momentum in the eight 3-cubes that bound that 4-cube. If this result is to come about automatically, via $\partial\partial = 0$, then the energy-momentum in any given 3-cube must in turn be the sum of contributions from each of the six 2-faces of that cube. This construction guarantees conservation. Thus every 2-face of one 3-cube will give a contribution which is equal in magnitude and opposite in sign to that of the abutting 2-face of another one of the eight 3-cubes.

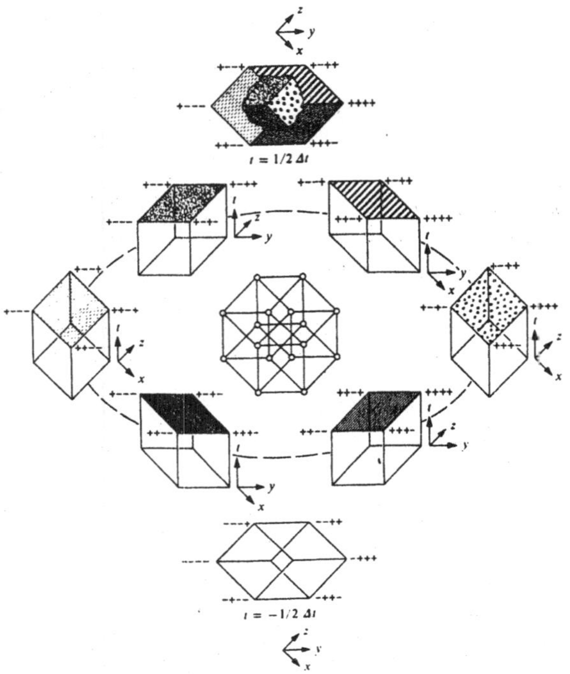

Fig. 4. A 4-cube of dimensions Δx, Δy, Δx, Δt and its eight 3-cube faces depicted as "exploded off" of it. The content of stress-energy tensor in the 3-cube at the upper right (shaded only on its "top," and with diagonal stripes) measures the amount of momentum-energy that passes out of the 4-cube across a surface of area ΔxΔy, located at $z + \frac{1}{2}\Delta z$, during the time Δt. To demand conservation of momentum and energy is to demand that the amounts of momentum and energy in all eight cubes, taken with due regard to sign, shall add to zero. This demand is fulfilled by representing the amount of momentum-energy in the given 3-cube (a vector or, in its dual representation, a 3-index tensor) as the sum of the "moments of rotation" associated with all six 2-dimensional faces of that 3-cube. When we add up these contributions for all eight 3-cubes we have $6 \times 8 = 48$ 2-dimensional faces to deal with, but their contributions all cancel out in pairs, as required. Thus the contribution of the striped face just considered is cancelled by the contribution of the striped face of the "top" or ΔxΔyΔz 3-cube. This is the payoff of the principle that the "2-dimensional boundary of the 3-dimensional boundary of a 4-cube" is zero. This is how spacetime grips mass so as to ensure and enforce the law of conservation of momentum and energy.

Élie Cartan seems to have been the first to apply $\partial\partial = 0$ to general relativity. He, Kibble, Sciama and Hehl have, in this connection, considered generalizations of Einstein's theory which call upon the geometrical notion of "torsion" as well as curvature. However, no one has yet proposed an experiment adequate in sensitivity to distinguish theories with torsion from Einstein's standard 1915 torsion-free theory.

Given curvature as our only building material and given one face of a 3-cube, we calculate only the rotation associated with that face. That rotation is no good for constructing the energy-momentum content of the 3-cube. (1) Added up for all six faces of the 3-cube it always gives a zero sum. (2) It is the wrong kind of geometric quantity. It has two indices rather than one. We turn to elementary statics for an idea.

For a body in static equilibrium the vector sum of the forces on it has to vanish. This granted, the sum of the moments of these forces is independent of the arbitrary point about which we choose to calculate this moment:

$$M = \Sigma(r_i + a) \times F_i = \Sigma r_i \times F_i \quad .$$

Guided by the example of statics, we select some arbitrary point, P, inside or near the 3-cube we are considering (Fig. 5). We take the distance, r_i, from that point to the center of the ith face of

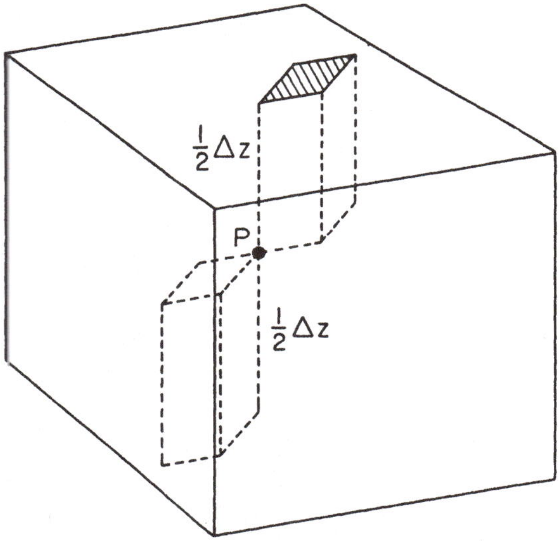

Fig. 5. Moments of rotation associated with the top and bottom faces of the 3-cube $\Delta x \Delta y \Delta z$, represented as parallelepipeds. The combined 3-volume of these two parallelepipeds is independent of the location of the point P. The sum of the volumes of the parallelepipeds associated with all six faces tells how much energy there is in the 3-cube.

the 3-cube. We evaluate the "moment of rotation," $r_i \times (\text{rotation})_i$, associated with that face. We add this moment of rotation up for all six faces. We identify this with the 8π times the amount of energy-momentum 4-vector in that 3-cube. The sum of the rotations for the six faces is zero, as we have already seen. Therefore, the sum of the moments of these rotations is independent of the arbitrary point about which we choose to calculate the moments. Moreover, this moment of rotation is a two-plus-one-equals-three-index quantity. It is "dual" (in the sense of 4-dimensional geometry) to a one-index quantity, a vector.

The vector representation of moment of rotation we identify (up to the conventional factor 8π) with energy-momentum because it is automatically conserved. That is how we make our source, energy-momentum, the slave of our field, spacetime geometry:

$$\begin{Bmatrix}\text{dual of moment}\\ \text{of rotation summed}\\ \text{over the six faces}\\ \text{of 3-cube}\end{Bmatrix} = 8\pi \begin{Bmatrix}\text{amount of}\\ \text{energy-momentum}\\ \text{in that 3-cube}\\ \end{Bmatrix}$$

This is the true content and meaning of Einstein's field equation.

We can spell out this concept of moment of rotation more fully for any of the eight 3-cubes that bound our 4-cube. However, it is simplest to take as example a time-directed region, a 3-cube of dimension Δx, Δy, Δz centered at the point x, y, z, t (all components of $d^3\Sigma_\beta$ zero except $d^3\Sigma_0$) and to inquire after only one component, the energy or time component, of what is contained in this region,

$$8\pi T^0{}_0 = \begin{Bmatrix}\text{dual of}\\ \text{moment of}\\ \text{rotation summed}\\ \text{over the six faces}\\ \text{of the 3-cube}\end{Bmatrix} .$$

To ask for the time component alone of the energy-momentum is to ask only for the xyz component of its dual, the moment of rotation itself. That means that we can disregard the time components of any of the six rotations. It is enough by way of illustration to look at the comtribution of two opposing faces of the 3-cube (Fig. 5). We take the arbitrary point P at the center of the 3-cube. We represent the moment of rotation as a parallelepiped. The moment of rotation associated with the top face of the 3-cube is:

$$\begin{Bmatrix}\text{moment of}\\ \text{rotation}\end{Bmatrix}_{\text{top face}} = -\underbrace{(\underbrace{R^{12}{}_{12}}_{\substack{\text{plane}\\ \text{of}\\ \text{rotation}}} \underbrace{\Delta x\Delta y}_{\text{area}})}_{\text{rotation}} \underbrace{\frac{\Delta z}{2}}_{\substack{\text{lever}\\ \text{arm}}} .$$

When we add these moments of rotation for all six faces and divide by the volume of the 3-cube we get the moment of rotation per unit volume, and at last the physical interpretation of the Einstein tensor,

$$G^{0}{}_{0} = \begin{pmatrix}\text{moment of}\\ \text{rotation per}\\ \text{unit volume}\end{pmatrix} = -(R^{12}{}_{12} + R^{23}{}_{23} + R^{31}{}_{31}) \quad .$$

If only the principle $\partial\partial = 0$ came into play we might say that we had got everything out of nothing. However, we needed more--duality--which, in essence, means the idea of metric itself. Without that we could not have obtained the one-index vector of energy-momentum out of the three-index moment of rotation. That means that there are important elements in physics which still elude the principle of "law without law."

To use the statement $\partial\partial = 0$ allows us also to deduce, even more simply, almost all of classical electrodynamics from almost nothing; and, with a few more complications, almost all of chromodynamics from almost nothing.

The almost trivial identity of algebraic geometry, that the boundary of a boundary is zero, pervades the breadth and depth of the field of physics of our day. This circumstance gives us renewed occasion to take seriously the hypothesis, that almost all physics is built on almost nothing.

RECAPITULATION

In physics at the quantum level our equipment asks the question and nature gives the answer, in the shape of an elementary quantum phenomenon. On such Bohr phenomena all experience is built, directly or indirectly. Does that mean that all of existence --including particles, fields of force, and spacetime itself--is built on the higgledy-piggledy of elementary quantum phenomena and nothing more? On this hypothesis it is far from possible to reach any definitive conclusion today, but in support of it are two striking circumstances: first, fields and geometry let themselves be defined by the quantum-mechanical phase changes around closed loops; and second, each of the three great laws of physics of our time contains as central element the trivial--but great--identity of algebraic topology that the boundary of a boundary is zero, and this twice over, once at the 1-2-3-dimensional level, and again at the 2-3-4-dimensional level.

Preparation of this report for publication was assisted by the University of Texas Center for Theoretical Physics and by NSF Grant No. 8205717.

REFERENCES

[1] W. Heisenberg, Zeits. f. Physik 43, 172-198 (1927).
[2] J. A. Wheeler and W. H. Zurek, eds., "Quantum Theory and Measurement," Princeton University Press, Princeton, New Jersey (1983), may be referred to for a fuller account of a few of the issues taken up in the present report. It includes reprints or translations of several of the items cited in the text.
[3] N. Bohr, Nature 121, 580-590 (1928).
[4] N. Bohr, "Atomic Theory and the Description of Nature," Cambridge University Press, Cambridge, U.K. (1934), p. 10.
[5] N. Bohr, "Atomic Physics and Human Knowledge," Wiley, New York (1958), p. 90.
[6] W. K. Wootters and W. H. Zurek, Phys. Rev. D19, 473-484 (1979).
[7] A. Einstein, B. Podolsky and W. Rosen, Phys. Rev. 47, 777-780 (1935).
[8] D. Bohm, "Quantum Theory," Prentice-Hall, Englewood Cliffs, New Jersey (1951), Chap. 22, sections 15-19; reprinted in ref. 2.
[9] J. A. Wheeler, in A. R. Marlowe, ed.,"Mathematical Foundations of Quantum Theory," Academic Press, New York (1978), pp. 9-48.
[10] W. Miller and J. A. Wheeler, in "Proceedings of the August 29-31, 1983 Tokyo International Symposium on the Foundations of Quantum Mechanics," scheduled to appear in Proc. Phys. Soc. Japan, 1984.
[11] J. A. Wheeler, Am. J. Phys. 51, 398-404 (1983).
[12] Y. Aharonov and D. Bohm, Phys. Rev. 115, 485 (1959).
[13] J. Anandan, Phys. Rev. D15, 1448-1457 (1977).

INDEX